T0260405

# IoT and Analytics in Renewable Energy Systems (Volume 2)

Smart cities emanate from a smart renewable-energy-aided power grid. The smart grid technologies offer an array of benefits like reliability, availability and resiliency. Smart grids phenomenally contribute to facilitating cities reaching those sustainability goals over time. Digital technologies such as the Internet of Things (IoT), automation, artificial intelligence (AI) and machine learning (ML) significantly contribute to the two-way communication between utilities and customers in smart cities.

Five salient features of this book are as follows:

- Smart grid to the smart customer
- Intelligent computing for smart grid applications
- Novel designs of IoT systems such as smart healthcare, smart transportation, smart home, smart agriculture, smart manufacturing, smart grid, smart education, smart government and smart traffic management systems
- Innovations in using IoT and AI in improving resilience of smart energy infrastructure
- Challenges and future research directions of smart city applications

# IoT and Analytics in Renewable Energy Systems (Volume 2)

## AI, ML and IoT Deployment in Sustainable Smart Cities

Edited by
O. V. Gnana Swathika, K. Karthikeyan, and
Sanjeevikumar Padmanaban

CRC Press
Taylor & Francis Group
Boca Raton London New York

CRC Press is an imprint of the
Taylor & Francis Group, an **informa** business

MATLAB® is a trademark of The MathWorks, Inc. and is used with permission. The MathWorks does not warrant the accuracy of the text or exercises in this book. This book's use or discussion of MATLAB® software or related products does not constitute endorsement or sponsorship by The MathWorks of a particular pedagogical approach or particular use of the MATLAB® software.

Cover image © Shutterstock

First edition published 2024
by CRC Press
2385 Executive Center Drive, Suite 320, Boca Raton FL 33431

and by CRC Press
4 Park Square, Milton Park, Abingdon, Oxon, OX14 4RN

ISBN: 9781032448282 (hbk)
ISBN: 9781032448299 (pbk)
ISBN: 9781003374121 (ebk)

DOI: 10.1201/9781003374121

Typeset in Times
by codeMantra

# Contents

Editors ................................................................................................................................ ix

Contributors ........................................................................................................................ xi

**Chapter 1**   Efficient Solutions from Smart Healthcare Ecosystem in the 21st Century – A
Brief Study ...................................................................................................... 1

*Luke Gerard Christie and Deepa Cherian*

**Chapter 2**   IoT-Based Vehicle Monitoring System ............................................................ 9

*R. Sricharan, E. Karthikeyan, K. Kaamesh, O.V. Gnana Swathika,
and V. Berlin Hency*

**Chapter 3**   G-GET: A Donation App to Reduce Poverty ................................................. 17

*K. Karthik, V. Meenakshi Sundaram, and J. Ranjani*

**Chapter 4**   Enhanced *K*-Means with Automated *k* Estimation and Outlier Detection Using
Auto-Encoder Neural Network ..................................................................... 29

*P. Illavenil, J. Kiron, S. Ramnath, and G. Hannah Grace*

**Chapter 5**   LPG Leakage Detector .................................................................................. 41

*Zaina Nasreen, D. Subbulekshmi, S. Angalaeswari, and T. Deepa*

**Chapter 6**   IoT-Based Intelligent Garbage Monitoring Management System to Catalyse
Farming ......................................................................................................... 51

*B.V.A.N.S.S. Prabhakar Rao and Rabindra Kumar Singh*

**Chapter 7**   IoT-Based Wildfire Detection and Monitoring System Using
Predictive Analytics ..................................................................................... 65

*Chimata Shriya, Varsha Jayaprakash, C.N. Rachel, O.V. Gnana Swathika,
and V. Berlin Hency*

**Chapter 8**   Rainfall Prediction Model Using Artificial Intelligence Techniques ............. 75

*Souvik Laskar and A. Menaka Pushpa*

**Chapter 9**   Intelligent Coconut Harvesting System ......................................................... 97

*B.V.A.N.S.S. Prabhakar Rao and Ram Prasad Reddy Sadi*

**Chapter 10**  IoT-Based Live Ambulance: Management and Tracking System ......................... 111

  *G. Gayathri, Sarada Manaswini Upadhyayula, O.V. Gnana Swathika,
  and V. Berlin Hency*

**Chapter 11**  Smart Robot Car for Industrial Internet of Things ................................. 123

  *Ayushi Chakrabarty, H.R. Deekshetha, S. Reshma, O.V. Gnana Swathika,
  and V. Berlin Hency*

**Chapter 12**  IoT-Based Monitoring System with Machine Learning Analytics of
  Transformer: Mini Review ..................................................... 143

  *Aadi Ashutosh Chauhan, Rohan Bhojwani, Rahul Pal, O.V. Gnana Swathika,
  Aayush Karthikeyan, and K.T.M.U. Hemapala*

**Chapter 13**  Design of Earthquake Alarm ................................................. 151

  *K. Lokeswar, D. Subbulekshmi, T. Deepa, and S. Angalaeswari*

**Chapter 14**  Energy Demand and Flexibility of Energy Supply: A Case Study ......................... 157

  *R. Rajapriya, Varun Gopalakrishnan, Milind Shrinivas Dangate,
  and Nasrin I. Shaikh*

**Chapter 15**  Internet of Things-Based Toddler Security Monitoring and
  Management System ........................................................ 171

  *Karmel Arockiasamy, G. Kanimozhi, and Dheep Singh*

**Chapter 16**  Adaptive Traffic Control ................................................... 181

  *Yashashwini Dixit, Gyanadipta Mohanty, and Karmel Arockiasamy*

**Chapter 17**  Genuine Investments for Economic Energy Outputs ................................ 189

  *Saravanan Chinnusamy, R. Rajapriya, Milind Shrinivas Dangate,
  and Nasrin I. Shaikh*

**Chapter 18**  IoT-Based Power Theft Detection: Mini Review ..................................... 205

  *Ameesh Singh, Harsh Gupta, O.V. Gnana Swathika, Aayush Karthikeyan,
  and Akhtar Kalam*

**Chapter 19**  Design and Implementation of Bluetooth-Enabled Home Automation System ....... 215

  *Nagavindhya Nagavindhya, Krithikka Jayamurthi, V. Berlin Hency,
  O.V. Gnana Swathika, Aayush Karthikeyan, and K.T.M.U. Hemapala*

**Chapter 20**  IoT-Based Smart Electricity Management ........................................ 231

  *R. Sricharan, E. Karthikeyan, K. Sethu Narayanan, O.V. Gnana Swathika,
  and V. Berlin Hency*

**Chapter 21** IoT-Based COVID-19 Patient Monitoring System ....................................................243

*S. Charan, K. Kaamesh, B. Aswin, O.V. Gnana Swathika, and V. Berlin Hency*

**Chapter 22** Interleaved Cubic Boost Converter ..........................................................257

*C. Sankar Ram, Aditya Basawaraj Shiggavi, A. Adhvaidh Maharaajan, R. Atul Thiyagarajan, and M. Prabhakar*

**Chapter 23** Emerging Role of AI, ML and IoT in Modern Sustainable Energy Management ...........................................................273

*Arpan Tewary, Chandan Upadhyay, and A.K. Singh*

**Chapter 24** Automated Water Dispenser – A Hygiene Solution for Pandemic...........................289

*G.G. Raja Sekhar, D. Kalyan, R. Ramkumar, and M. Lakshmi*

**Chapter 25** Review of IoT-Based Smart Waste Management Systems........................................299

*T. Sakthi Ram, S. Vetriashwath, V. Mruthunjay, Rahul Srikanth, Yuvan Shankar, L. Yogesh, O.V. Gnana Swathika, and Aayush Karthikeyan*

**Chapter 26** Cyber Security in Smart Energy Networks...............................................309

*Sanjeevikumar Padmanaban, Mostafa Azimi Nasab, Tina Samavat, Mohammad Zand, Morteza Azimi Nasab, and Erfan Hashemi*

**Index**....................................................................................327

# Editors

**O. V. Gnana Swathika** (Member '11–Senior Member '20, IEEE) earned a BE in Electrical and Electronics Engineering from Madras University, Chennai, Tamil Nadu, India, in 2000; an MS in Electrical Engineering from Wayne State University, Detroit, MI, USA, in 2004; and a PhD in Electrical Engineering from VIT University, Chennai, Tamil Nadu, India, in 2017. She completed her postdoc at the University of Moratuwa, Sri Lanka in 2019. Her current research interests include microgrid protection, power system optimization, embedded systems and photovoltaic systems. She is currently serving as Associate Professor Senior, Centre for Smart Grid Technologies, Vellore Institute of Technology, Chennai, India.

**K. Karthikeyan** is an electrical and electronics engineering graduate with a master's in personnel management from the University of Madras. With two decades of rich experience in electrical design, he has immensely contributed toward the building services sector comprising airports, Information Technology Office Space (ITOS), tall statues, railway stations/depots, hospitals, educational institutional buildings, residential buildings, hotels, steel plants and automobile plants in India and abroad (Sri Lanka, Dubai and the UK). Currently, he is Chief Engineering Manager – Electrical Designs for Larsen & Toubro (L&T) Construction, an Indian multinational Engineering Procurement Construction (EPC) contracting company. Also, he has worked at Voltas, ABB and Apex Knowledge Technology Private Limited. His primary role involved the preparation and review of complete electrical system designs up to 110 kV. Perform detailed engineering stage which includes various electrical design calculations, design basis reports, suitable for construction drawings, and Mechanical Electrical Plumbing (MEP) design coordination. He is the point of contact for both client and internal project team, leads and manages a team of design and divisional personnel, engages in day-to-day interaction with clients, peer review of project progress, manages project deadlines and project time estimation, and assists in staff appraisals, training and recruiting.

**Sanjeevikumar Padmanaban** (Member '12–Senior Member '15, IEEE) received a PhD degree in electrical engineering from the University of Bologna, Bologna, Italy, in 2012. He was an Associate Professor at VIT University from 2012 to 2013. In 2013, he joined the National Institute of Technology, India, as a Faculty Member. In 2014, he was invited as a Visiting Researcher at the Department of Electrical Engineering, Qatar University, Doha, Qatar, funded by the Qatar National Research Foundation (Government of Qatar). He continued his research activities with the Dublin Institute of Technology, Dublin, Ireland, in 2014.

Further, he served as an Associate Professor with the Department of Electrical and Electronics Engineering, University of Johannesburg, Johannesburg, South Africa, from 2016 to 2018. From March 2018 to February 2021, he has been an Assistant Professor in the Department of Energy Technology, Aalborg University, Esbjerg, Denmark. He continued his activities from March 2021 as an Associate Professor in the CTiF Global Capsule (CGC) Laboratory, Department of Business Development and Technology, Aarhus University, Herning, Denmark. Presently, he is a Full Professor in Electrical Power Engineering in the Department of Electrical

Engineering, Information Technology and Cybernetics, University of South-Eastern Norway, Norway.

S. Padmanaban has authored over 750+ scientific papers and received the Best Paper cum Most Excellence Research Paper Award from IET-SEISCON'13, IET-CEAT'16, IEEE-EECSI'19 and IEEE-CENCON'19, and five best paper awards from ETAEERE'16-sponsored Lecture Notes in Electrical Engineering, Springer book. He is a Fellow of the Institution of Engineers, India, the Institution of Electronics and Telecommunication Engineers, India, and the Institution of Engineering and Technology, UK. He received a lifetime achievement award from Marquis Who's Who – USA 2017 for contributing to power electronics and renewable energy research. He is listed among the world's top 2 scientists (from 2019) by Stanford University, USA. He is an Editor/Associate Editor/Editorial Board for refereed journals, in particular the *IEEE Systems Journal*, *IEEE Transaction on Industry Applications*, *IEEE ACCESS*, *IET Power Electronics*, *IET Electronics Letters* and *Wiley-International Transactions on Electrical Energy Systems*; Subject Editorial Board Member—*Energy Sources—Energies Journal*, MDPI; and the Subject Editor for the *IET Renewable Power Generation*, *IET Generation, Transmission and Distribution* and *FACETS* Journal (Canada).

# Contributors

**S. Angalaeswari**
School of Electrical Engineering
Vellore Institute of Technology
Chennai, India

**Karmel Arockiasamy**
School of Computer Science and Engineering
Vellore Institute of Technology
Chennai, India

**B. Aswin**
School of Electronics Engineering
Vellore Institute of Technology
Chennai, India

**Rohan Bhojwani**
School of Electrical Engineering
Vellore Institute of Technology
Chennai, India

**Ayushi Chakrabarty**
School of Electronics
Vellore Institute of Technology
Chennai, India

**S. Charan**
School of Electrical Engineering
Vellore Institute of Technology
Chennai, India

**Aadi Ashutosh Chauhan**
School of Electrical Engineering
Vellore Institute of Technology
Chennai, India

**Deepa Cherian**
Indian School of Business
Hyderabad, India

**Saravanan Chinnusamy**
R & D Division, Medicinal Chemistry
Chemveda Life Sciences India Pvt. Ltd.
Hyderabad, India

**Luke Gerard Christie**
School of Social Sciences and Languages
Vellore Institute of Technology
Chennai, India

**Milind Shrinivas Dangate**
Chemistry Division, School of Advanced
    Sciences
Vellore Institute of Technology
Chennai, India

**H.R. Deekshetha**
School of Electronics
Vellore Institute of Technology
Chennai, India

**T. Deepa**
School of Electrical Engineering
Vellore Institute of Technology
Chennai, India

**Yashashwini Dixit**
Vellore Institute of Technology
Chennai, India

**G. Gayathri**
School of Electronics
Vellore Institute of Technology
Chennai, India

**Varun Gopalakrishnan**
School of Mechanical Engineering
Vellore Institute of Technology
Chennai, India

**G. Hannah Grace**
School of Advanced Sciences
Vellore Institute of Technology
Chennai, India

**Harsh Gupta**
School of Electrical Engineering
Vellore Institute of Technology
Chennai, India

**Erfan Hashemi**
Department of Business Development and
    Technology
CTiF Global Capsule
Herning, Denmark

**K.T.M.U. Hemapala**
Department of Electrical Engineering
University of Moratuwa
Moratuwa, Sri Lanka

**V. Berlin Hency**
School of Electronics Engineering
Vellore Institute of Technology
Chennai, India

**P. Illavenil**
School of Computer Science and Engineering
Vellore Institute of Technology
Chennai, India

**Krithikka Jayamurthi**
School of Electrical Engineering
Vellore Institute of Technology
Chennai, India

**Varsha Jayaprakash**
School of Electronics Engineering
Vellore Institute of Technology
Chennai, India

**K. Kaamesh**
School of Electronics Engineering
Vellore Institute of Technology
Chennai, India

**Akhtar Kalam**
Smart Energy Research Unit, College of
    Engineering and Science
Victoria University
Victoria, Australia

**D. Kalyan**
Department of EEE
Koneru Lakshmaiah Education Foundation
Vaddeswaram, India

**G. Kanimozhi**
Centre for Smart Grid Technologies
Vellore Institute of Technology
Chennai, India

**K. Karthik**
Sri Sairam Engineering College
Chennai, India

**Aayush Karthikeyan**
Schulich School of Engineering
University of Calgary
Calgary, Canada

**E. Karthikeyan**
School of Electronics Engineering
Vellore Institute of Technology
Chennai, India

**J. Kiron**
School of Computer Science and Engineering
Vellore Institute of Technology
Chennai, India

**M. Lakshmi**
Department of EEE
Dhanalakshmi Srinivasan University
Trichy, India

**Souvik Laskar**
Vellore Institute of Technology
Chennai, India

**K. Lokeswar**
Vellore Institute of Technology
Chennai, India

**A. Adhvaidh Maharaajan**
School of Electrical Engineering
Vellore Institute of Technology
Chennai, India

**V. Meenakshi Sundaram**
Sri Sairam Engineering College
Chennai, India

**Gyanadipta Mohanty**
Vellore Institute of Technology
Chennai, India

**V. Mruthunjay**
School of Electrical Engineering
Vellore Institute of Technology
Chennai, India

**Nagavindhya Nagavindhya**
School of Electrical Engineering
Vellore Institute of Technology
Chennai, India

**K. Sethu Narayanan**
School of Electronics Engineering
Vellore Institute of Technology
Chennai, India

**Morteza Azimi Nasab**
Department of Business Development and
    Technology
CTiF Global Capsule
Herning, Denmark

**Mostafa Azimi Nasab**
Department of Business Development and
    Technology
CTiF Global Capsule
Herning, Denmark

**Zaina Nasreen**
School of Electrical Engineering
Vellore Institute of Technology
Chennai, India

**Sanjeevikumar Padmanaban**
Department of Business Development and
    Technology
CTiF Global Capsule
Herning, Denmark

**Rahul Pal**
School of Electrical Engineering
Vellore Institute of Technology
Chennai, India

**M. Prabhakar**
Centre for Smart Grid Technologies
Vellore Institute of Technology
Chennai, India

**A. Menaka Pushpa**
Vellore Institute of Technology
Chennai, India

**C.N. Rachel**
School of Electronics Engineering
Vellore Institute of Technology
Chennai, India

**R. Rajapriya**
School of Mechanical Engineering
Chemistry Division, School of Advanced
    Sciences
Vellore Institute of Technology
Chennai, India

**C. Sankar Ram**
School of Electrical Engineering
Vellore Institute of Technology
Chennai, India

**T. Sakthi Ram**
School of Electrical Engineering
Vellore Institute of Technology
Chennai, India

**R. Ramkumar**
Department of EEE
Dhanalakshmi Srinivasan University
Trichy, India

**S. Ramnath**
School of Computer Science and Engineering
Vellore Institute of Technology
Chennai, India

**J. Ranjani**
Sri Sairam Engineering College
Chennai, India

**B.V.A.N.S.S. Prabhakar Rao**
School of Computer Science and Engineering
Vellore Institute of Technology
Chennai, India

**S. Reshma**
School of Electronics
Vellore Institute of Technology
Chennai, India

**Tina Samavat**
Department of Business Development and
    Technology
CTiF Global Capsule
Herning, Denmark

**Ram Prasad Reddy Sadi**
Department of Information Technology
Anil Neerukonda Institute of Technology and
    Sciences
Visakhapatnam, India

**G.G. Raja Sekhar**
Department of EEE
Koneru Lakshmaiah Education Foundation
Vaddeswaram, India

**Nasrin I. Shaikh**
Department of Chemistry
Nowrosjee Wadia College
Pune, India
and
Department of Chemistry
Abdul Karim Ali Shayad Faculty of
    Engineering and Polytechnic, MMANTC
Malegaon, India

**Yuvan Shankar**
School of Electrical Engineering
Vellore Institute of Technology
Chennai, India

**Aditya Basawaraj Shiggavi**
School of Electrical Engineering
Vellore Institute of Technology
Chennai, India

**Chimata Shriya**
School of Electronics Engineering
Vellore Institute of Technology
Chennai, India

**Ameesh Singh**
School of Electrical Engineering
Vellore Institute of Technology
Chennai, India

**A.K. Singh**
Department of Sciences
R. P. P. G. College Sultanpur
India

**Dheep Singh**
Vellore Institute of Technology
Chennai, India

**Rabindra Kumar Singh**
School of Computer Science and Engineering
Vellore Institute of Technology
Chennai, India

**R. Sricharan**
School of Electronics Engineering
Vellore Institute of Technology
Chennai, India

**Rahul Srikanth**
School of Electrical Engineering
Vellore Institute of Technology
Chennai, India

**D. Subbulekshmi**
School of Electrical Engineering
Vellore Institute of Technology
Chennai, India

**V. Meenakshi Sundaram**
Sri Sairam Engineering College
Chennai, India

**O.V. Gnana Swathika**
Centre for Smart Grid Technologies, School of
    Electrical Engineering
Vellore Institute of Technology
Chennai, India

**Arpan Tewary**
Department of Sciences
Central University of Jharkhand
Ranchi, India

**R. Atul Thiyagarajan**
School of Electrical Engineering
Vellore Institute of Technology
Chennai, India

**Chandan Upadhyay**
Department of Sciences
Dr. R. M. L. Avadh University
Ayodhya, India

**Sarada Manaswini Upadhyayula**
School of Electronics
Vellore Institute of Technology
Chennai, India

**S. Vetriashwath**
School of Electrical Engineering
Vellore Institute of Technology
Chennai, India

**L. Yogesh**
School of Electrical Engineering
Vellore Institute of Technology
Chennai, India

**Mohammad Zand**
Department of Business Development and
   Technology
CTiF Global Capsule
Herning, Denmark

# 1 Efficient Solutions from Smart Healthcare Ecosystem in the 21st Century – A Brief Study

*Luke Gerard Christie*
Vellore Institute of Technology

*Deepa Cherian*
Indian School of Business

## CONTENTS

1.1  Introduction ........................................................................................................2
1.2  Healthcare Devices............................................................................................3
1.3  Influence of IoT, AI and Big Data....................................................................4
1.4  Data Protection and Privacy .............................................................................5
1.5  Conclusion .........................................................................................................5
Bibliography ................................................................................................................6

**ABSTRACT**

Today's ecosystem is awash for efficient solutions in healthcare. With newer disease outbreaks, global warming and more people falling sick often, stakeholders in healthcare facilities must aim to build resilient and energy-efficient solutions in healthcare. The influx of emergent technologies creates opportunities to build systems that are robust and sophisticated to save energy and costs. Hospitals in Global South countries must address energy crisis due to the shifting from fossil fuels to electricity. The fundamental questions we have to ask are the preparation to build systems of energy consumption that are smart and efficient for the future. Policy makers, foremost, must be highly sensitive to a new age and must be calculative as their policies do affect society for the better or for the worse. This chapter aims to discuss efficient energy solutions with new technologies to save life and create a more effective and safer healthcare ecosystem that is informed and intelligent with Internet of Things and artificial intelligence. Global South countries that aim to be less carbon-dependent must also aim to be holistic unlike the previous decades of poor-informed unsustainable growth and avoid the lacuna for future generations to tackle as the crisis can be beyond revitalization in an already-sensitive ecology.

## KEYWORDS

Smart Healthcare; Pandemics; Emergent Technologies; Sensitive Environments; Ecological Protection

DOI: 10.1201/9781003374121-1

## 1.1  INTRODUCTION

*The future of healthcare will be digital. Digitising healthcare is the key enabler for expanding preci-*
*sion medicine, transforming care delivery, and improving patient experience, allowing healthcare*
*providers to increase value through better outcomes.*

*Dileep Mangsuli, Siemens Healthlineers*

India has conceptualized a smart healthcare system on a global standard after bitter experiences with the COVID-19 pandemic. The goal that the government has is in line with policies that support the initiative where all citizens of the land will have quality access to personalized healthcare with capability of cyber physical health systems and immediate diagnosis of disease. Smartphones becoming a commonality only amplifies the prospects for citizens to use web-based apps with or without zero remuneration when consulting healthcare specialists online. The current ecosystem is embedding artificial intelligence (AI) and data mining in planning healthcare and policies associated around healthcare to create an atmosphere of accessibility for all. In the Global North countries, the healthcare environment has naturally encouraged startups, new businesses and technology startups and innovators to integrate with the ecosystem, ensuring accessibility and affordability. With India's nerve centre shifting towards going digital and with the doctor–patient ration of 1:1,500, it makes access through smart technology more easily available, reducing the burden on doctors and an overworked healthcare environment. In case, of pandemics and with newer predictions due to global warming and climate change, that poor countries or Global South countries, India being the largest and one of them will face tremendous challenges owing to its population. A pandemic such as Covid-19 witnessed painful moments due to population being the over-riding factor and where social distancing is an impossibility. To add to the bureaucratic struggle, healthcare becomes largely inaccessible in an offline atmosphere. The most practical way out of this conundrum is through the smart ecosystem approach of consultation and early diagnosis online and further tests from our private places (hospitalized only when we realize that one needs more healthcare support or more surgeries).

The earliest initiative for all citizens is the Digital Health ID that has been generated for all citizens of the country with a massive boost to investment in healthcare spending especially in a country that has no history of Universal Healthcare coverage – health and accessibility for all. The other major crisis is the rapid increase in the population, in particular senior citizens who will not be able to access or consult with a physician in an offline mode, and for them, the online space is more private and more convenient. An increasing older-age population for any country only proves a huge increase in chronic diseases that require not only frequent visits to healthcare providers, but also hospitalization needs that will keep increasing over time. Though richer countries have access to new-age technologies embedded into the healthcare system, the challenges are equally many due to the cost of living and the increase in the population of senior citizens. The healthcare environment in the digital space has become more affordable and accessible with the advent of Electronic Health Records (EHRs), a system which records patient's every visit to a healthcare facility, diseases' diagnoses, and medication taken in earlier times in a digital format. This helps the healthcare fraternity to address a health concern immediately and avoid wasting time or running multiple tests to address the existing concern due to a poor maintenance of data upkeep. EHRs can be easily shared amongst healthcare providers and accessed from a central database, as the entire health history of a patient (previous or current medications, immunizations, laboratory test results, current or previous diagnosis, doctors consulted) is stored, which makes it easier for a patient to handle.

We will find that with AI and data mining, there is more flexibility and convenience in healthcare management due to the continued access and connections being linked to Wi-Fi, smartphones, and mobile applications that reveal to a patient his blood pressure or the strength of heart beat or his breathing patterns depending on the number of times, doctors ask their patients for a check-up. The new digital era in healthcare technologies is more convenient and guarantees more transparency in the process and confidentiality between doctor and patient where the shift from connected health has evolved to smart health.

The following devices are used extensively in smart healthcare ecosystem:

i. Wearable medical devices (such as smart phones, blood pressure monitors, glucometers, smart watches, and smart contact lenses)
ii. Internet of Things (IoT) gadgets (such as implantable or ingestible sensors) to enable continuous patient monitoring and treatment even when patients are at their homes or at office.

## 1.2 HEALTHCARE DEVICES

Healthcare devices remain in the 21st century the swiftest growing sector using IoT, and economic analysts project that this sector on a global scale has the potential to reach $178 billion by early 2026. IoT, due to its massive booming in medical field, is being defined as Internet of Medical Things (IoMT), and this proves to be the most rapidly growing sector with emergent technologies.

The successes scripted by IoMT are as follows:

i. **IoT robotic surgery**: Most of us are aware of robotic surgeries being used in multi-lateral hospital across the globe. Robotic surgeries are performed with cameras inserted deep into the body of the patient helping surgeons perform complex surgeries. It has also been observed that micro-incisions in the human body are better accomplished with IoT nano-sized devices that lead to less intrusive process and quicker healing.
ii. **IoT Parkinson's disease monitoring**: With less technology, it becomes a challenge for healthcare providers to monitor the severity of a disease, but in this case, IoT sensors help monitor the fluctuations of body movements in a day to help understand the severity of the disease. This makes it possible for patients to save time rather than spending time at a healthcare facility for observation.
iii. **IoT heart rate monitoring**: IoT devices guard patients from rapid fluctuations in heart rates allowing more autonomy to move freely around, guaranteeing a 90% accuracy.
iv. **IoT hand hygiene monitoring**: A device that monitors hand hygiene that ensures providers and patients have clean hands minimizing contagion.
v. **IoT depression and mood monitoring**: The challenge to chronicle mood swings is immense or to collect depression symptoms is difficult. In a normal traditional setup, healthcare providers acquire information based on prompts and responses from their patients where macrocosmic information is gained, leaving out the microcosmic details. IoT here comes with Mood-Aware Devices that address these challenges by analysing heart and blood pressure that infer a person's mental state or even where devices track movement of a person's eyes. However, it is to be understood that exactness can betray the diagnosis outcome but have proved to be less infringing on a patient who is unwilling to open up.
vi. **IoT-connected inhalers**: Asthma attacks or chronic obstructive pulmonary disorder can be diagnosed with IoT-connected inhalers which help patients track frequency of respiratory problems or heavy breathing and are also used to collect information of the environment that causes respiratory illnesses. IoT-connected inhalers inform patients when they misplace their inhalers elsewhere.
vii. **IoT-ingestible sensors**: It is difficult to insert a probe to collect information of the digestive tract and is also challenging and complex to collect data for diagnosis. IoT-ingestible sensors help collecting information in a less invasive way by offering insights on stomach pH levels, or to help find the source of internal bleeding.

viii. **IoT glucose monitoring**: Healthcare providers find it challenging to monitor glucose levels of patients, as after a test is taken, the unpredictability is the constant fluctuation. IoT devices offer a consistent and constant monitoring of glucose levels and eliminate the need to keep tabs or records by intimating the patients when levels of glucose are haphazard.

ix. **IoT-connected contact lenses**: These help collect information in a less invasive way and have possible microcameras that help the wearer take pictures with blinking of eyelids. Google, so far, has patented the connected contact lenses.

x. **IoT remote patient monitoring**: This IoT device is a common application in healthcare collecting immediate information of blood pressure, heart rate, temperature and other requirements remotely without being physically present in the healthcare facility. Sophisticated data collected are passed to the healthcare provider or the healthcare provider where all involved can have access to information. The possibility of sophisticated algorithms generate information or send out alerts when there's evidence of minor or major fluctuation that help the healthcare provider or facility to immediately intervene preventing a crisis for the patient.

IoT devices are small, and require less effort in taking care without interfering in the normalcy of a patient's life and less electricity. However, it all depends on how meticulous and careful is the patient handling their devices.

## 1.3 INFLUENCE OF IoT, AI AND BIG DATA

The generic idea behind smart health is not only accessibility but timely treatment, i.e., continuous monitoring of patient with the inclusion of AI or IoT equipment where information can be transmitted to anywhere in the world, making the environment of healthcare safe and sustainable. All the sensors with the aid of AI or IoT serve as data collecting points based on physiology of the patient. Further the data is transmitted with mobile telephony or Wi-Fi which is common to a healthcare facility of to the consulting doctor. All information of the patient is regularly updated and is mobile all of the time reducing healthcare challenges for the doctor and allowing him to focus on his strategic accountability to his patient. The smart healthcare ecosystem makes healthcare information relevant and accessible to both doctors and patients and even can be studied across geographies to prevent a disease outbreak. The data or information with the advent or influx of new-age technologies are mined constantly and can be studied extensively to take informed decisions and proactively work on efficient healthcare management. Big Data will aggregate from the convergence of different data sets by offering intelligent analysis and reducing the financial costs on the patient or a government on a large scale. Smart healthcare ecosystem prevents a disease outbreak and can be used to prevent mortality rates from increasing. The manner of knowledge based on specificity of disease guarantees type of treatment upon gathering of knowledge through the aid of Big Data mining complex data units.

It is to be observed that sophisticated computer applications and IoT tools can provide the basis for treatment or a rough abstract idea of underlying disease but it is proved to be less cost-effective and less time-consuming and avoid travel costs. IoT in healthcare also helps patient to self-educate themselves on their bodily conditions and the potential diagnosis of a disease. Simulated annealing, ant colony optimization, artificial neural network and genetic algorithms are most relevant algorithms used in the biggest healthcare facilities for data mining and analysis. The advancement of new sophisticated algorithms can help in solving and finding better solutions for an expansive healthcare ecosystem. The Hadoop MapReduce, Flink, Apache Spark and Apache Storm are also a few sophisticated computing platforms in data analytics for the purposes of data analysis and storage of large unstructured data sets. These applications and computing tools have created a robust and efficient atmosphere in the digital space, and the depth of new technologies is crowding the digital healthcare atmosphere bringing with innovation and a high level of sophistication to standardize and help build standardization in for global solutions on the Internet-integrated meticulously and updated as the healthcare management evolves.

## 1.4  DATA PROTECTION AND PRIVACY

With the many opportunities that the smart healthcare system offers, there are a myriad of challenges too. The primary aspects are privacy and data protection of patients whose health care data can be easily accessed, as data are shared and spread over the Internet. India is still fraught with technology issues and an awkward Data Protection Bill that needs to be fit into the legal framework is indeed a work in progress. The positives do outweigh the negatives but the challenge for individual patients and healthcare facilities and doctors or surgeons needs to guarantee each other that all information shared on the Internet is safe and there a transparency clause signed upon by the doctors or surgeons as we see with the existence of a confidentiality clause between doctor and patient. The challenge in the 21st century is to ensure that no data are exposed to a third party or where data breaches are contained to prevent discussion of a disease or possible identification problems. It is also a known fact that with IoT and web-based applications using AI and Big Data, there's autonomy on the world-wide space, and the autonomy can be a technological nightmare for patients or doctors if there's a data breach due to miscalculation of data protection or improper digital protection mechanisms in place. Technological eavesdropping by hacking computer systems or in the case of IoT web applications due to wireless communications makes both the doctor and patient vulnerable. IoT has been observed to have low computing power and capacity, which makes it harder to implement complex algorithms for protection. Although there are benefits of quicker accessibility and cost-benefits, governments and healthcare facilities have to work on ensuring trust, transparency, efficient and effective dense mechanism of data protection, and privacy. All stakeholders have to do a full blown analysis on data protection and guarantee security, as this can become an infringement on patient or individual privacy. It is also to be understood that that there has to be a distinguishing line in specifying a diagnosis or prognosis of disease as technology can over-ride sensitive information, creating a confusion to the healthcare provider and the patient, and further leading to a data storm due to lack of specifications. We must also not forget that data collected can come from different sources if the patient has visited multiple healthcare providers and doctors have to ensure that they abide by specifications. It is important to understand that sophisticated algorithms must be written keeping specificity and precision in mind and to be discriminating of sensitive information. A recent insight in healthcare is to ensure that more than subjective analyses of the said information, doctors have to be more objective in their data standardization that makes accessibility easier and healthcare information easier to comprehend.

## 1.5  CONCLUSION

Emerging technologies are advancing at an exponential rate with new players entering the fray for business purposes. As discussed, the deployment of new technologies that aid the patient and doctor is encouraging, creating an affordability scope for patients but the challenges in poor or poorer countries are immense as most will prefer meeting a doctor in person and are not very technology-sound. The governments in developing countries must consider campaigning for a digital society where technology is accessible to all guaranteeing that the people who belong to the bottom of the socio-economic pyramid can have the same facilities as those in the top of the economic pyramid as new-age technology can be an effective equalizer. Health is the determinant for the success of a country, and technology can help bridge many divisions. At the same time, governments must work hard with newer ideas to ensure that laws are in place and technology companies must work with the healthcare ecosystem to prevent and protect data breaches. If these problems remain, they will serve as painful bottlenecks in effective healthcare management, though we observe that the healthcare systems in Global North and Global South countries have adopted intelligent technology and Big

Data making the infrastructure more efficient and robust. We also witness that with more advancement, systems can be vulnerable to bad actors or to hacking and for that governments and healthcare facilities must ensure an in-house team that looks to protect infringement and data breaches. In the 21st century, the obvious truth is that global warming and climate change will be the major challenge for an environment with economic distress and green technology can reduce the effects if utilized properly shifting economies to being low carbon and amplifying the scope in the technology diversification in healthcare too.

## BIBLIOGRAPHY

1. World Health Organization. "Density of Physicians (Total Number per 1000 Population, Latest Available Year), Global Health Observatory (GHO) Data", Situation and Trends, available at: http://www.who.int/gho/health_workforce/physicians_density/en/ (accessed 10 August 2018).
2. Ministry of AYUSH, "Government of India", available at: http://www.ayush.gov.in (accessed 4 August 2022).
3. "Rural Health Statistics 2017-1: Ministry of Health and Family Welfare, Government of India", available at: https://main.mohfw.gov.in/sites/default/files/Final%20RHS%202018-19_0.pdf.
4. Anagnostopoulos, I., Zeadally, S. and Exposito, E. (2016), "Handling big data: research challenges and future directions", *Journal of Supercomputing*, Vol. 72, No. 4, pp. 1494–1516.
5. BroadbandCommission (2017), "Digital health: a call for government leadership and cooperation between ICT and health", available at: www.broadbandcommission.org/Documents/publications/WorkingGroupHealthReport-2017.pdf (accessed August 2022).
6. Broda (2007), "Managing Trust in e-Health with Federated Identity Management", eHealth Workshop, Konolfingen.
7. Chaudhury, S., Paul, D., Mukherjee, R. and Haldar, S. (2017), "Internet of thing based HealthCare monitoring system", The 8th IEEE Annual Conference on Industrial Automation and Electromechanical Engineering, Bangkok.
8. Cisco (2019), "Making connected health a reality", Cisco Systems, available at: www.cisco.com/c/dam/en_us/solutions/industries/docs/healthcare/connected_health_brochure.pdf (accessed August 2022).
9. Dimitrov, D.V. (2016), "Medical internet of things and big data in healthcare", *Healthcare Informatics Research*, Vol. 22, No. 3, pp. 145–170, available at: www.ncbi.nlm.nih.gov/pmc/articles/PMC4981575/
10. Dwivedi, S., Kasliwal, P. and Soni, S. (2016), "Comprehensive study of data analytics tools (RapidMiner, Weka, R tool, Knime)", IEEE Symposium on Colossal Data Analysis and Networking (CDAN), Indore.
11. eMarketer (2017), "Wearables still far from mass adoption", available at: www.emarketer.com/content/wearables-still-far-from-mass-adoption (accessed August 2019).
12. Firouzi, F., Farahani, B., Ibrahim, M. and Chakrabarty, K. (2018), "From EDA to IoT eHealth: promise, challenges, and solutions", *IEEE Transactions on Computer-Aided Design of Integrated Circuits and Systems*, Vol. 37 No. 12, pp. 2955–2987.
13. Garg, V. (2015), "Optimization of multiple queries for big data with Apache Hadoop/Hive", IEEE International Conference on Computational Intelligence and Communication Networks, Jabalpur.
14. Gawanmeh, A. (2016), "Open issues in reliability, safety, and efficiency of connected health", First IEEE Conference on Connected Health: Applications, Systems and Engineering Technologies, Washington, DC.
15. Gopi, P. and Hwang, T. (2016), "BSN-care: a secure IoT-based modern healthcare system using body sensor network", *IEEE Sensors Journal*, Vol. 16 No. 5, pp. 1358–1376.
16. Haluza, D. and Jungwirth, D. (2014), "ICT and the future of healthcare: aspects of doctor-patient communication", *International Journal of Technology Assessment in Health Care*, Vol. 30 No. 3, pp. 298–305.
17. HealthAffairs (2019), "National health expenditure projections, 2018–27: economic and demographic trends drive spending and enrollment growth", available at: www.healthaffairs.org/doi/full/10.1377/hlthaff.2018.05499 (accessed August 2022).
18. intelliPaat (2019), "What is data analytics", available at: https://intellipaat.com/blog/what-is-data-analytics/ (accessed August 2019).
19. Jain, P. and Mayrya, J.P. (2017), "Comparative analysis using hive and pig on consumers data", *International Journal of Computer Science and Information Technologies*, Vol. 8, No. 2, pp. 275–287.
20. Johri, P., Singh, T., Das, S. and Anand, S. (2017), "Vitality of big data analytics in healthcare department", IEEE International Conference on Infocom Technologies and Unmanned Systems (Trends and Future Directions), Dubai.

21. Loiselle, C.G. and Ahmed, S. (2017), "Is connected health contributing to a healthier population?", *Journal of Medical Internet Research*, Vol. 19, No. 11, p. e386.
22. Oasis (2005), "Security and privacy considerations for the OASIS security assertion markup language (SAML) V2.0", OASIS Standard.
23. Zeadally, S., Isaac, J.T. and Baig, Z. (2016), "Security attacks and solutions in electronic health (E-health) systems", *Journal of Medical Systems*, Vol. 40, No. 12, p. 263.

# 2 IoT-Based Vehicle Monitoring System

*R. Sricharan, E. Karthikeyan, K. Kaamesh,*
*O.V. Gnana Swathika, and V. Berlin Hency*
Vellore Institute of Technology

## CONTENTS

2.1 Introduction ............................................................................................................. 9
2.2 Related Work ......................................................................................................... 10
2.3 Proposed System.................................................................................................... 11
    2.3.1 App Creation through MIT App Inventor ................................................. 11
    2.3.2 Firebase Realtime Database Code for Storing Username and Password for the
    Application........................................................................................................ 12
2.4 Results.................................................................................................................... 12
    2.4.1 Thingspeak Output ................................................................................... 12
    2.4.2 App Outputs.............................................................................................. 13
    2.4.3 Firebase Realtime Database Storage for Username and Password ........... 14
2.5 Conclusion and Future Works................................................................................ 14
References......................................................................................................................... 16

### ABSTRACT

In the current times, vehicles play a major role in our lives. They have become an essential/vital part of our lives. Vehicles help us to reach our destinations on time. Vehicles are used as transport for goods and stocks. Vehicles are used in agriculture, military, healthcare, and in every industry that we think of. So, continuous monitoring of vehicles will help us in the long run. Several parameters can be monitored to improve vehicle performance. Thus, this paper aims to help users to monitor their vehicles for its better performance.

## KEYWORDS

Vehicle Monitoring System; IoT; Performance

## 2.1 INTRODUCTION

The proposed system has the capability to measure the temperature of the engine, detect abnormal vibrations, monitor fuel tank levels and constantly detect for smoke. This system can be used in mostly all cars, since usually these services are given to high-end luxury cars. Without more technological help, these can also be used in bikes and other vehicles. Our study has features like realtime data collection and storage in cloud, anomaly detection like vibration monitoring, monitoring of data from Web-Application Programming Interface (API) like Thingspeak, Monitoring of data from mobile application and analysing old data regarding the usage of the vehicle to indicate the trends.

DOI: 10.1201/9781003374121-2

The objective of our study is to create a simple and efficient system where users can monitor their vehicle's parameters like Vibration, Heat Produced in Engine, Fuel Tank Capacity and Smoke Detection. Users can monitor these parameters through Web-API or An App which shows the parameter values. The data gathered from the sensors are also used for analysis of the user's usage of the vehicle.

## 2.2 RELATED WORK

The discussion begins with an idea which deals with the Global Positioning System (GPS) module which updates the live and destination location. A local database is designed to store the records [1]. To monitor other parameters as well, another paper deals with the GPS/ Global System for Mobile communication (GSM)/General Packet Radio Services (GPRS) module which updates the live location and also monitors the fuel consumption, vehicle speed and engine compartment [2]. Internet of Things (IoT) has made our life easier, as it allows remote access. Therefore, this paper deals with the GPS module which updates the live location. A local database is designed to store the records with the help of cloud computing. In addition, vehicle speed, fuel level and driver's condition are monitored. Remote access with Web Portal feature is discussed in [3]. GSM technology is also used in ideas proposed in this paper which deals with fuel indication systems. With the use of GSM, a message can be sent to the vehicle owner on the status of the fuel tank [4]. Other technologies like the Fuzzy logic system with feedback for fire accidents and timer circuits for the extinguishing process are also proposed [5]. Parameters like speed of vehicle are also important, so this paper also deals with the speed module which updates the live speed in the device. A small local database is designed to store the records for image processing. Vehicle speed is determined using image processing [6].

Vehicles range in various sizes. A previous study deals with the GPS module which updates the live and destination location in the aircraft system. The study concludes that the system should be more accurate as it deals with the real life remotely piloted aircraft system. Error detection and correction option is also present in the GPS system [7]. One idea is IoT networking using NodeMCU and Blynk app. The transfer of sensor data from the microcontroller to the IoT cloud can be done wirelessly using Blynk App in smartphones [8]. Sometimes, parameters become vulnerable, and this topic is discussed in a paper which deals with the GPS module and its accuracy and challenges because of the factors like weather, environment around the vehicle and GPS receiver. The practical implementation of this paper will help in realizing a cheap, effective, reliable system for security [9]. While some proposals are still in the developmental phase, MultEYE is a traffic monitoring system that can detect, track and estimate the velocity of vehicles in a sequence of aerial images [10]. Accident prevention is monitored by using this model. Accidents will be detected immediately with severity level. Whereabouts will be intimated to authorities without any delay. It will also help the user to find and control the stolen vehicles. It will update the parameters like acceleration, fuel and impact sensors in the constructed website [11]. A research study by the author Srinivasan proposes two separate models, one being to develop Bluetooth low-energy technology as the communication module for data transmission to the cloud database, and the other being 4G Dongle that can be used for transmitting data directly to the cloud and the mobile application from Raspberry pi. It further sends data like pressure, location, fuel, etc. to the user [12].

There are different platforms to monitor these parameters, such as Blynk IoT platform which also uses MPX5700AP sensor and Wemos D1 mini microcontroller. A prototype is built to read the tyre pressure, and the pressure data is analysed in the microcontroller. If it is lower than a threshold, a notification is sent to the owner's mobile via the Blynk platform. Other than this, generally, all the data are recorded in the Thingspeak Cloud [13]. Another proposed platform is the Node-RED platform, which is discussed in detail in a paper which mainly provides us insights on how to connect sensors with the Node-RED platform. The sensor data are injected into the Node-RED circuit, and based on the user's choice, the analysed result is viewed either as a display in the dashboard or published as a HTTP out link [14]. Wireless technology is applied so that vehicle owners are able to enter and protect their automobiles with more passive involvement. The GPS receiver on the kit will locate the latitude and longitude of the vehicle using the satellite service. The ignition is also controlled as per user's voice. It is simulated using Proteus software to get a better idea [15]. This concludes the discussion of important aspects and challenges in Vehicle Monitoring and several ideas are analysed in implementing them.

## 2.3   PROPOSED SYSTEM

The design of the proposed system uses sensors to capture the parameters of the vehicle and record the readings continuously and store them in the cloud. The data collection, storage and process are done in real time. The sensors used in this system include temperature sensor (LM35), vibration sensor (SW-420), MQ2 smoke detection sensor and liquid-level depth detection sensor. Threshold values for some parameters are fixed, and if the detected values exceed the threshold values, the anomaly can be seen through the widgets present in the Thingspeak Cloud. Users can also access the mobile application for parameters monitoring and user data analysis.

Figure 2.1 explains the concept of the entire Vehicle Monitoring System. The Arduino UNO collects data from the four sensors, namely vibration, temperature, MQ2 and liquid-level detector that is connected to it. The Arduino UNO sends that collected data to the NodeMCU via serial communication. The NodeMCU sends the received data from the Arduino UNO to a Web-API (Thingspeak). Thingspeak allows us to enable realtime monitoring. A mobile application is created for users/customers for realtime monitoring and implementing automated analysis of the status of the vehicle in the IoT-Based Vehicle Monitoring System.

### 2.3.1   App Creation through MIT App Inventor

**1. Blocks**

Figure 2.2 shows the app being built using MIT App Inventor. The layout for both the login and main page is designed and the block codes are assigned to each component such as buttons, text boxes and labels in their respective layout.

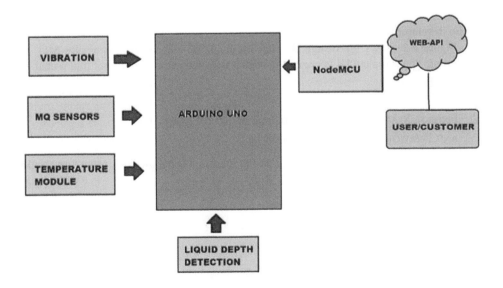

**FIGURE 2.1**   Block diagram consisting of important components.

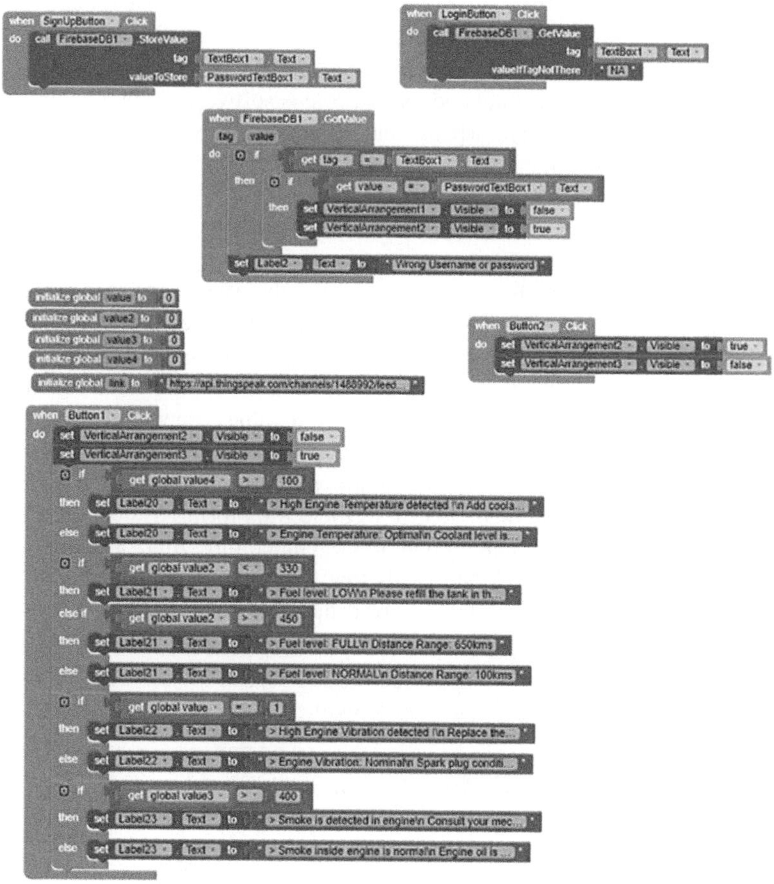

**FIGURE 2.2**    Block codes through MIT App Inventor.

### 2. App development screen

Figure 2.3 shows the design for the layout of the app created using the MIT App Inventor. The complete UI is created in such a way that the user does not find it challenging to operate.

### 2.3.2 FIREBASE REALTIME DATABASE CODE FOR STORING USERNAME AND PASSWORD FOR THE APPLICATION

The read and write rules are changed to true in the Realtime Database for storing and checking of the login credentials as shown in Figure 2.4.

## 2.4 RESULTS

### 2.4.1 THINGSPEAK OUTPUT

Figure 2.5 shows the output of Thingspeak. Here, each field shows us the value obtained from ESP8266 Wi-Fi Module, and some of the sensory devices like temperature sensor, vibration sensor, liquid-level detector and $CO_2$ detector, and the output is displayed both graphically and numerically.

**FIGURE 2.3**    App development screen in MIT App Inventor.

**FIGURE 2.4**    Firebase database.

## 2.4.2 App Outputs

The login page is made user-friendly with a clear description of the text boxes to enter their username and password. The Login and Sign Up button is highlighted to give a clear view. The main page begins with a login status message followed by four sections, Engine temperature, fuel level,

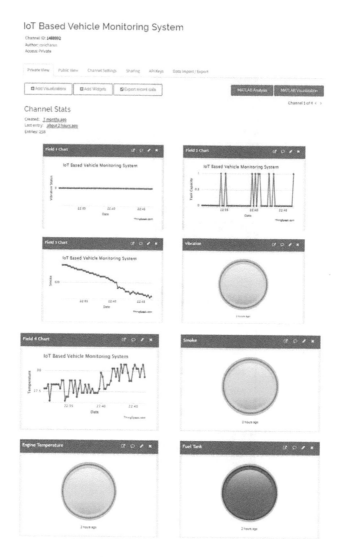

**FIGURE 2.5**   Thingspeak output.

vibration and $CO_2$ emission. The Analysis page gives a detailed explanation of the current status for the vehicle. Figure 2.6 represents the sample layout of the login and main page of the app.

### 2.4.3   Firebase Realtime Database Storage for Username and Password

To do the task of storing for the usernames and passwords, Firebase is used. In the sign up process, the username and password are sent to the Realtime Database in Firebase and stored in the format as shown in Figure 2.7.

## 2.5   CONCLUSION AND FUTURE WORKS

The proposed design of the IoT-Based Vehicle Monitoring System showed promising results on gathering the data of the MQ-2 sensor that has successfully detected any smoke in the car. The liquid-level depth detection sensor has successfully detected the fuel level. The SW-420 sensor has

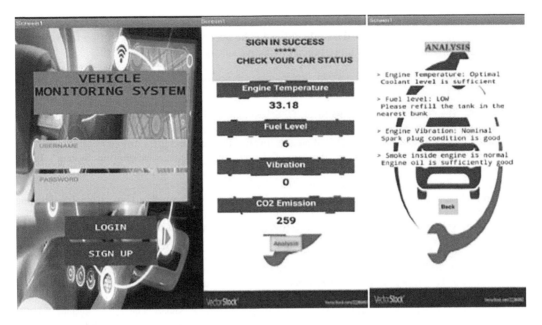

**FIGURE 2.6** App Outputs.

**FIGURE 2.7** Firebase Realtime Database.

successfully detected engine vibration. The LM-32 sensor has successfully gathered the temperature from the engine.

All the gathered data are updated in the Thingspeak website and stored successfully for the analysis using the Vehicle Monitoring System app. The Vehicle Monitoring System app performs all the analyses and displays to the user the current status of their own vehicle successfully.

The future work should include a GPS component that can be added for tracking the car. A GSM-based system can be implemented for sending any alert SMS to the user mobile directly. The database for the storage can be upgraded to premium for good service on the storage of data. More sensor elements can be added to the main system. The accuracy of the data can be improved by updating the current sensors to a more precise/sensitive sensor for better results.

## REFERENCES

[1] Kannan Hemachandran, Shubham Tayal, G. Sai Kumar, Vamshikrishna Boddu, Swathi Mudigonda, and Muralikrishna Emudapuram, "A Technical Paper Review on Vehicle Tracking System", Proceeding of the International Conference on Computer Networks, Big Data and IoT, (2019).

[2] Mayuresh Desai, Arati Phadke, "Internet of Things Based Vehicle Monitoring System", IEEE, Fourteenth International Conference on Wireless and Optical Communications Networks (WOCN), pp. 1–3, (2017), doi: 10.1109/WOCN.2017.8065840.

[3] Dimil Jose, Sanath Prasad, V. G. Sridhar, "Intelligent Vehicle Monitoring Using Global Positioning System and Cloud Computing", *Procedia Computer Science*, Vol. 50, pp. 440–446 (2015).

[4] Rajesh Krishnasamy, Ramkumar Aathi, Booma Jayapalan, K. Karthikeyen, Mohamed Nowfa, "Automatic Fuel Monitoring System", *International Journal of Recent Technology and Engineering (IJRTE)*, Vol. 8, pp. 348–352 (2019).

[5] Kwame O. Robert Sowah, Abdul Ofoli Ampadu, Koudjo Koumadi, Godfrey A. Mills, Joseph Nortey, "Design and Implementation of a Fire Detection and Control System for Automobiles Using Fuzzy Logic", IEEE Industry Applications Society Annual Meeting, pp. 1–8, (2016), doi: 10.1109/IAS.2016.7731880.

[6] Chomtip Pornpanomchai, Kaweepap Kongkittisan, "Vehicle Speed Detection System", IEEE International Conference on Signal and Image Processing Applications, pp. 135–139, (2009), doi: 10.1109/ICSIPA.2009.5478629.

[7] Attila Boer, Marius Luculescu, Luciana Cristea, Sorin Zamfira, "Comparative Study between Global Positioning Systems Used on Remotely Piloted Aircraft Systems", Scientific Research And Education In The Air Force, AFASES, (2016), doi: 10.19062/2247-3173.2016.18.1.16.

[8] Nabilah Mazalan, "Application of Wireless Internet in Networking using NodeMCU and Blynk App", Seminar LIS, (2019).

[9] Sonali Kumari, Simran Ghai, Bharti Kushwaha, "Vehicle and Object Tracking Based on GPS and GSM", *International Journal of Novel Research in Computer Science and Software Engineering*, Vol. 3, No. 1, pp. 255–259 (2016).

[10] Navaneeth Balamuralidhar, Sofia Tilon and Francesco Nex, "MultEYE: Monitoring System for Real-Time Vehicle Detection, Tracking and Speed Estimation from UAV Imagery on Edge-Computing Platforms", *Remote Sensing*, (2021), doi: 10.3390/rs13040573.

[11] S. Kumar Reddy Mallidi, V. V. Vineela, "IoT Based Smart Vehicle Monitoring System", *International Journal of Advanced Research in Computer Science*, Vol. 9, No. 2, pp. 738–741 (2018).

[12] A. Srinivasan, "IoT Cloud Based Real Time Automobile Monitoring System", IEEE, 2018 3rd IEEE International Conference on Intelligent Transportation Engineering (ICITE), (2018).

[13] Jacquline Morlav S. Waworundeng, Deva Fernando Tiwow, Lothar Mark Tulangi, "Air Pressure Detection System on Motorized Vehicle Tires Based on IoT Platform", 2019 1st International Conference on Cybernetics and Intelligent System (ICORIS), pp. 251–256, (2019).

[14] Milica Lekić, Gordana Gardašević, "IoT Sensor Integration to Node-RED Platform", IEEE, 17th International Symposium Infoteh-Jahorina (INFOTEH), pp. 1–5, (2018), doi: 10.1109/INFOTEH.2018.8345544.

[15] Pravada P. Wankhade, S. O. Dahad, "Real Time Vehicle Locking and Tracking System Using GSM and GPS Technology-An Anti-Theft System", *International Journal of Technology and Engineering System (IJTES)*, Vol. 2, No. 3, pp. 272–275 (2011).

# 3 G-GET
## *A Donation App to Reduce Poverty*

*K. Karthik, V. Meenakshi Sundaram, and J. Ranjani*
Sri Sairam Engineering College

## CONTENTS

3.1 Introduction ........................................................................................................ 17
3.2 Literature Review ............................................................................................... 17
3.3 Proposed Solution .............................................................................................. 18
3.4 Working ............................................................................................................... 19
    3.4.1 WANT HELP ............................................................................................ 20
    3.4.2 WANT TO HELP ...................................................................................... 21
    3.4.3 System Architecture ................................................................................ 26
3.5 Technologies Used .............................................................................................. 26
3.6 Future Work ........................................................................................................ 27
3.7 Conclusion .......................................................................................................... 27

### ABSTRACT

Poverty stands as a barrier in one's life to achieving their dream and making life hard. Our application is designed to help the people who are struggling in their life due to poverty and many other issues. This app focuses on forming a bridge between the needy and those who are willing to help, so that both are satisfied. Basic needs of a person like food, medical care and education can be resolved by using this app. The application has two major divisions: WANT HELP and WANT TO HELP. Those who are willing to help others by providing food (e.g. for orphanages), donating blood, donating money, etc. come under WANT TO HELP, and those who get help from the donors come under WANT HELP. The donors and needy can get connected by registering in this app.

## 3.1 INTRODUCTION

India stands in the second position in terms of population. Here, poverty plays a crucial role. Due to poverty, most of Indian population struggle to fulfil their daily needs. Those who suffer from poverty do not get sufficient food. Further, when these people get ill, they do not have enough money to visit a doctor or buy medicine. Such poor people often die due to prolonged illness. Also they cannot afford to do their higher studies or any education-related needs. Due to these conditions, they are unable to achieve their dream or have a peaceful life. These issues can be resolved to an extent through donation.

## 3.2 LITERATURE REVIEW

The current working application is designed specifically for either food donations or education-related donations or health-related donations. None of existing working applications had incorporated all the three options in the same app. In most of the current working applications, donations are made through online portals like crowd-funding applications, which can lead to a number of scams and the people who are helping won't get positive feedback, since there is no real interaction

DOI: 10.1201/9781003374121-3

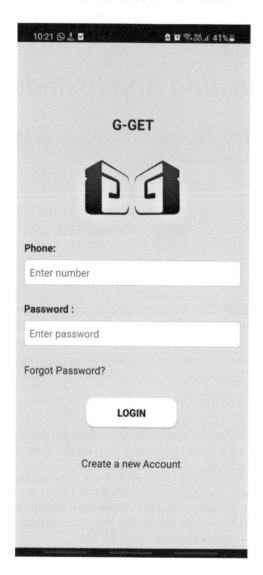

**FIGURE 3.1**   Login page.

between the donor and the needy. These applications didn't have a place where people who get help can show their gratitude to the people who provide help, for example, a page where people who got help can say thanks and rate them or a page where they can give a negative review about the person if he/she tried to scam them. These applications didn't have a rank list page where people can see the top donors. This ranking of donors will make them feel good and make them donate more.

## 3.3   PROPOSED SOLUTION

In our application, all the three donation options are incorporated in a single application, so the user can utilize all the three donations in the same app. In this app, we only display the information about the request raised by the needy and do not entertain payment to be done through the app. So if a donor wants to help, he/she has to go in person and donate or contact him with the details specified and get his/her account information and donate through online mode. By this method many

**FIGURE 3.2**    Choosing page.

scams can be avoided and the donor can get positive feedback on his donation. In this application, the receiver can give a feedback or write a review about the donor and can rate for his/her donation. This app also includes a page displaying donor rank list where different donors can view their current position in the rank list based on the number of donations and also based on the ratings given.

## 3.4  WORKING

After opening the application, the login page will be displayed as shown in Figure 3.1.

If you're already registered, then you can enter your details and login or if you're a new user, you have to click the "create new account" and register. In the choosing page, there will be two tabs "WANT HELP" and "WANT TO HELP". If you want help from others (needy), you have to click "WANT HELP" and register. If you're willing to help others, then you have to click the "WANT TO HELP" tab and register as shown in Figure 3.2.

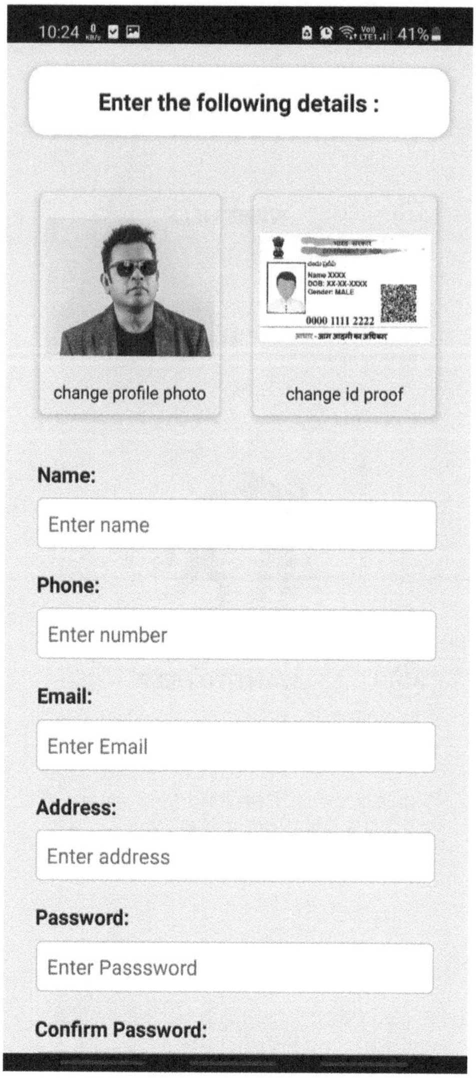

**FIGURE 3.3** Registration page.

### 3.4.1 WANT HELP

After choosing, you have to fill in your details and upload your profile photo and ID proof as shown in Figure 3.3.

After registration, the home page will be displayed as shown in Figure 3.4.

In the home page, three help tabs and a thanksgiving tab will be displayed. In the top left corner, a menu bar will be displayed; if you click on the menu bar, your profile and previous requests will be displayed as shown in Figure 3.5.

In the home page out of the three tabs, you can choose one in which you want help. After clicking on the tab, you have to fill in the details asked and place a request as shown in Figure 3.6. In the thanksgiving tab, you can give feedback about the donor and rate him/her by giving your rating.

**FIGURE 3.4**   Homepage of WANT HELP.

### 3.4.2   WANT TO HELP

After choosing and filling your details same as in the "WANT HELP" tab, the home page will be displayed as shown in Figure 3.7.

In the home page, three help tabs and a ranklist tab will be displayed. In the top left corner, a menu bar will be displayed; if you click on the menu bar, your profile and previous donations will be displayed.

In the home page, out of the three tabs, you can choose one in which you want to help. After clicking on the tab, all the requests raised by the needy will be displayed and you can choose one by seeing their details as shown in Figure 3.8.

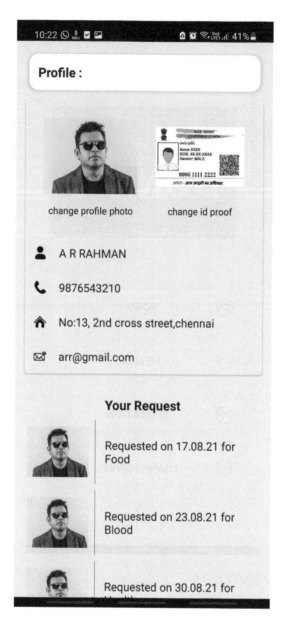

**FIGURE 3.5**   Profile page.

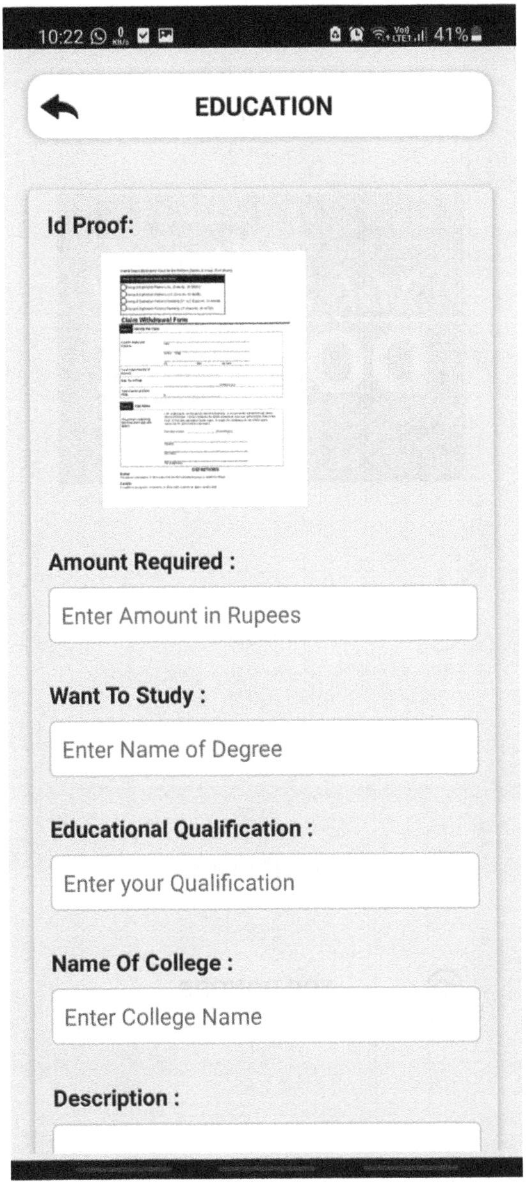

**FIGURE 3.6**   Requesting page.

**FIGURE 3.7**　Homepage of WANT TO HELP.

**FIGURE 3.8** Request list.

**FIGURE 3.9**  Receiver details.

After choosing the request which you wish to help, you can contact the person with the details specified as shown in Figure 3.9. On the ranklist tab you can see your rank, top donors and their details in the list.

### 3.4.3  System Architecture

The working of the application is shown in Figure 3.10.

## 3.5  TECHNOLOGIES USED

A. **Android studio**: For developing the front end of the application using xml and developing the backend using java.
B. **Figma**: For designing the front end of the application.
C. **Firebase**: A database to store user details and information of the requests raised by needy.

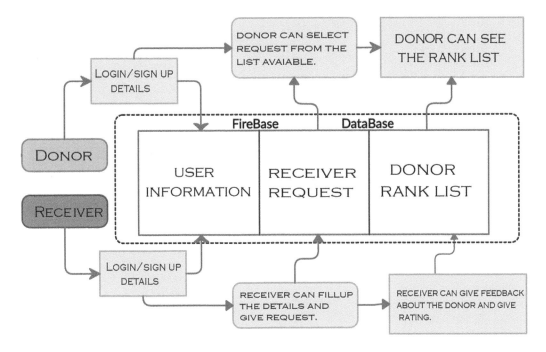

**FIGURE 3.10** Application System architecture.

## 3.6 FUTURE WORK

1. The requests in the "WANT TO HELP" tab can be sorted using map API, so that people can filter the requests to find a nearby location like in the case of food donation.
2. Integrating AI software that checks the user details and proofs and verifies by itself, so that fake profiles can be identified.

## 3.7 CONCLUSION

In this fast moving world, no one has time to notice the struggles of the poor people. It's our responsibility to make everyone lead a happy life by helping them and spreading positivity with the people around us. Till now, there is no application which has all the mentioned three donation options integrated in a single app that also enables users to give ratings. Through our app, the discussed problems can be overcome and donation can be made easier.

# 4 Enhanced *K*-Means with Automated *k* Estimation and Outlier Detection Using Auto-Encoder Neural Network

*P. Illavenil, J. Kiron, S. Ramnath, and G. Hannah Grace*
Vellore Institute of Technology

## CONTENTS

4.1 Introduction: Background and Driving Forces ................................................................29
    4.1.1 Within-Cluster Sum of Squares ..............................................................................30
    4.1.2 The Traditional "Elbow" Method ...........................................................................30
    4.1.3 Auto-Encoders .........................................................................................................30
    4.1.4 Silhouette Score .......................................................................................................30
4.2 Related Work ......................................................................................................................31
4.3 Proposed Algorithm ..........................................................................................................32
    4.3.1 Naïve-Sharding Initialization ..................................................................................33
    4.3.2 Automating the Process of Detecting Optimal *k*-Value ........................................34
    4.3.3 Noise Detection Using Auto-Encoders ...................................................................34
        4.3.3.1 Auto-Encoders .........................................................................................34
        4.3.3.2 Anomaly Detection ..................................................................................35
4.4 Experiment Results ............................................................................................................35
4.5 Conclusion .........................................................................................................................38
References ..................................................................................................................................39

**ABSTRACT**

Given a data set with vast amounts of data, it can become complicated to study unstructured data. The clustering technique helps in finding structure and groups a set of data in such a way that the items in that group are more comparable to each other than the other elements. One such clustering method is *K*-means, which maximizes inter-cluster distance while minimizing intra-cluster distance. Still, the major drawback of this algorithm is the process of choosing the right *K*-value and its inability to detect noise. To overcome the disadvantages mentioned above, this chapter discusses various techniques—the use of naïve-sharding centroid initialization to detect the initial centroids, which improves the efficiency of the method drastically. A more effective elbow method has been proposed, which helps in identifying the optimal number of clusters. Finally, anomaly detection using auto-encoder neural networks is used to filter out outliers/noise. Recent research demonstrates that forecasting techniques using *k*-means clustering algorithms are used to boost the final forecast due to the increased integration of renewable energy resources into the energy grid.

## 4.1 INTRODUCTION: BACKGROUND AND DRIVING FORCES

In today's scenario, the world is driven by information, which makes it essential to make informed decisions. Many companies now understand the value of extracting relevant information from their

DOI: 10.1201/9781003374121-4

vast data. Clustering is the grouping of similar data into a cluster/group. These clusters of data are identical to each other in one way or another, and this relevance can be used to study them. There are many clustering algorithms like Mean-Shift Clustering, DBSCAN [1], Agglomerative Hierarchical Clustering, etc. But $K$-means clustering is commonly used. This algorithm iteratively creates a "$K$" number of clusters with data in each cluster similar to one another on various attributes. Although famous, the algorithm has its fair share of flaws like inefficient initialization of centroids, etc. This chapter proposes some techniques to curtail those flaws with the help of Jupyter notebook, Python, and modules like sklearn, TensorFlow, Keras, and matplotlib. Our proposed algorithm discussed an enhanced "Elbow" method that automates the process of detecting $K$ and naïve-sharding strategy to initialize centroids [2]. Finally, anomaly detection using auto-encoder neural networks [3] has been implemented to remove unnecessary data points which do not belong to any cluster, which can make the above process incredibly efficient.

### 4.1.1 WITHIN-CLUSTER SUM OF SQUARES

Within-cluster sum of squares (WCSS) measure is developed within the ANOVA (Analysis of variance) framework. The distance between points in a cluster from its cluster centroids is measured as a sum of squares. Minimizing the WCSS (approx. 0) means that each cluster has precisely one data point, On the other hand, maximizing the WCSS implies that all data points are in the same cluster, which is of no use. We require our WCSS to be low value but not the minimum amount.

To calculate the WCSS, we need to find Euclidian distance of a point and its respective cluster centroid. Repeat the process for all the cluster points, estimate the sum, and divide by the number of points in the cluster. Repeat the steps for all clusters and find their average. The resulting value gives the average WCSS. The hybrid wind power forecasting method can also be used [4], where the wind power training data uses $K$-means with different initializations. The accuracy can also be evaluated using Root Mean Square Error (RMSE), mean absolute error (MAE) and relative mean absolute error (rMAE) when time series analysis is used.

### 4.1.2 THE TRADITIONAL "ELBOW" METHOD

Choosing the optimal hyper-parameter $k$ for $K$-means is one of its top disadvantages. To find the optimal $k$-value, WCSS (y-axis) vs. no of clusters (x-axis, $K = [1, N]$) is plotted. We choose the $k$-value at the "elbow" of the graph, i.e., the point after which the Inertia starts decreasing linearly or, in other words, the point where the WCSS is middle ground (low but not minimum).

Thus, by visualizing the plot, we can identify the "elbow" point and decide the value of $k$. It isn't logical and efficient to visualize every scenario and then determine the value of $k$, as it might be ambiguous for a lot of cases where our naked eye cannot process or estimate the optimal $K$.

### 4.1.3 AUTO-ENCODERS

Auto-encoder neural network is an unsupervised learning technique used for dimensionality reduction, outlier detection, etc.

### 4.1.4 SILHOUETTE SCORE

It is a method of validation of consistency within the cluster set. It is a measure of closeness of intra-cluster points compared with the separation of inter-cluster points. The silhouette score ranges from −1 to 1, a higher value of silhouette score implies that the data points within the cluster have high similarity, and similarity is weak when compared with other cluster data points.

The silhouette coefficient $S(i)$:

$$S(i) = \left( b(i) - a(i) \right) / \left( \max \left\{ \left( a(i), b(i) \right) \right\} \right), \tag{4.1}$$

where $a(i)$ represents the average dissimilarity of the $i$th data point with respect to all other data points in the cluster, and $b(i)$ represents the average dissimilarity of the $i$th data point with respect to all data points in the neighbouring cluster.

## 4.2 RELATED WORK

Ashutosh Mahesh Pednekar [2] proposed a method which improves the efficiency of the *K*-means algorithm by reducing the number of iterations by around six times by initializing the centroids using the Particle Swarm Optimization (PSO) algorithm. Without the PSO, the centroids are assigned randomly, and it takes 24 iterations to converge in the test case that the paper shows finally, whereas if PSO is applied as proposed by Ashutosh Mahesh Padnekar, the clustering converges in just six iterations, thus rendering it more accurate and useful.

Krishna, B. L., P. Jhansi Lakshmi, and P. Satya Prakash [5] proposed an algorithm to solve the disadvantages of both *K*-means clustering and DBSCAN. This is achieved by combining both the algorithms of *K*-means and DBSCAN to solve each other's problems. First, DBSCAN is implemented to the data which clusters the data sets into m groups, and then *K*-means is applied with $k$ as input. If $k > m$, the groups are partitioned into $k$ groups, and if $k < m$, the groups are combined to form $k$ groups. Now, the drawbacks of both algorithms are solved. For example, clusters formed by DBSCAN may overlap each other, which is undesired. By combining the *K*-means algorithm, the clusters no longer overlap.

A proposed approach on *K*-value selection techniques of the *K*-means clustering algorithm was made by Yuan, Chunhui, and Haitao Yang [6]. The Elbow Method, Gap Statistic, Silhouette Coefficient, and Canopy *K*-value selection methods are examined, along with experimental verification. To cluster the Iris data set and determine the *K*-value and clustering outcome of the data set, these four types of *K*-value selection algorithms are utilized. The experimental results include the obtained *K*-value along with the execution time of each algorithm, which then compared to meet the requirements.

The *K*-means algorithm is described in depth by Manoharan, J. James and Dr. S. Hari Ganesh [7], where the ultimate clustering outcome of the process heavily depends upon the starting centroids. It claims that it is important to choose the appropriate number of clusters in conventional *K*-means clustering. The majority of prior methods required inputs like threshold settings for the quantity of data points in a data set. In order to address the issues of locating initial centroids and assigning data items to appropriate clusters, this paper suggests the "divide and conquer" strategy. Initially, the cluster centres are obtained, followed by *K*-means to give optimal cluster centres. This method aims to increase the execution speed and also to reduce the complexity.

Xie, Yiqun, and Shashi Shekhar [1] proposed an algorithm incorporating statistical significance in DBSCAN clustering. Since DBSCAN forms clusters with spurious patterns, individual companies don't find it useful. To address this issue, spatial scan statistics are used with DBSCAN. The study also proposes another method called baseline Monte Carlo method to see the importance of clusters and a Dual-Convergence algorithm to speed-up the computation. Cluster detection of varying cluster densities has also been discussed. The improved algorithm eliminates chance patterns, thus giving better solutions. These tweaks make the algorithm more productive and reduce execution time.

Syakur, M. A. [8] deals with the combination of the *K*-means method and the elbow method to identify the best cluster of customer profiles. The study uses the elbow method to determine the number of clusters that produce the same amount of clusters *K* on different data. Initially, customer segmentation is carried out by performing clustering analysis so that each segment can contain customers with similar characteristics. Then, the *K*-means clustering algorithm partitions the data into clusters. Optimal value of *K* is determined using the above-mentioned best technique.

Lithio A and Maitra R [9] proposed a method to solve the clustering of data sets that have incomplete records. Traditional $K$-means clustering wouldn't be able to cluster those data sets. The Hartigan–Wong $K$-means clustering technique has been implemented for the data sets that have incomplete records. The paper discusses the modifications to the initialization method. To estimate the number of clusters, the jump statistic method has been used. The test results, when examined with astronomical data, showed promising outcomes in line with expected results.

Karami, Amin, and Manel Guerrero-Zapata [10] proposed to secure private content more effectively by using an anomaly detection system. Although traditional $K$-means is a famous anomaly detection method, it is sensitive to the initialization of centroids. Hence, a novel fuzzy anomaly detection system is proposed. This system has two phases—a training phase where $K$-means and PSO hybridization happens to find the number of well-distanced clusters and their centroids, and a detection phase where a fuzzy approach is employed by combining two distance-based methods. This proposed detection system seems to create well-separated clusters while also decreasing anomalies greatly.

The paper by Lyudchik and Olga [11] aims to detect outliers (data points that deviate from other points' behaviour so much that they become undesirable) from a set of observations. This is implemented by using deep auto-encoder which is an artificial neural network that consists of encoding and decoding networks. Reference [12] makes use of the MNIST data set delivered by Keras which contains 60,000 digits from 0 to 9. Although there were fluctuations in the outcome when compared to the expected results, the proposed technique seems promising. It might give better results upon retraining the models several more times.

## 4.3  PROPOSED ALGORITHM

The idea behind our proposed algorithm tackles the significant drawbacks of the existing $K$-means algorithm, both efficiently and effectively. Our algorithm as shown in Figure 4.1 consists of the combination of three key steps: centroid initialization using the naïve-sharding technique, finding the optimal $k$-value automatically using a proposed method that improves the existing elbow method, and finally filtering out the noise/outliers which are detected using the auto-encoder neural networks. The upcoming divisions will elaborate on the above-mentioned three-step process.

**Algorithm**:

1. $K=[1,N]$
2. For $k$ in range of $K$:
   i.   Cluster_centroids=naïve_sharding(dataframe, $k$).
   ii.  Perform $K$-means iteratively for each $k$-value ranging from 1 to $N$ by initializing the cluster centres with cluster centroids calculated in (i).
   iii. Calculate the WCSS also known as inertia.
        End for loop.
3. Construct a graph with $K$ as the $x$-axis and WCSS as the $y$-axis.
4. Join the first point and the last point of the $K$ values.
5. Calculate the perpendicular distance for each point of elbow curve with the line that connects the first and last $K$ values.
6. The point with the maximum distance from the elbow curve to the line constructed in Step 3 is the optimum $k$-value.
7. Perform $K$-means for the estimated optimal $k$-value and customized cluster centroid corresponding to the optimal $k$.
8. Split each cluster and store each in an array.
9. Perform outlier detection using auto-encoders neural networks for each cluster.
10. Combine the filtered clusters.

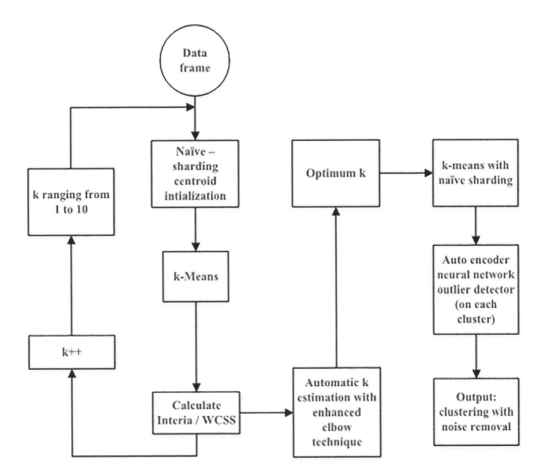

**FIGURE 4.1**    Flowchart of the proposed algorithm.

### 4.3.1 Naïve-Sharding Initialization

Cluster initialization is one of the significant fallbacks of the *K*-means algorithm. The existing *K*-means uses random centroid initialization; thus, a weak starting point could shoot up the number of iterations required for convergence, increasing the overall run-time, resulting in a less-efficient method. The naïve-sharding technique is a straightforward, efficient, and time-saving method to initialize centroids that don't depend on random initialization. The principal objective of this algorithm is to initialize cluster centres close to the actual optimal centres.

The procedure to find the optimal cluster centre is quite direct; it mainly depends on calculating the composite sum of each attribute value of a particular instance, and the data set is sorted for the calculated composite sum. The sorted data set is split into *k* shards/pieces. At last, the original attributes of each shard are summed up, and their mean is computed. The calculated mean values are the centroids used for initialization.

**Algorithm**:

1. Calculating the attribute sum for each row of the data set.
2. Sort the data set with respect to the attribute sum.
3. Horizontally cut the sorted data set into *k* equal shards.
4. For each shard, sum the original attribute column and compute its mean value.

5. For each shard, the calculated mean value of each attribute is the coordinates of the cluster centroids.

### 4.3.2 Automating the Process of Detecting Optimal *k*-Value

A simple tweak in the WCSS vs. no. of clusters graph mentioned in Section 2.2 and a deterministic distance calculation is all that requires to automate the whole process.

**Algorithm**:

1. Connect the two endpoints of the "elbow graph" and connect the points ($k=1$, WCSS OF $K=1$) and ($k=N-1$, WCSS of $K=N-1$).
2. Using the distance measure, determine the distance of the point from the line drawn in Step 1 for all values of $K$. The distance from a line to a point ($x0$, $y0$) is given in the situation of a line in the plane that is given by the equation $ax+by+c=0$, where $a$, $b$, and $c$ are real constants and $a$ and $b$ are not both zero as shown in Figure 4.2.

$$\text{distance}\left(ax + by + c = 0, \left(x_0, y_0\right)\right) = \frac{\left|ax_0 + by_0 + c\right|}{\sqrt{a^2 + b^2}} \tag{4.2}$$

The point on this line which is closest to ($x_0, y_0$) has coordinates:

$$x = \frac{b\left(bx_0 - ay_0\right) - ac}{a^2 + b^2}. \tag{4.3}$$

$$y = \frac{a\left(-bx_0 + ay_0\right) - bc}{a^2 + b^2}. \tag{4.4}$$

3. Plot a graph with the distance of the point from the line as the $Y$ coordinate and no. of clusters as the $X$ coordinate.
4. Find the maximum distance using the max function.
5. The *k*-value with the maximum distance is the line which has the optimal value of $K$. 4.

### 4.3.3 Noise Detection Using Auto-Encoders

#### 4.3.3.1 Auto-Encoders

Auto-encoders follow a three-step methodology: first collects the input, encodes the information into a compressed format, and finally, the compressed version is reconstructed—evaluation of the auto-encoder based on closeness, the reconstructed value, and the inputs.

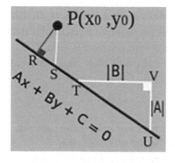

**FIGURE 4.2**   Finding distance.

#### 4.3.3.2   Anomaly Detection

There are several machine learning algorithms developed to detect outliers such as Isolation Forest, elliptical envelopes, one-class support vector machines (SVM), etc. But the reason we choose deep learning and auto-encoders to identify anomalies is due to the ability to perform better on scalable/large data sets, and its improved analysis of unstructured data.

The auto-encoder uses two fundamental components, namely the encoder and the decoder. The encoder accepts the input data and compresses it, and the decoder reconstructs the compressed data. When training the auto-encoder, we measure the mean squared error (MSE) between the input and the reconstructed data. The MSE is a statistical measure of deviation from the estimated to the actual value.

$$\text{MSE} = \frac{1}{n}\Sigma\left(y - \hat{y}\right)^2, \tag{4.5}$$

where $\left(y - \hat{y}\right)^2$ is the square of the difference between actual and predicted.

0.99 quantile of the normally distributed MSE values is set as the threshold. Therefore, If MSE is higher than a set threshold value, the data point is considered an outlier.

**Algorithm**:

1. Perform *K*-means with the estimated *k*-value.
2. Split all the clusters and store them in a separate array.
3. Build an auto-encoder function using the Keras module.
4. For each cluster, model the training set of the respective cluster with auto-encoder function.
5. For each cluster, estimate the decoded value with the test set using the auto-encoder's predict function and calculate the reconstruction error.
6. For each cluster, compare the reconstruction error/MSE with the calculated threshold.
7. For each cluster, if the corresponding MSE is greater than the threshold, consider the point as an outlier, and filter it out.
8. Combine all the filtered clusters.

### 4.4   EXPERIMENT RESULTS

Here are a few snapshots illustrating the implementation of our proposed method. Our proposed algorithm is tested using the random make-blob data set (generates isotropic Gaussian blobs for clustering) using the sci-kit learn library as shown in Figures 4.3 and 4.4. On comparing the time

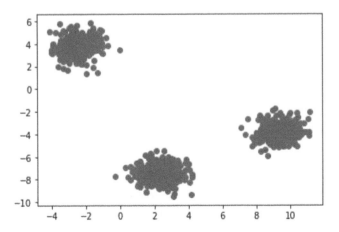

**FIGURE 4.3**   Thousand samples clustered into three sets.

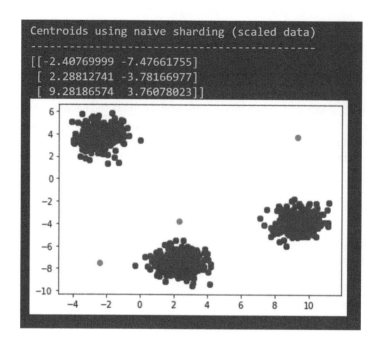

**FIGURE 4.4**   Initialization of centroid position with the aid of naïve-sharding technique.

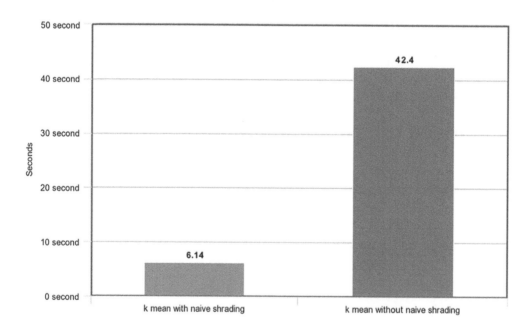

**FIGURE 4.5**   Time comparison chart of *K*-means with and without naïve-sharding.

elapsed to converge for ten iterations of *K*-means between random initialization and naïve-sharded initialization notifies us about the drastic increase in efficiency when we use naïve-sharded technique as seen in Figure 4.5. Figure 4.6 shows the plot between the WCSS and number of clusters to find the optimal *k*-value. Figure 4.7 is the plot between the distance of points from the line in Figure 4.8 and the number of clusters. The optimal *k*-value is the point with maximum distance

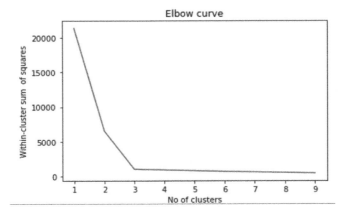

**FIGURE 4.6**    Elbow curve.

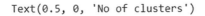

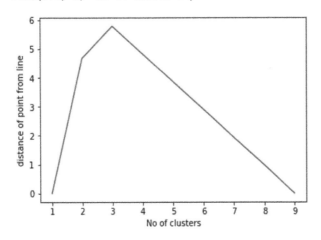

**FIGURE 4.7**    Distance of points from line.

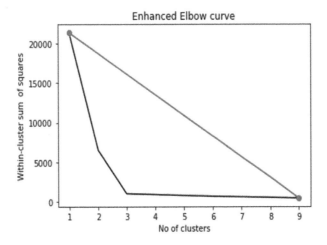

**FIGURE 4.8**    Automation process curve of *k*-detection.

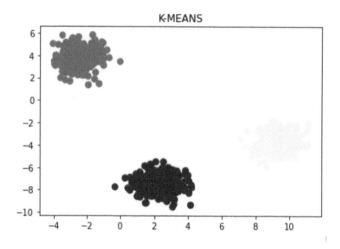

**FIGURE 4.9**   Performing $K$-means for optimal $K$ with naïve-sharding.

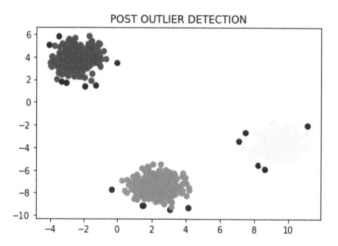

**FIGURE 4.10**   K-means after detecting outliers using auto-encoders (black dots represent outliers).

from the line. The optimal value of $k$ is 3. In Figure 4.9, we can see output of $K$-means with naïve-sharding, and in Figure 4.10, the black dots represent outliers.

Finally, we validate our test results using the average silhouette score, which can be seen clearly from Figures 4.11 and 4.12.

## 4.5   CONCLUSION

In this chapter, three significant techniques (Naïve-Sharding Initialization, Automated Elbow Method, and Anomaly Detection Using Auto-encoder Neural Networks) have been proposed, which eradicates all the disadvantages of $K$-means clustering. While it took 42.4 seconds just to initialize the centroids without naïve-sharding method, using it reduced the initialization time to 6.14 seconds. Thus, the naïve-sharding process proves to be one of the effective ways to initialize centroids. Finding the optimal $k$-value is usually done by the elbow method. But the $k$-value has to be assigned by the user after visualizing the elbow curve. Automating that process makes the clustering algorithm more user-friendly and solves the issue of occurrence of any human error. Removing any

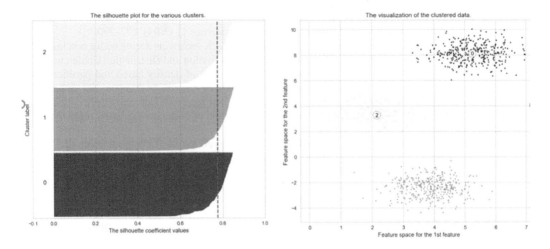

**FIGURE 4.11**    Silhouette analysis for *K*-means clustering on sample data.

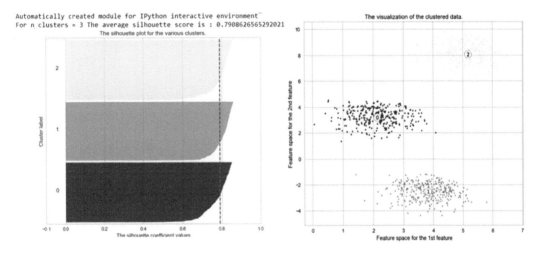

**FIGURE 4.12**    Silhouette analysis for enhanced *K*-means clustering on sample data with *n*_clusters = 3.

outliers while analysing data is a definite advantage and the paper proposes to use auto-encoders to do the same. As shown in Figure 4.10, the anomalies have been effectively detected and removed from the clusters. Upon performing the silhouette analysis, it's clearly seen that the proposed method increases the validity of the algorithm significantly. These techniques prove to be vital while analysing large sets of data, since it's more efficient than the traditional *K*-means algorithm. As a future work, wind power data sets with diverse characteristics from different wind farms located in the United States will be used to determine the accuracy of renewable sources where the proposed enhanced *k*-means based on different initialization techniques will be adopted.

## REFERENCES

[1] Xie, Y., & Shekhar, S. "Significant DBSCAN towards Statistically Robust Clustering." Proceedings of the 16th International Symposium on Spatial and Temporal Databases (2019).
[2] Pednekar, A. M. "Optimal initialization of K-means using Particle Swarm Optimization." arXiv preprint arXiv:1904.09098 (2019).

[3] Delavar, A. G., & Mohebpour, G. H. "ANR: An algorithm to recommend initial cluster centers for k-means algorithm." *Journal of Mathematics and Computer Science* 11 (2014): 277–290.

[4] Ghofrani, M., de Rezende, M., Azimi, R., & Ghayekhloo, M. "K-means clustering with a new initialization approach for wind power forecasting." IEEE/PES Transmission and Distribution Conferences.

[5] Krishna, B. L., Lakshmi, P. J., & Prakash, P. S. "Combination of density based and partition based clustering algorithm DBK means." *International Journal of Computer Science and Information Technologies* 3.1 (2012): 4491–4494.

[6] Yuan, C., & Yang, H. "Research on K-value selection method of K-means clustering algorithm." *J—Multidisciplinary Scientific Journal* 2.2 (2019): 226–235.

[7] Manoharan, J. J., & Ganesh, S. H. "Initialization of optimized K-means centroids using divide-and-conquer method." *ARPN Journal of Engineering and Applied Sciences* 11.2 (2016): 1086–1091.

[8] Syakur, M. A., Khotimah, B. K., Rochman, E. M. S., & Satoto, B. D. "Integration K-means clustering method and elbow method for identification of the best customer profile cluster." IOP Conference Series: Materials Science and Engineering. Vol. 336. No. 1. IOP Publishing, 2018.

[9] Lithio A, & Maitra, R. "An efficient k-means-type algorithm for clustering data-sets with incomplete records." *Statistical Analysis and Data Mining: The ASA Data Science Journal* 11 (2018): 296–311.

[10] Karami, A., & Guerrero-Zapata, M. "A fuzzy anomaly detection system based on hybrid PSO-Kmeans algorithm in content-centric networks." *Neurocomputing* 149 (2015): 1253–1269.

[11] Lyudchik, O. "Outlier detection using autoencoders." No. CERN-STUDENTS-Note-2016-079. 2016.

[12] Chen, J., Sathe, S., Aggarwal, C., & Turaga, D. "Outlier detection with autoencoder ensembles." Proceedings of the 2017 SIAM International Conference on Data Mining. Society for Industrial and Applied Mathematics, 2017.

# 5 LPG Leakage Detector

*Zaina Nasreen, D. Subbulekshmi,*
*S. Angalaeswari, and T. Deepa*
Vellore Institute of Technology

## CONTENTS

5.1 Introduction ..................................................................................................... 41
5.2 Motivation........................................................................................................ 42
5.3 Description....................................................................................................... 43
5.4 Applications .................................................................................................... 43
5.5 Circuit Diagram ............................................................................................. 43
5.6 Components Explanation.............................................................................. 44
5.7 Working Principle of Our Study ................................................................. 47
5.8 Code................................................................................................................. 47
5.9 Result and Inference ..................................................................................... 48
5.10 Conclusion ...................................................................................................... 48
5.11 Future Scope................................................................................................... 49
Bibliography ............................................................................................................ 49

**ABSTRACT**

The primary goal of this research is to create a system that can detect gas leaks. The purpose of this system is to detect the presence of liquefied petroleum gas (LPG). When a leak is detected, the gadget should also take the appropriate actions, such as sounding an alarm and shutting off the valve automatically.

Petroleum liquid is the most common type of fuel abbreviated as LPG. It releases less pollution than firewood and charcoal when consumed and hence is considered as clean fuel. Gas leakage has been hazardous to both industrial and residential areas. Because of the rising number of accidents caused by gas leaks, home security has become a serious concern. To prevent gas leakage-related accidents, we plan to develop a gas leakage detecting system that will sound an alarm when a leak is discovered. This system includes a buzzer which will alert the users and a sensor which can efficiently detect the gas.

LPG is highly inflammable and can cause serious injuries. Accidents in residential area and home fires are taking place frequently, which is increasing the threat to human lives and property. Some common causes of fire incidents include poor rubber quality in the tube or the regulator not being turned off after usage. As a result, there is a pressing need to design a reliable gas leak detecting system. As a result, this study demonstrates an autonomous gas leak detection system with an alerting mechanism.

Large gas leakage detection systems are installed in industrial area for monitoring, especially in oil rigs, paper-manufacturing factories, etc.

This device can provide a much efficient step towards the protection and safety of people in residence where least protection is available.

## 5.1 INTRODUCTION

A serious problem which is observed widely in recent times is gas leakage. It goes without saying that gas leaks result in dangerous incidents. Gas sensors are utilized in a variety of applications, including safety, health, and instrumentation. Alarms for explosive or hazardous gases can be found

DOI: 10.1201/9781003374121-5

in both household and commercial settings, as well as in automotive applications. Gas sensors like these are often used in air quality control systems.

Liquefied petroleum gas (LPG) is a combustible mixture of hydrocarbon gases used as a fuel in a wide range of items, including homes, hotels, factories, automobiles, and trucks. It's because of its beneficial properties, which include less smoke, reduced soot, and a lower environmental impact, as well as its high calorific value. LPG is highly flammable and can spark a fire even if the source of the leak is far away. This is owing to the fact that it contains highly flammable chemical components such as propane and butane. In most homes, LPG is mostly used for cooking. If there is a leak, the released gases could create an explosion. Gas leaks can result in both property and human injury. Explosion, fire, and asphyxia hazards are determined by the physical properties of gas, such as toxicity, flammability, and suffocation. A gas leak can result in a catastrophic incident, such as the Bhopal gas disaster. Such explosions are caused by old valves, inferior cylinders, a lack of frequent checking of gas cylinders, worn-out regulators, and a lack of awareness of how to handle gas cylinders. As a result, gas leaks must be identified and managed in order to safeguard individuals from harm. LPG contains an odorant, such as ethanethiol, which allows most people to identify leaks quickly. Nevertheless, it's not like a keen sense of smell will be able to rely on this safety feature, which is where gas sensors come in handy. Different gas-detecting technologies are employed in the current technique. This chapter presents and analyses a low-cost gas leakage detection, warning, and control system based on sensors. The system is simple to operate, transportable, dependable, and inexpensive. It will cost only 950 Indian rupees and the one available in the market costs around 2,000–3,500 Indian rupees. The purpose of this study is to find leaks of LPG/Compressed Natural Gas (CNG) in household appliances. If a leak is detected, the device will sound an alarm.

## 5.2  MOTIVATION

LPG is extremely flammable and can catch fire even if the source of the leak is far away. A malfunctioning rubber tube or a regulator that is not turned off while not in use is the most common cause of fires. As a result, developing a gas leak detection system is essential. As a result of this research, a gas leakage alarm system has been developed, which can detect gas leaks and alert passengers on board.

According to official figures, 1,500–2,000 people die each year as a result of LPG gas leaks in homes, with many more severely affected mainly in India.

Early-warning devices, such as gas detectors, are an important part of most industries' safety plans for reducing risks to workers and equipment. These can help give you additional time to take corrective or preventative action. They can also be employed as part of an industrial plant's overall monitoring and safety system. The rapid expansion of the oil and gas industry has resulted in major and dangerous gas leakage accidents. Gas leaks result in a financial loss, so solutions must be found at the very least to limit the effects of these accidents. The challenges involve not just creating a prototype of the device that can detect leakage but also responding to it automatically when it occurs. When it is not adequately monitored, such an accident may occur. LPG is used as one of the heating options in four-season nations like Russia to keep inhabitants' homes warm throughout the winter. At the time of the explosion, no one was in the house. "LPG is also one of the gases that is difficult to detect due to human sense limitations. Cook claimed that natural gas caused the explosion, but he was unable to detect any gas smells since he had lost his sense of smell due to a working mishap. When the house blew up, he was on his way to pick up his daughter from school".

The condition of pipeline systems deteriorates over time. Corrosion speeds up over time, and long-term corrosion raises the risk of failure (fatigue cracking). Regularly monitoring only the "junk" section of pipelines leads to a pipeline system with uncertain integrity. The level of trust in honesty will fall below acceptable limits. Inspection of the pipeline system's currently uninspected parts becomes necessary. The purpose of this chapter is to provide information about "gas leak detection".

## 5.3 DESCRIPTION

A gasoline detector monitors the presence of gases in a specific area and is typically used as part of a safety equipment. This system can detect a gas leak or other pollution and communicate with a control unit to automatically turn off a procedure. Operators in the region where the leak is occurring can be alerted by a gasoline detector, allowing them to evacuate. Many gases can be harmful to organic living, including humans and animals; hence, this technology is necessary.

The technology of detecting potentially harmful gas leaks with the help of sensors is known as gasoline leak detection. When a toxic fuel is detected, these sensors usually emit an audible warning to warn people. Workers may be exposed to harmful gases while painting, fumigating, gas filling, creating, excavating contaminated soils, landfill operations, entering prohibited areas, and other tasks. Common sensors include combustible gas sensors, photoionization detectors, ultrasonic sensors, electrochemical gas sensors, and semiconductor sensors. These days, infrared imaging sensors are being employed more frequently. These sensors can be found in a variety of places, including commercial enterprises, refineries, pharmaceutical manufacturing, fumigation facilities, aviation and shipbuilding facilities, and more.

This chapter demonstrates how a microcontroller can update a large number of external components while also adding functionality at a cost equivalent to a simple integrated comparator. The prototype's hardware and microprocessor firmware have been optimized to impose a sophisticated LPG gasoline alert for automobiles running on LPG/CNG, so that it can raise alarm before any fatal catastrophe occurs.

It is made up of additives that are easily available on the market, and the circuit production is similarly simple.

To detect LPG gasoline, an LPG gasoline sensor module is utilized. When it detects LPG gasoline leakage, it sends an excessive pulse to its DO pin, which Arduino constantly reads.

When the LPG gasoline sensor module generates an excessive pulse, Arduino displays the "LPG fuel Leakage Alert" message on a 16×2 liquid crystal display and activates the buzzer, which sounds repeatedly until the gas detector module detects no gas in the vicinity.

## 5.4 APPLICATIONS

a. Protection from any gas leakage in residences.
b. For safety from gas leakage in heating gas-fired appliances like boilers, domestic water heaters, etc.
c. Used in small-scale industries which require gas for their production.
d. For safety from gas leakage in cooking gas-fired appliances like ovens, stoves, etc.
e. Protection from any gas leakage in vehicles.
f. It also detects alcohol, so it is used as liquor tester.
g. The sensor has excellent sensitivity combined with a quick fast response time.
h. The system is highly reliable, tamper-proof, and secure.
i. In the long run, the maintenance cost is very less when compared to the present systems.
j. It is possible to get instantaneous results and with high accuracy.

## 5.5 CIRCUIT DIAGRAM

Figure 5.1 shows the circuit diagram of proposed leakage detection system.

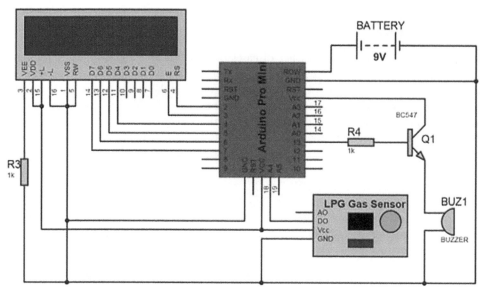

**FIGURE 5.1**    Circuit diagram of the proposed leakage detection system.

## 5.6    COMPONENTS EXPLANATION

Components used are as follows:

- 12 V high-gain siren/buzzer
- MQ-6 LPG sensor
- Arduino Uno
- LCD (LM016L)
- Logic toggle
- NPN Darlington transistor
- 5 V voltage regulator
- Two resistors of 1-kilo-ohm
- **PZ1**: 12 V high-gain siren/buzzer.

  A siren is a loud noise-making device. Most fire sirens are single tone and mechanically driven by electric motors with a rotor attached to the shaft. Some newer sirens are electronically driven speakers. Fire sirens are often called "fire whistles", "fire alarms", or "fire horns".

  Electronic sirens use circuits like oscillators, modulators, and amplifiers to create a specific siren tone (wail, yelp, pierce/priority/phaser, manual, etc.) that is delivered through external speakers.

  It's unusual to see an emergency vehicle with two types of sirens, especially in modern fire engines.

  To help draw attention, police sirens frequently use a tritone interval. Ronald H. Chapman and Charles W. Stephens of Motorola invented the first electronic siren that sounded like a mechanic siren in 1965.

- **GS1**: MQ-6 LPG sensor

  The MQ-6 gas sensor can detect gases like LPG and butane. The MQ-6 sensor module has a digital pin which helps this sensor to operate without using a microcontroller. Analog pin is used when measuring gases in ppm, which is Transistor-Transistor Logic (TTL)-driven and works on 5 V, and therefore it can be used with almost any of the common microcontrollers.

The MQ-6 sensor simplifies gas detection. Either the digital or analogue pin can be used to do this. The power LED illuminates when the module is powered up to 5 V, while the output LED remains dark when no gas is detected, indicating that the digital output pin is at 0 V. Before these sensors can be used, they must first be pre-heated. When the sensor is activated, the output LED should light up together with the digital pin when exposed to the gas that needs to be detected; if not, use the potentiometer to raise the output. The digital pin on this sensor will either go high (5 V) or stay low (−5 V) when exposed to this gas at this concentration (0 V).

The analogue pin can also be utilized for the same purpose. Using a microcontroller, read the analogue values (0–5 V), which will be exactly proportional to the gas concentration detected by the sensor. Experiment with these settings to see how the sensor reacts to various gas concentrations. Observe how the experiment goes and then develop your program accordingly.

For simulation purpose we use an extra pin which is known as test pin, that is not present in actual MQ-6 sensor. So, if we get output as 0 means, no gas detected, and if get it as 1 means gas detected.

- Features:
  - The operating voltage is +5 V
  - LPG or butane gas can be detected with it
  - **Digital output voltage**: ZERO V to FIVE V (TTL Logic)
  - **Analog output voltage**: ZERO V to FIVE V
  - **Preheat duration**: Twenty seconds
  - Can be used as a Digital or analog sensor
  - The digital pin's sensitivity can be modified with a potentiometer.
- Logic toggle

  The toggle flip-flop is another type of bistable sequential logic circuit based around the previous clocked JK flip-flop circuit. The toggle flip-flop can be used as a basic digital element for storing one bit of information, as a divide-by-two divider or as a counter. Toggle flip-flops have a single input and one or two complementary outputs of Q and Q which change state on the positive edge (rising edge) or negative edge (falling edge) of an input clock signal or pulse.

  In our study, it is used to toggle the input in MQ-6 sensor, which will be 0 if no gas is detected and 1 if gas is detected.
- Arduino UNO

  The Arduuino.cc created the Arduino UNO. It's an open-source microcontroller board based on the ATmega328P. The board's digital and analogue input/output (I/O) pins allow it to be interfaced with a variety of expansion boards and circuits. Using an Arduino Integrated Development Environment and a type B USB connector, the board may be programmed. There are 141 digital I/O pins and six analogue I/O pins on it.

  The Arduino UNO is the first release of the Arduino Software, and the name is derived from the Italian word "uno", which means "one". The bootloader in the ATmega328 is pre-programmed, allowing fresh code to be uploaded without the necessity of an external hardware programmer. The UNO board is the first of a series of USB-based Arduino boards.
- LCD (LM016L)

  Liquid crystal display, abbreviated as LCD, is a flat panel display which uses liquid crystal for its operation. Light-emitting diode, abbreviated as LED, is commonly used in smartphones, computers, television, and instrument panels.

  LEDs and gas plasma displays were replaced by LCDs, as it was huge reform of technology. It consumes less power than LEDs and gas plasma display as it works on the principle of blocking light rather than emitting it. LCDs have much thinner displays compared to cathode ray tube technology. In a LCD, liquid crystals produce image using blacklight.

- NPN Darlington transistor

    Darlington is a pair NPN transistor. It functions like a normal NPN transistor, but since it has a Darlington pair inside, it has a good collector current rating of about 5 A and a gain of about 1,000. It can also withstand about 100 V across its Collector-Emitter, and hence, can be used to drive heavy loads.

    This transistor is known for its high current gain (hfe = 1,000) and high collector current (IC = 5 A); hence, it is normally used to control loads with high current or in applications where high amplification is required. This transistor has a low Base-Emitter Voltage of only 5 V, and hence, can be easily controlled by a logic device like microcontrollers. However, care has to be taken to check if the logic device can source up to 120 mA.

    - Features:
        - It is a Darlington Medium-Power NPN Transistor.
        - The high DC Current Gain is generally thousand.
        - The continuous Collector Current is 5 A.
        - The Collector-Emitter Voltage is 100 V.
        - The Collector-Base Voltage is 100 V.
        - The Emitter-Base Voltage is 5 V.
        - The Base Current is 120 mA.

- 7805, 5 V voltage regulator

    A voltage regulator (VR) is a device that keeps the output voltage constant despite changes in the input voltage or load conditions. VRs keep the voltages from a power source within a range that is compatible with the other electrical components. While VRs are most commonly used to convert DC to DC electricity, some may also convert AC to AC or AC to DC. This study will concentrate on DC/DC VRs. They provide a consistent output voltage for a wide range of input voltages. The 7805 IC is a well-known regulator IC that is employed in a wide range of projects in our situation. The number 7805 has two meanings: "78" refers to a positive VR, and "05" refers to a 5 V output. As a result, our 7805 will produce a +5 V signal.

    - Features:
        - It is a Positive Voltage Regulator with a voltage of 5 V.
        - Input voltage must be at least 7 V.
        - The input voltage is capped at 25 V.
        - The current required for operation is 5 mA.
        - Thermal overload and short circuit current limiting protection are offered on the inside.

- Resistors

    A resistor is a passive electrical component that introduces resistance to current flow. They're present in almost all electrical networks and electronic circuits. Resistors are used in electronic circuits to reduce current flow, alter signal levels, divide voltages, bias active devices, and terminate transmission lines, among other things.

    Motor controllers, power distribution systems, and generator test loads all use high-power resistors, which can dissipate hundreds of watts of electrical power as heat. Fixed resistors' resistance varies only slightly as a function of temperature, time, or operating voltage. Variable resistors can be used as temperature, light, and humidity sensors, as well as to change circuit elements. Resistors are widespread in electronic equipment and are common components of electrical networks and electronic circuits. As discrete components, practical resistors can be made up of a variety of compounds and shapes. Integrated circuits employ resistors as well.

    Resistors are utilized in a variety of applications. Delimit electric current, voltage division, heat generation, matching and loading circuits, control gain, and establish time constants are only a few examples. They have resistance levels spanning more than nine orders of magnitude and are commercially available in stores. They can be used to disperse kinetic energy from trains as electric brakes, or they can be small.

The resistor has various applications; therefore, the properties are varied accordingly. The main parameter is the value of the resistance, as its key principle is to hinder the flow of current. The resistance tolerance is given in percentage to define is manufacturing accuracy.

Temperature coefficient and long-term stability are few parameters which affect the resistance and hence are specified. Temperature coefficient is determined by both resistive material and mechanical design, which is usually specified in high-precision applications.

For high-power applications, the power rating is crucial. It specifies the maximum power that the component can handle without causing damage. It is specified in room temperature and pressure. Heat sinks are required by large power rating devices, and it also demands a considerable size. Maximum voltage, pulse stability, etc. also play important roles in design specification. In this work, two 1-kilo-ohm resistors are used.

## 5.7  WORKING PRINCIPLE OF OUR STUDY

Figure 5.2 shows the simulation diagram of proposed detector.

An LPG gas sensor module is used to detect LPG. When a gas leak is detected, Arduino sends a single pulse to the DO pin, which it retains indefinitely. When Arduino receives a one pulse from the gas sensor, it shows the message "LPG detected" on the LCD and activates the buzzer, which continues to buzz until the gas detector detects no gas in the surroundings.

The LCD will display the "No LPG gas" message, if Arduino receives a LOW pulse from the gas detector system.

Arduino controls the entire system: reading the output of the LPG gas sensor module, sending a message to the LCD, and activating the buzzer. The included potentiometer can be used to alter the sensitivity of this sensor module.

## 5.8  CODE

Figure 5.3 shows the coding part.

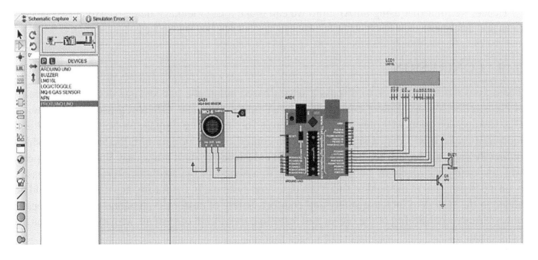

**FIGURE 5.2**  Proposed simulation circuit.

```
LPG_gas_detector
#include <LiquidCrystal.h>

LiquidCrystal lcd(7, 6, 5, 4, 3, 2);

#define MQPin A0
#define buzzer 1

void setup() {
   lcd.begin(16, 2);
   pinMode(MQPin, INPUT_PULLUP);
   pinMode(buzzer, OUTPUT);
    lcd.setCursor(5, 0);
    lcd.print("GAS");
    lcd.setCursor(3, 1);
    lcd.print("DETECTOR");
    delay(1000);
    lcd.clear();
}
```

```
void loop() {

int gas_value = digitalRead(MQPin);

if(gas_value==HIGH)
{
   digitalWrite(buzzer, HIGH);
   lcd.setCursor(6, 0);
   lcd.print("GAS");
    lcd.setCursor(3, 1);
   lcd.print("DETECTED");
   delay(200);
   lcd.clear();
   delay(200);

}
else
{
  lcd.clear();
  digitalWrite(buzzer, LOW);
}
```

**FIGURE 5.3**   Proposed coding for the detection system.

## 5.9   RESULT AND INFERENCE

When the Arduino receives a pulse from the gas sensor, it displays "LPG detected" on the LCD and activates the buzzer, which continues to buzz until the gas detector detects no gas in the area.

For software simulation we have used a test pin which initiates the buzzer if input is given as 1, when the gas is detected, and it doesn't gives any output if the input given is 0, which means no gas is detected.

Hence, we infer that using an Arduino we can make a LPG leakage detector which is very efficient and affordable.

## 5.10   CONCLUSION

The purpose of this study is to discuss a sensor-based automatic gas leakage detection system. It is a cost-effective, less power consumption, easy and safe to use, efficient gas-detecting device. Gas leakage detection will play an important role in controlling damage in both residential and industrial areas. It will play a significant role in controlling the pollution in the atmosphere which indirectly causes a hazard for human health. The designed system costs only 950 Indian rupees, which is easily affordable by people. According to a review of the available literature, there hasn't been much progress in the field of smart gas detection systems. In the future, new functions could be added to the system to increase safety and comfort. Safety of customers, tenants, and employees is now the key responsibility of the owner and installing these kind of systems with more specification which ensure safety will help them grow their market.

Gas leaks cause serious mishaps, resulting in both property loss and human injuries. The main cause of gas leaks is poor equipment maintenance and a lack of public awareness. Inhalation of toxic gases causes severe long-term diseases to human beings. Hence, detection of LPG leakage is important to protect both material and human lives. Hence, in this chapter, a LPG leakage detection system is proposed. When a leak is detected, it activates the buzzer and displays a warning on the LCD. This approach is both straightforward and dependable.

## 5.11  FUTURE SCOPE

One such theoretical advancement that is well feasible with present technical advances in this generation is temperature displays that last for long periods of time when no message buffers are empty. Another intriguing and significant advancement could be to house many receiver MODEMS in different locations throughout the geographical area, each with its own SIM card. Another delivery option within the project could be a multilingual display. It is possible to add audio output to make it more user-friendly.

The group of this study would like to suggest to the researchers that they continue to build this prototype tool in order to figure out how to combine the manipulation of the LPG tank hand wheel as well as something else to help with the suggested business.

## BIBLIOGRAPHY

Hasibuan, Muhammad Siddik, Iswandi Idris. Intelligent LPG Gas Leak Detection Tool with SMS Notification. *Journal of Physics: Conference Series*, Vol. 1424, No. 1, p. 012020, 2019.

Khan, Mohammad Monirujjaman. Sensor-Based Gas Leakage Detector System. *Engineering Proceedings*, Vol. 2, No. 1, p. 28, 2020.

Leavline, E. Jebamalar, D. Asir Antony Gnana Singh, B. Abinaya, H. Deepika. LPG Gas Leakage Detection and Alert System. *International Journal of Electronics Engineering Research*, Vol. 9, No. 7, pp. 1095–1097, 2017.

Paculanan, Rhonnel S., Israel Carino. LPG Leakage Detector Using Arduino with SMS Alert and Sound Alarm. *International Journal of Innovative Technology and Exploring Engineering (IJITEE)*, Vol. 8, No. 6C2, April 2019.

Rawat, Hitendra, Ashish Kushwah, Khyati Asthana, Akanksha Shivhare. LPG Gas Leakage Detection & Control System. In *National Conference on Synergetic Trends in Engineering and Technology*, 2014.

Shahewaz, Syeda Bushra, Ch. Rajendra Prasad. Gas Leakage Detection and Alerting System Using Arduino UNO. *Global Journal of Engineering and Technology Advances*, Vol. 5, No. 3, pp. 29–35, 2020.

Siddika, Ayesha, Imam Hossain. LPG Gas Leakage Monitoring and Alert System Using Arduino. *International Journal of Science and Research (IJSR)*, Vol. 9, No. 1, pp. 1734–1737, 2020.

Yadav, Vasudev, Akhilesh Shukla, Sofiya Bandra, Vipin Kumar, Ubais Ansari, Suraj Khanna. A Review on Microcontroller Based LPG Gas Leakage Detector. *Journal of VLSI Design and Signal Processing*, Vol. 2, No. 3, pp. 1–10, 2016.

# 6 IoT-Based Intelligent Garbage Monitoring Management System to Catalyse Farming

*B.V.A.N.S.S. Prabhakar Rao and Rabindra Kumar Singh*
Vellore Institute of Technology

## CONTENTS

6.1   Introduction ....................................................................................................... 52
6.2   Need..................................................................................................................... 52
6.3   Background........................................................................................................... 52
6.4   Literature Survey ................................................................................................ 53
6.5   Objectives ............................................................................................................ 53
6.6   Motivation............................................................................................................ 55
6.7   Goals.................................................................................................................... 55
6.8   Implementation ................................................................................................... 55
6.9   Description........................................................................................................... 58
6.10  Results and Discussions...................................................................................... 58
6.11  Summary ............................................................................................................. 60
6.12  Conclusion .......................................................................................................... 61
References.................................................................................................................... 62

**ABSTRACT**

The central idea of this chapter is to design and develop an Internet of Things (IoT)-based smart garbage monitoring system for proper waste segregation and disposal. In this system, an vigilant coordination has been designed to send an alert to the attendant for instant cleaning of the wastebasket according to the level of garbage filling based on the bin's capacity in terms of its height and weight as well as segregation based on the type of solid waste as identified by the sensors deployed in the bin. It is associated to the community-based garbage which is generated at regular intervals based on the activities performed and tasks carried out. Also message should be communicated to the concern authorities for necessary action at the earliest to cater the needs of farmers cultivated type of commodity with available type of bins, thereby reducing the need for manual verification. This system will not only help in creating a cleaner environment for the common man but also help in the prevention of diseases caused by the accumulation of waste in public areas. Hence, this system is intended to be extensively used by the common man without much hassle to take the time to decide which bin to put the garbage in, to keep the city clean. The proposed developed system will provide a complete mechanism to process the waste that amounts at different locations to deliver the desired droppings based on the need. The solution currently focuses to implement the same using IoT models. By employing this effort will avoid overflowing of debris from the basin, along with managing different kinds of waste in solid/liquid forms, in residential as well as public areas which were earlier segregated physically with the assistance of man-power. There is a squeezing requirement for feasible methodologies. Steps are being taken by the administration, yet, at the same time, an increasingly methodical methodology is required alongside the use of the most recent and savvy advances at different conceivable dimensions. The segregated waste materials can be used as recyclable products. Some of the products such as waste paper can be

DOI: 10.1201/9781003374121-6

reused. With the help of mechanical techniques, the other waste products can be used in the formation of manure, thereby supporting (catalysing) the development process of farming.

## 6.1  INTRODUCTION

India is a developing country. It faces many environmental problems associated with the waste generation and inadequate waste collection and treatment. In a developing country like India, community material surplus is one of the foremost areas of concern. The waste generated by the people living in a jurisdiction refers to the trash or the garbage discarded by the people. Solid waste consists of organic as well as inorganic waste. In a developing country like India, the increase in municipal solid waste is due to the urban expansion and increase in population. A piece of waste fluctuates through numerous variables. Because of urban blast, the civil strong waste has expanded on a higher rate, and the insufficient and informal treatment of the metropolitan strong waste (MSW) builds the well-being danger in the nation. The current Indian administration has a track on diverse work towards environmental hygiene. Indian cities namely Delhi, Kolkata Ahmedabad, Hyderabad, Chennai, Bangalore and Mumbai experience forceful regulatory expansion and high per capita waste accumulation. Basic disputes and complications must be considered like available land space for temporary storage of garbage till it processing whole gathering as per the economic feasible in terms of transport and time taken for full utilization as per the crop based on the season and so on. Consequently, there is need to analyse all the existing system related to smart garbage disposal system and understand the possible applications that can come out of these systems. Different methods used to implement such systems will give out different results with respect to the applications of the system. The strategic concern in the unused managing is sometimes that the amount of waste is so high that in spite of availability of bins, people dump the garbage near the bins in cases when they are full. Even after efforts put in by the government to keep our streets clean, by installations of separate bins for different kinds of waste, it has been frequently seen that due to lazy and carefree attitude of some citizens, proper garbage disposal is not practiced.

## 6.2  NEED

According to a report issued in 2017 by the renowned newspaper agency "The Times of India" on the basis of statements given by minister of state of Science and Technology, around 1,00,000 metric tons of waste is generated per day in our country. Composition of waste may vary due to different factors such as poor living standards, climatic conditions and socio-economic factors. Due to urban explosion, the municipal solid waste has increased on a higher rate and the inadequate and unscientific handling of the municipal solid waste increases the health hazard in the country. Daily scheduled intelligent phases are deployed for the collection and disposal process of garbage with complete optimization. Hence, this chapter mainly focuses on the study of currently developed/-under-development solutions to this problem and introduction to a new approach in order to cover almost all the aspects of efficient garbage segregation and disposal by applying different concepts of Computing Science and Engineering.

## 6.3  BACKGROUND

With the world developing and growing day by day and the living conditions of the people improving every day, the amount of garbage disposal has also increased. Garbage makes the environment unsuitable for the living beings to live in. With the ever-growing garbage and improper garbage disposal system, the quality of living conditions has degraded. This may lead to proliferation of numerous diseases as well. Hence, the importance of efficient waste disposal process (especially in our country) cannot be overemphasized. Human exercises make squander, and these squanders are taken care of, put away, gathered and discarded, which can present dangers to the earth and to general well-being. Quick urbanization and industrialization in India have brought about overemphasizing of urban foundation administrations, including civil strong waste (MSW) administrations. MSW incorporates family unit junk and garbage, road clearing, development and pulverization flotsam and jetsam, sanitation deposits, exchange and non-unsafe mechanical debris and treated bio-medicinal strong waste. City bodies are confronting significant challenges in giving sufficient administrations, for example, supply of water, power, streets, instruction and open sanitation, including Municipal Solid Waste Management (MSWM). In developing nations, the per capita age of the strong squanders in urban neighbourhoods is considerably less contrasted with the developed nations.

## 6.4 LITERATURE SURVEY

As per the current day needs, there is huge demand and immediate attention is required for developing sustainable disruptive intelligent machines [1–3].

Proper technology-based system puts into use the mass property of various things for separating [4] (Table 6.1).

Some of the existing approaches of smart garbage disposal system have different application depending on the different methodologies used to implement the system [5,6].

In Sweden, "the automated vacuum waste collection system is able to collect only four types of waste: general waste, organic waste, recyclable paper and recyclable cardboard" [7,8].

"An Auto Recycle work of Columbia University built a prototype which allowed the waste material to get segregated into the respective compartment of the dustbin after identification" [9] (Figure 6.1).

Different implementation methods will lead to different applications and alternative Internet of Things (IoT) implementation methods [10].

In yet another application, the proposed system has been designed for actual programmed separation of unused at homes, thereby reducing the physical efforts. The system uses concepts of Machine Learning, Image Processing and IoT [11].

Table 6.2 presents the above two approaches used for the segregation of garbage.

## 6.5 OBJECTIVES

This framework is to design an intelligent debris vigilant system, which is used for an appropriate waste segregation and disposal.

- We propose a smart alert by sending an alert to attendant for instant cleaning according to the level on the basis of the bin's capacity in terms of its height and weight as well as segregation on the basis of the type of solid waste as identified by the sensors deployed in the bin.
- This application is fully designed based on the prerequisite of the remote monitoring of the bins, thereby reducing the need of manual verification at regular intervals.

**TABLE 6.1**
**Various Physical Separation Processes**

| Name | Basis of Separation | Method |
|------|---------------------|--------|
| Trommel separator | Size | Perforations separating small and big object |
| Eddy current separator | Metallic, non-metallic | Electromagnetic method |
| Near-infrared sensor | Reflectance property | Measuring reflectance value of an object within a particular range |
| X-ray | Density | Identifies density of object to distinguish |

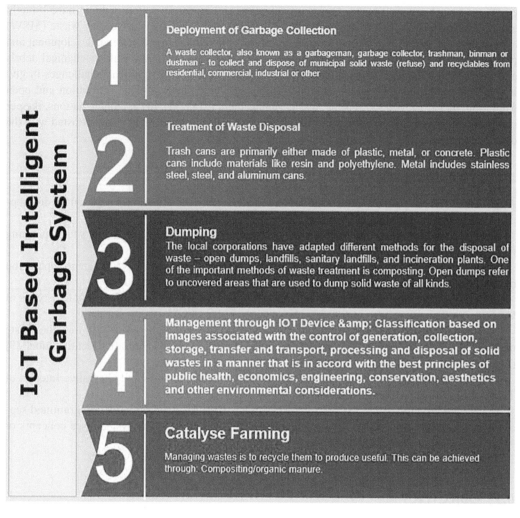

**IoT Based Intelligent Garbage System**

**1 Deployment of Garbage Collection**

A waste collector, also known as a garbageman, garbage collector, trashman, binman or dustman - to collect and dispose of municipal solid waste (refuse) and recyclables from residential, commercial, industrial or other

**2 Treatment of Waste Disposal**

Trash cans are primarily either made of plastic, metal, or concrete. Plastic cans include materials like resin and polyethylene. Metal includes stainless steel, steel, and aluminum cans.

**3 Dumping**

The local corporations have adapted different methods for the disposal of waste – open dumps, landfills, sanitary landfills, and incineration plants. One of the important methods of waste treatment is composting. Open dumps refer to uncovered areas that are used to dump solid waste of all kinds.

**4 Management through IOT Device &amp; Classification based on Images associated with the control of generation, collection, storage, transfer and transport, processing and disposal of solid wastes in a manner that is in accord with the best principles of public health, economics, engineering, conservation, aesthetics and other environmental considerations.**

**5 Catalyse Farming**

Managing wastes is to recycle them to produce useful. This can be achieved through: Compositing/organic manure.

**FIGURE 6.1** Flowchart.

**TABLE 6.2**

**Segregation of Garbage Using Different Approaches**

| Method/Approach | Category of Garbage Segregated | Algorithm Used |
|---|---|---|
| Approach 1 | Biodegradable, non-biodegradable | Convolutional neural network |
| Approach 2 | Metal, glass, paper, plastic | Convolutional neural network |

- This system will not only help in creating cleaner environment for the common man but also help in prevention of diseases caused by accumulation of waste in public areas.
- Since such smart bins are meant to be made accessible to as much area as possible, in order to increase waste management productivity, it'll be fairly easily accessible to everyone by making sure of cost reduction of the product to the maximum. Hence, this system is intended to be extensively used by the common man without much hassle to take the time to decide which bin to put the garbage in, in order to keep the city clean.

## 6.6  MOTIVATION

The following points motivated us to work on this framework:

1. Generally, in Indian urban communities, the formal handling and recuperation units are not set up.
2. Recovery and recyclable exercises are confined to little and medium.
3. Involvement of little youngsters and elderly individuals utilized for arranging and isolating waste.
4. No defensive dress/thought for cloth pickers/foragers.
5. In Indian urban communities, monetary ramifications of recuperation and reusing has not been examined or considered to utilize strong waste.
6. The commercial aspects are also to be considered.
7. Deploying a smart waste management system in the Indian cities on a commercial scale would be part of helping the plan of turning cities into Smart Cities. Hence, we thought of making a prototype which could work as a well-organized disposal and segregation organization in order to maintain a hygienic and clean surroundings in the modern era.

## 6.7  GOALS

The goals of the work include:

- Providing a smart garbage disposal system for the cleaner society.
- Providing a system which segregates the garbage into different categories.
- Providing an alert system so that the concerned workers can get the alert about the place where the garbage has been filled.
- An automatic system which can predict the areas where there will be more accumulation of garbage based on the previous data.

## 6.8  IMPLEMENTATION

Current statistics show that there is more than forty eight percentage growth in metropolitan residents in India (Figure 6.2).

- The main purpose of an efficient waste management is to ensure proper segregation of waste.

### Piling problem

Studies show that India's waste has more than doubled in the past 25 years

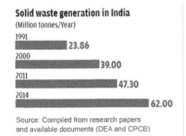

**FIGURE 6.2**   Statistics data on waste generation.

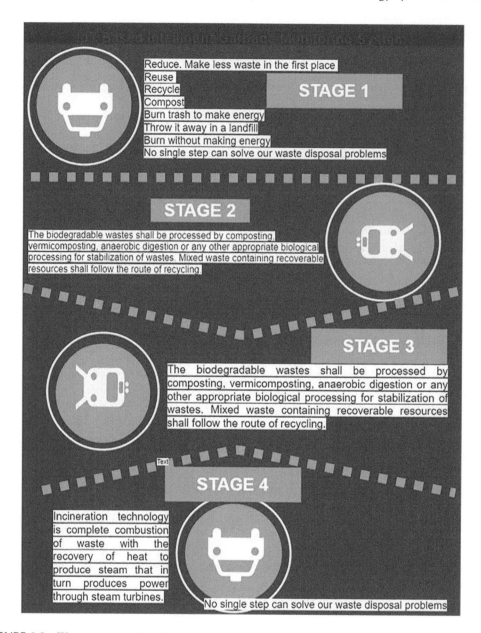

**FIGURE 6.3**  Waste to energy plant diagram.

- Deploying a smart waste management system in the Indian cities on a commercial scale would be part of helping the plan of turning cities into Smart Cities.
- After the appropriate segregation of waste products through the smart garbage system, it is essential to ensure that proper steps are taken to reuse or recycle the waste (Figure 6.3).

Different categories of Pyrolysis tools are shown in Table 6.3 [1].

- Recycling of Glass – the user throws cut-glass into salvage.
- Recycling of Paper – the paper is then taken to recycling plants where it is separated into types and grades (Figure 6.4).

**TABLE 6.3**
**Various Pyrolysis Technologies, Their Process Conditions and Major Products**

| Technology | Residence Time | Heating Rate | Temp. (°C) | Products |
|---|---|---|---|---|
| Carbonization | Hours-days | Very low | 300–500 | Charcoal |
| Pressure Carbonization | 15 minutes–2 hours | Medium | 450–550 | Charcoal |
| Conventional Pyrolysis | Hours | Low | 400–600 | Char, Oil, syn-gas |
| | 5–30 minutes | Medium | 700–900 | Char, syn-gas |
| Vacuum Pyrolysis | 2–30 seconds | Medium | 350–450 | Oil |
| Flash/Rapid Pyrolysis | 0.1–2 seconds | High | 450–650 | Oil |
| | <1 second | High | 650–900 | Oil, syn-gas |
| | <1 second | Very high | 1,000–3,000 | Syn-gas |

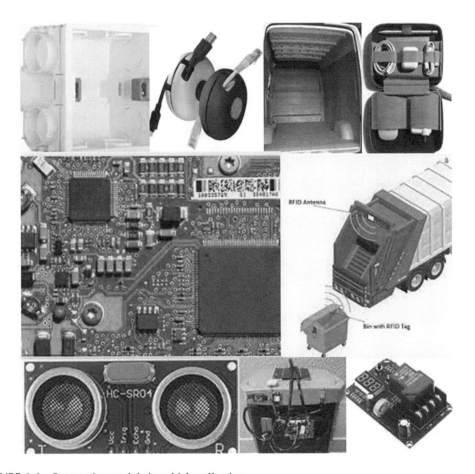

**FIGURE 6.4** Segregation module in vehicle collection.

In modern age, surplus has been a problematic to surroundings, although the usage of diverse approaches has been put in place to solve the problem due to the exponentially increasing amounts of waste but the total land area being a constant [12].

## 6.9   DESCRIPTION

The work has been designed keeping in mind the common man and his needs, and effective steps have been adopted in order to cut the cost along with increase in the efficiency and performance of our system.

- The main aim is to design this intelligent system for a suitable waste disposal combining the concepts for optimization of both garbage collection and disposal.
- The garbage remains placed on a platform attached to the waste bin where the segregation module would determine the kind of solid waste and with the help of a conveyor belt attached to the base of the platform, it would go through a series of smaller bins (fitted inside the main bin) in order to dispose it in the respective bin, i.e., metal/light weight/heavy weight.
- Metal sensor is used to detect metal waste first and separate it from the trash with the help of an induced magnet. The next step is to separate the paper waste with the help of a simple shaft to blow out this waste and separate it from the rest. The final step is to put the leftover waste in the heavy waste category.
- The height and weight of the waste in each separate bin are monitored using UV sensors and load sensors to check the percentage of bin that is full in terms of either height or weight uniquely for different bins.
- Once the bin is almost full, these sensors would send the data to a cloud server through a Wi-Fi module which would in turn send an alert to the local municipal garbage collection van driver with the details of the bin's location and the route (using Google Maps) to reach that location so that he/she can collect the waste and take it to the dumping site as soon as possible.
- The real-time monitoring of these bins is also possible through the same app, which would be used to send alerts to the garbage van drivers.
- After the segregation, time series algorithm is implemented for predicting the height and weight of the bins for the week (Figures 6.5 and 6.6).

## 6.10   RESULTS AND DISCUSSIONS

As we all know that any nation is basically depending on agriculture commodities for their livelihood, besides the income generated from the said product. But, nowadays, farmers are spending

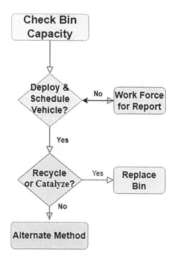

**FIGURE 6.5**   Web portal-based monitoring.

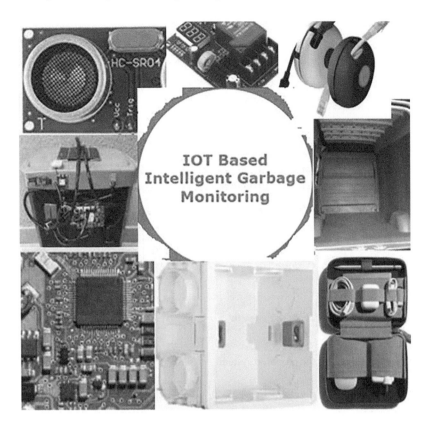

**FIGURE 6.6** Engineering design.

huge amount for buying fertilizers in cultivating an agricultural product. That causes to increase the product cost as well as bad impact on people's health. The main theme of the work is to develop a smart garbage management system for proper waste segregation and disposal that should accommodate the nearby farmers' needs in agriculture at an affordable cost. In this system, an alert system has been designed to send an alert signal to the municipal web server for instant cleaning of dustbins according to the level of garbage filling on the basis of the bins capacity in terms of its height and weight as well as segregation on the basis of the type of solid waste as identified by the sensors deployed in the bin. This system will not only help in creating cleaner environment for the common man but also help in prevention of diseases caused by accumulation of waste in public areas. Hence, this system is intended to be extensively used by the common man without much hassle to take the time to decide which bin to put the garbage in, in order to keep the city clean. India is a developing country. It faces many environmental problems associated with the waste generation and inadequate waste collection and treatment. Deployment of smart bins, tracking of garbage pickup trucks as well as the sanitation workers, route optimization for trucks to nearby farming lands for disposal, crosschecking of garbage weight, etc. can efficiently address the challenges of enforcement and transparency (Figure 6.7 and Table 6.4).

The cost incurred in the entire processing and transporting the waste is economically realistic (Figures 6.8–6.10).

Waste segregation suggests extrication unused into the different categories into which it might be classified; it can be heavy waste and light waste or wet and dry waste. Waterless waste includes wood and related items, metals and glass. In the entire recycling process, many technical and other staff may be involved to cater the needs of rural farmers, since they are putting lots of investment

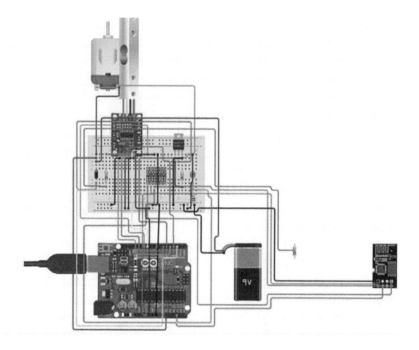

**FIGURE 6.7**  Circuit design for the proposed system.

**TABLE 6.4**
**Analysis of Data Collected for Different Locations in Chennai City**

| S. No. | Location | Day(s) with Maximum Number of Waste Items |
|---|---|---|
| 1 | Adyar | Thursday |
| 2 | Egmore | Friday and Saturday |
| 3 | Kelambakkam | Thursday |
| 4 | Navallur | Wednesday and Thursday |
| 5 | T Nagar | Thursday |

on fertilizers and chemicals in the processing of commodities as well as preserving the items till they reach the market [13].

## 6.11  SUMMARY

The main objective of this major work is to monitor waste segregation and disposal in terms of economic, technical and operational feasibility. In this alert system, the level of substantial debris on the basis of the bin's capacity in terms of its height and weight as well as segregation on the basis of the type of solid waste as identified by the sensors are deployed in the bin implemented in an effective manner [14]. This device is developed and interconnected to the public system to send alerts. This system will not only help in creating cleaner environment for the common man but also help in prevention of diseases caused by accumulation of waste in public areas. Hence, this system is intended to be extensively used by the common man without much hassle to take the time to decide which bin to put the garbage in, in order to keep the city clean.

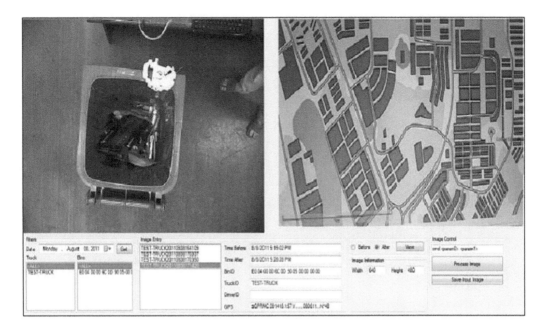

**FIGURE 6.8**   Global Information System (GIS)-based garbage collection.

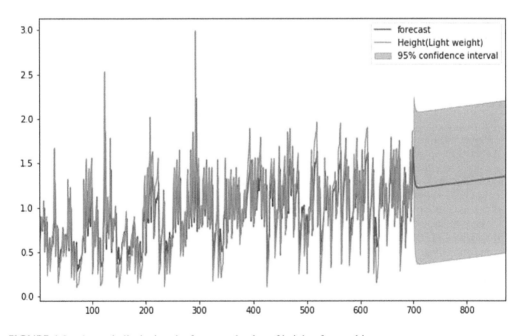

**FIGURE 6.9**   A graph displaying the forecasted value of height of waste bin.

## 6.12   CONCLUSION

The current systems available to manage garbage are clearly not sufficient and efficient enough and consume a lot of manual labour. The proposed method is used to implement different waste disposal and collection systems that have different applications namely segregation, collection and

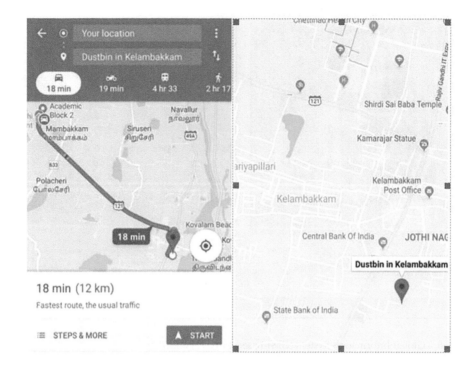

**FIGURE 6.10**   Map showing the route to the dustbin from the current location.

management which can further be implemented for better and more efficient waste collection process, giving rise to a healthier society to live in.

## REFERENCES

[1] Stringfellow, T.; Witherell, R. An independent engineering evaluation of waste to energy technologies. Renewable Energy World, 2014.

[2] Rao, B.P.; Singh, R.K. Disruptive intelligent system in engineering education for sustainable development. *Procedia Comput. Sci.* 2020, 172, 1059–1065.

[3] Deshmukh, S.; Deshmukh, C.N., Design wireless sensor network and IoT for smart city. *Int. J. Res. Advent Technol.* 2019, 7(5).

[4] Capel, C. Waste sorting - a look at the separation and sorting techniques in todays European Market. Waste Management World, 2008, 1–8.

[5] Lundin, A.C.; Ozkil, A.G.; Jensen, J.S. Smart cities: A case study in waste monitoring and management. In *Proceedings of the 50th Hawaii International Conference on System Sciences*, Waikoloa, HI, USA, 4–7 January 2017; pp. 1392–1401, 2017.

[6] Anagnostopoulos, T.; Kolomvatsos, K.; Anagnostopoulos, C.; Zaslavsky, A.; Hadjiefthymiades, S. Assessing dynamic models for high priority waste collection in smartcities. *J. Syst. Softw.* 2015, 110, 178–192.

[7] Larsson, V. Procurement of the vacuum waste collection systems. The Cases of Hammar by Sjöstad and Stockholm Royal Seaport; Stockholm University: Stockholm, Sweden, 2013.

[8] Nakou, D.; Benardos, A.; Kaliampakos, D. Assessing the financial and environmental performance of underground automated vacuum waste collection systems. *Tunn. Undergr. Space Technol.* 2014, 41, 263–271.

[9] Shi, H.; Bondarde, S.; Vishakh B. V., *AutoRecycle*, Columbia University, 2016.

[10] Desai, Y., Dalvi, A., Jadhav, P., Baphna, A. Waste segregation using machine learning. *Int. J. Res. Appl. Sci. Eng. Technol.* 2018, 6, 537–541.

[11] Hulyalkar, S.; Deshpande, R.; Makode, K.; Kajale, S. Implementation of smartbin using convolutional neural networks. *Int. J. Res. Appl. Sci. Eng. Technol.* 2018, 5(4), 3352–3358.

[12] Najaf, A.; Awais, M.; Muzammul, M.; Zafar, A. Intelligent system for garbage collection: IoT technology with ultrasonic sensor and Arduino mega. *Int. J. Comput. Sci. Netw. Secur.* 2018, 18(9), 102.

[13] Bircanoğlu, C.; Atay, M.; Beşer, F.; Genç, Ö.; Kızrak, M.A. RecycleNet: Intelligent waste sorting using deep neural networks. In *2018 Innovations in Intelligent Systems and Applications (INISTA)*, 2018.

[14] Sivajothi Kavitha, S.; Hemalatha, K.; Jamuna, V. A real-time smart dumpsters monitoring and garbage collection system using Iot. *Int. J. Recent Technol. Eng.* 2019, 8(2S5), 1–8.

# 7 IoT-Based Wildfire Detection and Monitoring System Using Predictive Analytics

*Chimata Shriya, Varsha Jayaprakash, C.N. Rachel, O.V. Gnana Swathika, and V. Berlin Hency*
Vellore Institute of Technology

## CONTENTS

7.1    Introduction ..................................................................................................65
7.2    Related Work ...............................................................................................66
7.3    Proposed System..........................................................................................67
     7.3.1    Hardware Architecture ....................................................................68
     7.3.2    Software Architecture ......................................................................68
     7.3.3    Predictive Analysis ..........................................................................68
7.4    Results and Discussion ................................................................................69
     7.4.1    Hardware System .............................................................................69
     7.4.2    Software System ..............................................................................70
     7.4.3    Predictive Analysis ..........................................................................70
7.5    Future Work.................................................................................................70
7.6    Conclusion ...................................................................................................72
References....................................................................................................73

### ABSTRACT

Wildfires listed among the most devastating natural disasters, wreak havoc on both wildlife and vegetation. This endangers forest's wealth and wipes out the entire flora and fauna regime, disrupts a region's biodiversity, ecology, and environment, and causes risk to domestic crops and inhabitants over there. As a result, predicting wildfires in advance and prevention of these fires from propagating in order to minimize the severity of the damage became an important research issue. This work describes design of an Internet of Things (IoT)-based system that collects data using sensor technology and applies predictive analysis method to predict occurrence of forest fires. Using this system, temperature and humidity data are gathered, evaluated, and transmitted to a cloud platform; logistic regression models are developed for predicting the risk in order to alert the inhabitants of the region. In case, an unplanned forest fire occurs, this IoT system detects and alerts the control room by specifying the location using a global positioning system module facilitating appropriate action by dispatching firefighters and informing homeowners via a manually controllable buzzer. This system also ensures a secure way to protect and preserve both wildlife and human beings from threatening forest fires.

## 7.1 INTRODUCTION

Wildfires are dangerous and unpredictable natural calamities that can inflict significant damage on both land and wildlife. With the advent of climatic change around the world, the number of forest fires has risen to several thousands per year, resulting in damage to over a million acres of land. This

DOI: 10.1201/9781003374121-7

causes the accumulation of gases such as carbon dioxide, which can alter the oxygen level in the atmosphere, causing asphyxia and respiratory problems in living beings as well as property damage [1]. To avoid the greatest amount of damage, fires must be recognized, predicted, and controlled as soon as possible. The rapid advancement of aviation technology and satellite communication has made it possible to patrol forest areas and hilly regions using satellite imaging, radar monitoring, aerial monitoring, near-ground monitoring, and other methods to monitor forest fires [2,3]. However, the major drawback with these systems is that they can only identify fires on a large scale, i.e., after significant damage to forest regions has already occurred. They often fail to detect the fires immediately after they start and before propagating into a disaster.

The use of Internet of Things (IoT) and wireless sensor networks (WSNs) to implement wildlife monitoring systems has assumed importance in recent years as a result of its user-friendliness, ability to record, and update data on a timely basis, which could aid in the timely prevention, detection, and tracking of wildfires [4]. They should be designed and distributed in such a way that they can cover a wider region while still providing a secure mode of communication. Due to the ambiguity of weather changes, it is important to modify the system by including a decision-making model, such as fuzzy logic and Analytic Hierarchy Process (AHP), to assess the data and reach a conclusion [5]. Machine learning techniques can be utilized to examine the spatial pattern or classify the occurrence of fire by gathering different datasets from wildfires around the world. Boosted regression tree, general linear model, support vector machine, artificial neural networks, Bayes network, multivariate logistic regression, and random forest are some of the techniques used by previous researchers in this field of study [6–8]. This research study focuses on the development of an IoT-based wildfire monitoring system that uses temperature, humidity, and flame sensors to monitor the environment and transmit data to a cloud database, as well as use a simple machine learning technique called logistic regression to predict the likelihood of a fire. The objective of this research is to develop economical and effective solutions based on simple techniques while ensuring safety through the use of sensors such as Flame Sensor, Temperature Sensor and Global Positioning System (GPS) Module. On detection, the system generates warnings and alerts neighboring residents via warning lights and manually controllable buzzers.

This paper is organized as follows: Section 7.1 gives an insight into the work carried out by previous researchers in the similar field. The proposed system and implementation are discussed in Section 7.2. Section 7.3 discusses the various results obtained, followed by the conclusion and future work in Section 7.4.

## 7.2   RELATED WORK

A system to detect forest fires by incorporating Zigbee and WSNs to read the intensity of the fire in forest areas by collecting data using temperature sensors was designed in Ref. [9]. Additionally, they also included a global system for mobile communications to alert residents and control room and a fire dynamics simulator to avoid false alarms, and a language program. Their results proved the ability of the system to detect fires accurately with minimal energy consumption. Support for forest fire prevention in reserve areas was proposed in Ref. [10], involving the concept of IoT to collect local meteorological and environmental data. The data were analyzed using spatial analysis techniques such as geographic information system and multi-criteria decision analysis to detect fires and for risk management purposes. Further, fuzzy AHP method along with Technique for Order Preference for Similarity to Ideal Solution was used for forest fire susceptibility zonation. Their system worked in the drastic conditions of Serbia and could be used in the national as well as the local levels to detect and coordinate emergency situations.

IoT-based information delivery system using sensors, wireless technology, etc. was discussed along with cloud computing to transmit the chances of fire occurrence along with latitude and longitude in Ref. [11]. They used charts and graphs to alert the user and authorities using service-oriented architecture. An unmanned aerial vehicle (UAV) was proposed in Ref. [12], which was capable of sensing and transporting data using sensors and multiple industrial IoT techniques. The response time to sense and transmit data was minimized using a learning-based cooperative particle swarm optimization algorithm along with Markov random field strategy in order to ensure optimal resource allocation. From the comparison of results from the simulation experiments of other state-of-the-art methods, it is concluded that the proposed method was superior in terms of quickest response time to monitor forest fires.

A novel IoT-based architecture that would consider meteorological data as well as images and perform a comparative study using various machine learning algorithms such as boosted decision trees, averaged perceptron, 2-Class Bayes Point Machine, and Binary Neural Network model was proposed in Ref. [13]. The boosted decision tree model proved to be more suitable to fire prediction system with the highest under area curve value of 0.78. An IoT-based fire detection and management system were developed in Ref. [14] using a flame sensor, temperature sensor, and Node Microcontroller Unit (MCU) to read the real-time values continuously and update the data on a database. The system was further developed by including techniques to generate notifications whenever the measured data crossed the assigned threshold values. An IoT-based system using temperature, color, and smoke sensors to capture data in forest areas and compare it with images obtained from satellites in order to find out if the particular area is prone to fire was proposed in Ref. [15]. Further, UAVs are employed to load water based on the intensity of fire using distributed adaptive routing to put out the fire in specified locations using drones.

A three-layer disaster management framework to predict the wild fires was proposed in Ref. [16]. They used various embedded sensors in order to collect the data and also location. Bayesian belief network and classification-based analysis methods were adopted at the fog layer for detection purposes and intimation to control room. It was further extended to a Web Services Interface (WSI)-based cloud layer in order to overcome the resource constraints of fog layer and to store the data more efficiently. A privacy-preserving IoT-based fire detector was designed in Ref. [17] using Raspberry Pi to capture forest scenarios using video cameras and transmit selective features in order to ensure privacy of data. Convolutional neural networks and Binary video descriptors were employed to extract features for classification respectively. Results showed that the proposed technique outperformed the other available techniques which used raw videos and provided a classification accuracy of 97.5%. A threefold methodology for fire safety in mountain regions was proposed in Ref. [18]. They used a swarm optimization model for containment of fire, developed a novel machine learning, principal component regression, and heuristic- based ensemble model, and applied IoT-based task orchestration approach to notify fire safety information to safety authorities. Their methodology adopted for predictive analytics of the fire turned out to be more accurate than other state-of-the-art methods.

An early fire detection model using required sensors and Raspberry Pi to read and store the data in a centralized server was proposed in Ref. [19]. The data were then analyzed using a feed-forward neural network model for prediction purposes and alerts were generated to users within the proximity. An alertness-adjustable cloud computing for monitoring of fires based on forest fire risk forecasting was developed in Ref. [20] that integrates the technologies of IoT, WSN, and cloud computing to build a hybrid network architecture, and its performance was evaluated based on the response time and ability to adjust the sensing behavior according to the criticality of the environmental conditions. It was shown that leveraging using fog computing aids in reducing delay with increased accuracy throughput. The system was also proven to augment data curation and visualization using the Chandler's burning index scoring methodology.

## 7.3 PROPOSED SYSTEM

A novel system was developed in this work for detection and prevention of wildfire. This system comprises a flame sensor, a temperature and humidity sensor, and a GPS module along with a buzzer. This system also includes a machine learning model for predictive analysis. Arduino open-source prototyping software, google collab (python), and Blynk app are used for software simulation. The framework of the system is shown in Figure 7.1. Various sub-systems, viz., (a) hardware system, (b) software system, and (c) predictive analysis model, are explained in this section.

### 7.3.1 HARDWARE ARCHITECTURE

As seen in Figure 7.1, a flame sensor is incorporated into the system so as to detect fire. This sensor is designed to detect light source in the wavelength range of 750–1,100 nm. It can detect the flame effectively at a distance of 100 cm. The output of this sensor is a digital signal. Similarly, a temperature and humidity sensor, DHT-11 is used to measure and constantly monitor the temperature of the affected region. This uses digital signal acquisition technique, includes a resistive-type humidity measuring component, and a Negative Temperature Coefficient (NTC)-type temperature measuring component. The output values of this sensor are stored in Blynk and can be graphically represented on a continuous basis so that any abrupt increase in temperature can be detected. The output of these sensors is used to activate the buzzer which is used to alert the authorities. This system also comprises a GPS module which constantly sends latitude and longitude data that would be updated in Blynk. This would be helpful as in case there are multiple systems of the same type installed in different locations; we can figure the exact spot by using the location data.

### 7.3.2 SOFTWARE ARCHITECTURE

For coding the entire system, Arduino open-source prototyping software was used. Additionally, data can also be observed on a serial monitor. For predictive analysis, google collab was used to write and execute Python code, as it is an excellent tool for machine learning tasks. Data obtained from sensors is displayed on Blynk dashboard, notifications are received on Blynk app. Blynk app is also used to control buzzer and shows GPS data.

### 7.3.3 PREDICTIVE ANALYSIS

Predictive analysis (Machine Learning Strategy) is used for predicting the probability of occurrence of fire in the forest at a given time. In predictive analysis, machine learning algorithm such as logistic regression is used. Temperature and Humidity data that are collected overtime from sensors are

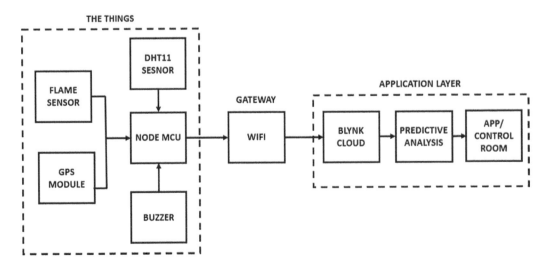

**FIGURE 7.1**  Architecture of wild fire detection and monitoring system.

fed into this model. Figure 7.2 demonstrates that the model is a development and training process which is used for predicting whether there will be a fire in the forest or not at a given point of time. Data from Blynk cloud will be fed into the model, and the model is trained using logistic regression based on temperature and humidity data. Predicted values are obtained after running the entire algorithm and hence, probability of fire is predicted based on that.

## 7.4   RESULTS AND DISCUSSION

Various results obtained during the implementation and execution of the proposed system using the required hardware components as well as software to display and analyze the data are discussed in the following subsections.

### 7.4.1   HARDWARE SYSTEM

The proposed system which is used to predict or detect wildfires is as shown in Figure 7.3. The temperature and humidity of the surroundings are continuously monitored by the DHT11 sensor. The

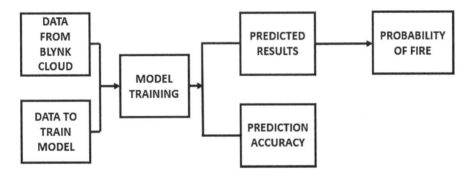

**FIGURE 7.2**   Predictive and analytics procedure.

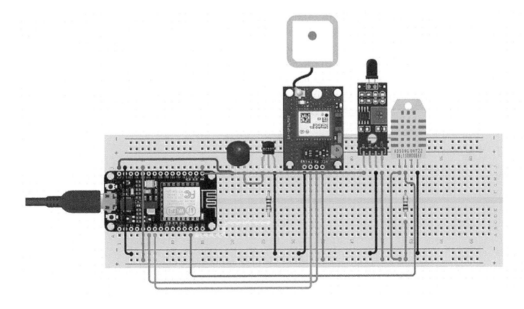

**FIGURE 7.3**   Proposed system.

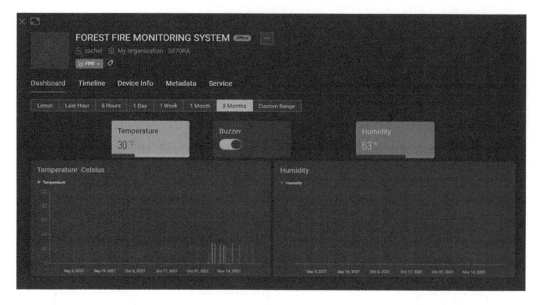

**FIGURE 7.4** Web dashboard.

surroundings are continuously monitored using the flame sensor to check for flames. The buzzer module, which is controlled using the Blynk dashboard, is used to alert residents in the surrounding regions in case a fire is detected or predicted. The GPS module is used to sense the location of the system and alert the control station of the exact location of the fire. Node MCU Wi-Fi development module is the development board used to control this system. The buzzer is controlled from the dashboard and LED (flash lights) is turned on every time a fire was detected.

### 7.4.2 Software System

The Node MCU board which controls all the sensors and modules in the system is connected to the internet through Wi-Fi, thus enabling it to transfer data to the cloud. Web and Mobile dashboards are created where temperature and humidity data can be observed and visualized. Figure 7.4 depicts the creation of web dashboard, while Figure 7.5 depicts the mobile dashboard. The web dashboard can also be used to control the buzzer module that is used to alert residents. In case a high temperature is detected at a particular region, pop-up notifications are sent to the registered users (control station). Figure 7.6 depicts the pop-up receival of notification.

### 7.4.3 Predictive Analysis

Predictive analysis is performed using logistic regression and an accuracy of 84.61% was obtained for the created model. Figure 7.7 shows a snippet of the table which consists of temperature and humidity values that are obtained from the sensor. The predictive analysis performed, helped analyze how prone a particular forest is to forest fires. Figure 7.8 shows the model accuracy obtained, while Figure 7.9 shows the probability of prediction of a fire using this model.

## 7.5  FUTURE WORK

Future researchers can make this system more foolproof by including an automatic system of predictive analytics inside the system. In the analysis part, artificial intelligence and deep learning

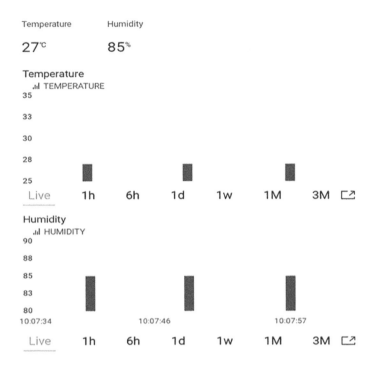

**FIGURE 7.5**   Mobile dashboard.

**FIGURE 7.6**   Pop-up notification received.

methods can be used to automate the system. In addition to identification or prediction of forest fires, a novel feature can be incorporated, which helps in extinguishing the fires in the immediate vicinity of this system automatically. Features such as encryption can be introduced to ensure the safe transmission of resourceful data. In a special case, for forests that are of national importance, passwords and highly complicated authentication features can be used for allowing accessibility to only authorized personnel.

| | A | B | C |
|---|---|---|---|
| 1 | Temperature | Humidity | |
| 2 | | | |
| 3 | 28.7 | 79% | |
| 4 | 28.7 | 79% | |
| 5 | 28.7 | 79.00% | |
| 6 | 29.4 | 80.40% | |
| 7 | 29.4 | 80.40% | |
| 8 | 29.4 | 80.40% | |
| 9 | 30 | 81.20% | |
| 10 | 30 | 81.20% | |
| 11 | 30 | 81.20% | |
| 12 | 31.2 | 81.20% | |
| 13 | 31 | 81.20% | |
| 14 | 31 | 81.20% | |
| 15 | 28.7 | 79% | |
| 16 | 28.7 | 79% | |
| 17 | 28.7 | 79% | |
| 18 | 29.4 | 80.40% | |
| 19 | 29.4 | 80.40% | |
| 20 | 28.7 | 79% | |

**FIGURE 7.7**    Snippet of table containing DHT11 sensor values.

```
Logistic Regression Training Accuracy:  0.8461538461538461
```

**FIGURE 7.8**    Model accuracy.

```
Your Forest is safe.
Probability of fire occuring is  0.009319618860452076
```

**FIGURE 7.9**    Probability of fire occurring.

## 7.6    CONCLUSION

This paper presents an approach for the comprehensive development of a novel system to detect and monitor forest fires. The temperature and humidity sensor values obtained using DHT11 sensors are highly precise with an accuracy of 95%–98%. The data received are displayed on a web and mobile dashboard. The GPS module will detect the location of the system in case a fire is detected, and warning notifications will be automatically generated. Flash lights at the location of the system are switched in order to warn residents and travelers without disturbing the wildlife. Manually controlled buzzers are connected to the system so that they can be turned in case a fire is detected during night times as residents might not notice the flash lights. Hence, it will also help in alerting them so that they move to a safer place. The predictive analytics model developed using logistic regression helps in predicting the probability of fire based on a sample data analyzed from forest fires all over the world. The accuracy obtained using logistic regression is 85%, which is greater than the mean reported accuracy of 79%.

## REFERENCES

1. S. E. Page, F. Siegert, J. O. Rieley, H.-D. V. Boehm, A. Jaya, and S. Limin, "The Amount of Carbon Released from Peat and Forest Fires in Indonesia During 1997," *Nature*, vol. 420, no. 6911, pp. 61–65, 2002.
2. K. Muhammad, S. Khan, M. Elhoseny, S. H. Ahmed, and S. W. Baik, "Efficient Fire Detection For Uncertain Surveillance Environment," *IEEE Transactions on Industrial Informatics*, vol. 15, no. 5, pp. 3113–3122, 2019.
3. L. Da Xu, W. He, and S. Li, "Internet of Things in Industries: A Survey," *IEEE Transactions on Industrial Informatics*, vol. 10, no. 4, pp. 2233–2243, 2014.
4. J. Toledo-Castro, I. Santos-González, P. Caballero-Gil, C. Hernández-Goya, N. Rodríguez-Pérez, and R. Aguasca-Colomo, "Fuzzy-Based Forest Fire Prevention and Detection by Wireless Sensor Networks." In: International Joint Conference SOCO'18-CISIS'18-ICEUTE'18. SOCO'18-CISIS'18-ICEUTE'18 2018. Advances in Intelligent Systems and Computing, vol. 771. Springer, Cham, 2019. https://doi.org/10.1007/978-3-319-94120-2_46.
5. J. Toledo-Castro, P. Caballero-Gil, N. Rodríguez-Pérez, I. Santos-González, C. Hernández-Goya, and R. Aguasca-Colomo, "Forest Fire Prevention, Detection, and Fighting Based on Fuzzy Logic and Wireless Sensor Networks", *Complexity*, vol. 2018, p. 1639715, 2018. https://doi.org/10.1155/2018/1639715.
6. H. R. Pourghasemi, A. Gayen, R. Lasaponara, and J. P. Tiefenbacher. "Application of Learning Vector Quantization and Different Machine Learning Techniques to Assessing Forest Fire Influence Factors and Spatial Modelling". *Environmental Research*, vol. 184, p. 109321, 2020.
7. B. T. Pham, A. Jaafari, M. Avand, N. Al-Ansari, T. Dinh Du, H. P. H. Yen, T. V. Phong, D. H. Nguyen, H. V. Le, D. Mafi-Gholami, I. Prakash, H. Thi Thuy, and T. T. Tuyen, "Performance Evaluation of Machine Learning Methods for Forest Fire Modeling and Prediction," *Symmetry*, vol. 12, no. 6, p. 1022, Jun. 2020. http://dx.doi.org/10.3390/sym12061022.
8. M.S. Tehrany, S. Jones, F. Shabani, F. Martínez-Álvarez, and D. Tien Bui. "A Novel Ensemble Modeling Approach for the Spatial Prediction of Tropical Forest Fire Susceptibility Using LogitBoost Machine Learning Classifier and Multi-Source Geospatial Data," *Theoretical and Applied Climatology*, vol. 137, pp. 637–653, 2019. https://doi.org/10.1007/s00704-018-2628-9.
9. F. Saeed, A. Paul, A. Rehman, W. Hong, and H. Seo, "IoT-Based Intelligent Modeling of Smart Home Environment for Fire Prevention and Safety," *Journal of Sensor and Actuator Networks*, vol. 7, no. 1, p. 11, 2018. http://dx.doi.org/10.3390/jsan7010011.
10. I. Novkovic, G. B. Markovic, D. Lukic, S. Dragicevic, M. Milosevic, S. Djurdjic, I. Samardzic, T. Lezaic, and M. Tadic, "GIS-Based Forest Fire Susceptibility Zonation with IoT Sensor Network Support, Case Study—Nature Park Golija, Serbia," *Sensors*, vol. 21, no. 19, p. 6520, 2021. http://dx.doi.org/10.3390/s21196520.
11. R. Tomar and R. Tiwari, "Information Delivery System for Early Forest Fire Detection Using Internet of Things". In: Singh M., Gupta P., Tyagi V., Flusser J., Ören T., Kashyap R. (eds) *Advances in Computing and Data Sciences. ICACDS 2019. Communications in Computer and Information Science*, vol 1045. Springer, Singapore, 2019. https://doi.org/10.1007/978-981-13-9939-8_42.
12. L. Sun, L. Wan, and X. Wang, "Learning-Based Resource Allocation Strategy for Industrial IoT in UAV-Enabled MEC Systems," *IEEE Transactions on Industrial Informatics*, vol. 17, no. 7, pp. 5031–5040, 2021. https://doi.org/10.1109/TII.2020.3024170.
13. Sharma, R., Rani, S., and Memon, I. "A Smart Approach for Fire Prediction under Uncertain Conditions Using Machine Learning". *Multimedia Tools and Applications* vol. 79, pp. 28155–28168 2020. https://doi.org/10.1007/s11042-020-09347-x.
14. S. Deepthi, S. G. Krishna, K. B. Sahana, H. R. Vandana, and M. Latha. "IOT Enabled Forest Fire Detection and Management," *International Journal of Engineering Research & Technology (IJERT)*, vol. 8, no. 11, pp. 196–199, 2020.
15. V. R. Karumanchi, S. H. Raju, S. Kavitha, V. L. Lalitha, and S. V. Krishna, "Fully Smart fire detection and prevention in the authorized forests". In: *2021 International Conference on Artificial Intelligence and Smart Systems (ICAIS)*, pp. 573–579, 2021. https://doi.org/10.1109/ICAIS50930.2021.9395969.
16. H. Kaur and S. K. Sood. "A Smart Disaster Management Framework for Wildfire Detection and Prediction," *The Computer Journal*, vol. 63, no. 11, pp. 1644–1657, 2020. https://doi.org/10.1093/comjnl/bxz091

17. A. H. Altowaijri, M. S. Alfaifi, T. A. Alshawi, A. B. Ibrahim, and S. A. Alshebeili. "A Privacy-Preserving IoT-Based Fire Detector," *IEEE Access* vol. 9, pp. 51393–51402, 2021. https://doi.org/10.1109/ACCESS.2021.3069588.

18. N. Iqbal, S. Ahmad, and D. H. Kim, "Towards Mountain Fire Safety Using Fire Spread Predictive Analytics and Mountain Fire Containment in IoT Environment," *Sustainability*, vol. 13, no. 5, p. 2461, 2021. http://dx.doi.org/10.3390/su13052461.

19. V. Dubey, P. Kumar, and N. Chauhan. "Forest Fire Detection System Using IoT and Artificial Neural Network", In: *International Conference on Innovative Computing and Communications: Proceedings of ICICC 2018*, vol. 1, pp. 323–337. Springer, Singapore, 2019.

20. A. Tsipis, A. Papamichail, I. Angelis, G. Koufoudakis, G. Tsoumanis, and K. Oikonomou. "An Alertness-Adjustable Cloud/Fog IoT Solution for Timely Environmental Monitoring Based on Wildfire Risk Forecasting," *Energies*, vol. 13, no. 14, p. 3693, 2020. http://dx.doi.org/10.3390/en13143693.

# 8 Rainfall Prediction Model Using Artificial Intelligence Techniques

*Souvik Laskar and A. Menaka Pushpa*
Vellore Institute of Technology

## CONTENTS

8.1 Introduction .................................................................................................. 76
8.2 Taxonomy of Rainfall Prediction Terms .................................................... 76
    8.2.1 Long-Term Prediction System ........................................................... 76
    8.2.2 Short-Term Prediction System ........................................................... 77
8.3 Technologies Adopted for Rainfall Prediction Model................................. 78
    8.3.1 Machine Learning-Based Models ....................................................... 78
        8.3.1.1 The ARIMA (Auto-Regressive Integrated Moving Rate) Model ............... 78
        8.3.1.2 Support Vector Machines ..................................................... 79
        8.3.1.3 Support Vector Regression.................................................... 79
        8.3.1.4 Random Forest ...................................................................... 79
    8.3.2 Neural Networks and Deep Learning-Based Predictions Models........... 79
        8.3.2.1 Artificial Neural Network..................................................... 79
        8.3.2.2 Back-Propagation Neural Network (BPNN)........................ 79
        8.3.2.3 Layer Recurrent Network...................................................... 80
        8.3.2.4 Multilayer Perceptron (MLP) ............................................... 80
        8.3.2.5 Convolutional Neural Networks (CNNs) ............................ 80
        8.3.2.6 Long Short-Term Memory (LSTM)...................................... 80
8.4 Preprocessing Techniques in Rainfall Prediction System........................... 81
    8.4.1 Downsampling..................................................................................... 81
    8.4.2 Principal Component Analysis ........................................................... 82
    8.4.3 Inverse Distance Weighting ............................................................... 82
8.5 Performance Measures in Rainfall Prediction Modules ............................. 82
    8.5.1 Root Mean Square Error (RMSE)....................................................... 82
    8.5.2 Normal Standard Error (NSE)............................................................ 82
    8.5.3 Threat Score (TS) ............................................................................... 83
8.6 Analysis of Existing Works ........................................................................ 83
    8.6.1 Publication Statistics........................................................................... 86
    8.6.2 Country-Wise Systems........................................................................ 87
8.7 Datasets for Rainfall Prediction Model ...................................................... 88
    8.7.1 Modules with Weather Numerical Dataset.......................................... 88
    8.7.2 Modules with Fusion of GIS and Weather Numerical Dataset ........... 93
8.8 Conclusion .................................................................................................. 93
Bibliography ......................................................................................................... 94

DOI: 10.1201/9781003374121-8

**ABSTRACT**

The process of predicting rainfall is critical to humankind, with drastic climate change patterns disrupting human life and agricultural output. In particular, thunderstorms and rainfall are difficult to handle without prior prediction, though such prediction is a challenge. Generally, machine learning (ML) algorithms are used to enhance rainfall prediction. The different types of ML algorithms include Auto-Regressive Integrated Moving Average (ARIMA), Support Vector Machine, Logistic Regression, and Artificial Neural Networks. This work provides clear insights into existing rainfall prediction models and their characteristics in terms of prediction terms and algorithms, preprocessing techniques, regions, datasets, performance metrics, and constraints. This work offers further directions for extended research in this field.

## 8.1   INTRODUCTION

The Indian economy is largely agriculture-driven, and success in agriculture is only possible with the judicious use of rainwater, which is where accurate rainfall forecasting plays a huge part. Rain-related information, when received sufficiently early, helps farmers manage their crop production better and boosts economic growth. Rainfall forecasting helps avert floods, and prevent loss of life, property, and infrastructure. In addition, it helps manage water sources. The variability of rainfall and the quantum received makes rainfall prediction a major challenge for meteorologists. The weather service is provided by the Department of Meteorology in every country, and it provides a national and local weather forecast across the world. The statistical methods applied earlier to predict rainfall were not wholly accurate because there was no process in place to train the system using historical data. In machine learning (ML), a huge amount of previous data can be used to train the model for accurate prediction. Consequently, ML algorithms and deep learning (DL) models are extensively used to predict rainfall. ML functions with a dataset(s) given to it and gradually offers better results over time. DL, on the other hand, is based on Artificial Neural Networks (ANNs). Despite the plethora of models that have been developed, research is critical to accurate prediction. Current research has seen the use of several ML algorithms applied to divergent datasets sourced from numerous countries. Of these, the ANNs, Random Forest (RF), Support Vector Machine (SVM), and Support Vector Regression (SVR) algorithms are the most widely used today in much of the existing work.

Section 8.2 of this chapter discusses the methodologies used for rainfall prediction. Section 8.3 discusses appropriate state-of-the-art preprocessing techniques adopted for rainfall prediction. Section 8.4 discusses the metrics used to evaluate the performance of existing weather prediction models. Section 8.5 describes the datasets employed in the previous work, while Section 8.6 discusses the advantages and disadvantages of each. Finally, Section 8.7 concludes the chapter.

## 8.2   TAXONOMY OF RAINFALL PREDICTION TERMS

Numerical weather prediction models were traditionally considered for rainfall prediction. Today, however, researchers worldwide have developed a range of ML and DL algorithms that predict rainfall most effectively. In terms of rainfall prediction over time, the existing body of literature on the subject may be broadly classified into two categories of prediction, short-term and long-term.

### 8.2.1   LONG-TERM PREDICTION SYSTEM

Long-term prediction models predict rainfall for the next 2 or 3 months, rather than for the following day or the day after or the following week. Lazri et al. [20] estimated precipitation from the Meteosat Second Generation (MSG) images for rainfall prediction for about a month. Diez-Sierra et al. [21] evaluated the performance of a few statistical methods and ML models for daily rainfall prediction. In 2020, a study by Chen et al. [22] proposed a data fusion framework to predict rainfall accurately. Hossain et al. [32] proposed a prediction model that outputs Australia's long-term spring rainfall. The papers discussed above, described in detail in Figure 8.1, worked on long-term prediction systems.

## 8.2.2 SHORT-TERM PREDICTION SYSTEM

Short-term prediction system models predict rainfall for the next 1 or 2 hours rather than the next 2 or 3 months. Maandhar et al. [16] advanced a technique that predicts rainfall about 30 minutes prior to the actual event. Khan et al. [17] worked on a multi-step, daily rainfall prediction method, with the relevant information presented 1–5 days in advance. In 2018, Zhang [18] propounded a system where regional short-term rainfall is predicted, while Moon et al. [19] proposed a model where warning signals are issued prior to extreme rainfall events. The papers discussed above, described in detail in Figure 8.1, worked on short-term prediction systems.

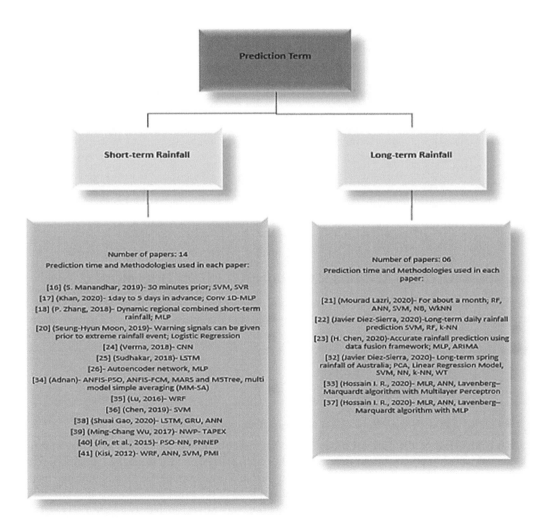

**FIGURE 8.1** Rainfall prediction terms.

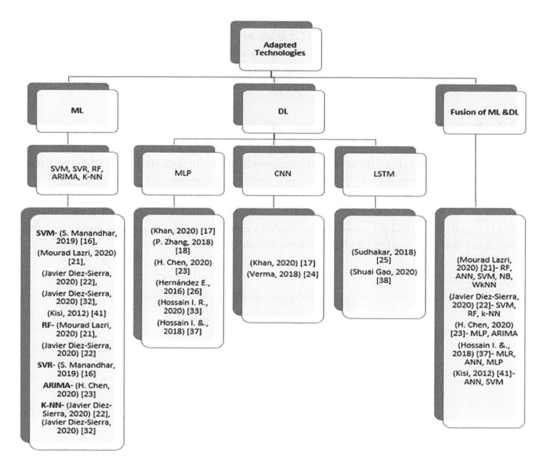

**FIGURE 8.2**    Methodologies in the rainfall prediction system.

## 8.3   TECHNOLOGIES ADOPTED FOR RAINFALL PREDICTION MODEL

Rainfall prediction systems are based on different technologies, ranging from ML and DL to a fusion of both ML and DL for maximal accuracy and minimal prediction errors. The papers discussed above are detailed in Figure 8.2 in terms of those that adopted ML algorithms, DL techniques, or an ensemble model of the two, ML and DL.

### 8.3.1   Machine Learning-Based Models

#### 8.3.1.1   The ARIMA (Auto-Regressive Integrated Moving Rate) Model

Karimi et al. [41] used the ARIMA model to predict monthly rainfall in the city of Urmia in Iran. The model, which predicts time series and handles both analysis and forecasting, comprises four phases. (a) In phase 1, a series of responses is identified and used to calculate the timeline and auto-correlations using the IDENTIFY statement. (b) In phase 2, the previous estimates and specified variables are ascertained, following which their parameters are estimated using the ESTIMATE statement. In phase 3, the results of the diagnostic tests, obtained using the variables and parameters above, are collected. In phase 4, the time series forecasts predictable future values, using the ARIMA model with the FORECAST statement.

### 8.3.1.2 Support Vector Machines

Manandhar et al. [16] used a SVM to build a model that predicts rainfall with less false alarm rates. Lazri et al. [20] developed a model using the SVM for precipitation estimation for satellite data. Diez-Sierra et al. [21] also used the SVM in their rainfall prediction model. SVM algorithms categorize patterns and non-linear regressions. They help create the best boundary line that splits n-dimensional spaces into classes. Further, SVMs are ideal for supervised learning because they generalize better than other neural network models. The solution provided by the SVM which, while identical to that of the others, is much more optimized. The SVM [15] chooses the extreme points or vectors that help form a hyperplane. The points closest to the hyperplane are termed support vectors, and the method is known as the SVM. The few researchers who used this method to predict rainfall obtained satisfactory results.

### 8.3.1.3 Support Vector Regression

Manandhar et al. [16] proposed this model, alongside the SVM, to predict rainfall and reduce false alarm rates. The SVR and SVM, though very similar, show a few key differences. The SVR has an adjustable parameter known as $\varepsilon$ (epsilon). The epsilon determines the breadth of the specified tube surrounding the anticipated function, which is called a hyperplane. The system does not penalize points falling under the tube because they are deemed to be accurate forecasts. Support vectors are positions that lie outside of the specified tube, rather than simply those at the boundary. In the final analysis, the "slack" variable counts the difference between points that are outside the tube and, accordingly, a regularization parameter is tuned. The SVR algorithm handles non-linear situations as well.

### 8.3.1.4 Random Forest

Lazri et al. [20] developed a multi-classifier model that incorporates the RF to predict precipitation estimates for the given satellite data. RF, a supervised ML technique generally applied for classification and regression, contains a number of decision trees. The RF model does not rely on a single decision tree and, instead, anticipates each tree's eventual output. There is increased accuracy if there is a corresponding increase in the number of trees. Random K number of data samples are selected from the training dataset to build the decision trees. RF takes less training time than other algorithms.

## 8.3.2 NEURAL NETWORKS AND DEEP LEARNING-BASED PREDICTIONS MODELS

### 8.3.2.1 Artificial Neural Network

Lazri et al. [20] developed a multi-classifier model that uses the ANN to predict precipitation estimation for the given data received from a satellite. They have adapted different techniques along with RF (ANN). The ANN is a computation model that replicates the human brain. It is made up of a large number of connected neurons [5] that function mostly side-by-side and are properly organized. Neural networks are classified into two, single-layer or multilayer, depending on the number of layers. There is a hidden layer between the input and output layers. A single-layer feed-forward (SLFF) neural network has an input layer and an output layer that are connected with weighted nodes. Adding hidden layers to an SLFF neural network [31] creates a multilayer feed-forward (MLFF) neural network.

### 8.3.2.2 Back-Propagation Neural Network (BPNN)

The BPNN is a single-input, hidden, and output layer-based MLFF neural network. The error, which is calculated from the difference in the predicted value of the output layer and the ground truth, may be reduced by the weight adjustment in each BPNN iteration. In the BPNN back-forward process, the error is dispatched backward in order to adjust the weight until it reaches a minimal value or zero. This involves the phases of (a) adjusting weights and nodes, (b) calculating the error, and (c) modifying weights.

### 8.3.2.3 Layer Recurrent Network

A directed cycle is generated by the connections between units in the layer recurrent neural network. Unlike typical feed-forward neural networks, the RNN uses its internal memory to handle arbitrary sequences of inputs. The hidden and functional layer outputs, from previous procedures, are delivered into the network as part of the input to the next hidden layer operations that follow.

### 8.3.2.4 Multilayer Perceptron (MLP)

Khan and Maity [17] proposed a hybrid Conv-1D MLP model, while Zhang et al. [18] advanced an MLP model to lower the dataset dimension in their work. Chen et al. [22] offered a unique ML -based model fusion framework to enhance satellite-based precipitation retrievals. Emilcy Hernandez et al. [25] forecast the cumulative daily precipitation in Manizales city in Colombia using this technique. ANNs are sometimes called neural networks or MLP networks, named after the most common sort of neural network. A single-neuron model, referred to as a perceptron, paved the way for larger neural networks. A neural network is like a branch of computing, in that it investigates how fundamental biological brain models may also be applied to difficult computational problems like ML and predictive modeling. The goal is to develop strong algorithms and data structures that represent complex scenarios, instead of building genuine brain models. Neural networks are powerful and consistently demonstrate the ability to match the representation in the training data to the output variable expected to be predicted.

In this sense, neural networks learn mapping. A universal approximation algorithm is needed for them to be able to learn a mapping function theoretically. The predictive potential of neural networks comes from their hierarchical or multi-layered structure. The data structure learns to recognize features of various sizes and resolutions and combines them to build higher-order features.

### 8.3.2.5 Convolutional Neural Networks (CNNs)

Mohd Imran Khan, Rajib Maity et al. [17] and Ali Haidar et al. [23] designed a system to predict rainfall using the CNN. Their system is a regularized version of the MLP. A convolutional layer, pooling layer, and fully linked layer are the three primary layers of the CNN [26]. The CNN's primary building component is a convolutional layer. It works in a hierarchical fashion, with each neuron connected to the layers before it. CNNs are commonly employed in classification applications. In this case, the pooling layer reduces the representations acquired from the convolutional layer, and subsequent layers are often employed to transform multidimensional sizes into one dimension.

Mohd Imran Khan, Rajib Maity et al. [17] and Ali Haidar et al. [23] used CNN as a prediction model in their proposed method. The one-dimensional CNN has uses in applications such as sequence prediction and series analysis. The CNN consists of an input layer and an output layer that are connected using an arbitrary number of hidden layers. These layers are linked by an activation function for input and output data flow. The input layer accepts 3D (input data) signals and dispatches them to the hidden layer. The model's computational engine is represented through the hidden layers. Depending on the requirements of the task, the CNN model may have one or more Conv1D layers, as well as a max-pooling dropout. Filters capture input characteristics, while kernels determine filter peaks. The model can extract sequences of hidden information from the training phase using a particular input dimension. Further, it reduces output complexity, owing to the possibility of overtraining.

### 8.3.2.6 Long Short-Term Memory (LSTM)

Swapna et al. [24] applied the LSTM technique to train their proposed system. In data science, the sequence prediction problem is perhaps the most difficult of all, though a DL model like the LSTM helps tackle the problem successfully. This is because it has a unique ability to selectively retain patterns for long durations, rendering it superior to others. Multiplication and addition are used to make minor changes to the data, and the information is passed via cell states by the LSTM. There are three dependencies, comprising the prior cell state, private concealed state, and current time step input. There are three gates in total: the forget gate keeps asides the information from the cell state, the input gate can give information to the cell state, and the output gate alters the data by adding or multiplying it.

## 8.4 PREPROCESSING TECHNIQUES IN RAINFALL PREDICTION SYSTEM

Few preprocessing techniques such as Principal Component Analysis (PCA) and Inverse Distance Weighting (IDW) have been adapted to improve the quality of rainfall prediction datasets in the existing systems. Based on such techniques, existing works are classified and shown in Figure 8.3.

### 8.4.1 DOWNSAMPLING

Downsampling reduces the number of training samples in the majority category since it helps balance target category counts. When the data gathered are deleted, key information is often lost [14], and this is where downsampling comes in, given that it balances the quantum of positive and negative labels. The downsampling approach is used to balance the amount of positive and negative labels. Manandhar et al. [16] took into account all of the examples from both the minority and majority scenarios, picked at random in order to have a balanced minority/majority ratio. Although the ratio 1:1 is commonly used, alternative ratios may be considered as well.

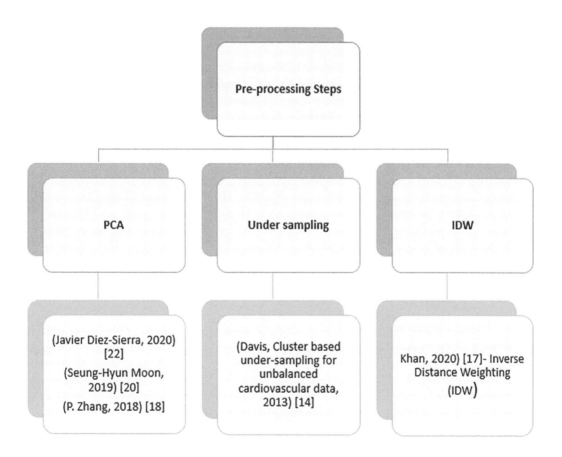

**FIGURE 8.3**  Data preprocessing techniques in the rainfall prediction systems.

### 8.4.2 Principal Component Analysis

In 2020, Diez-Sierra et al. [21] used rainfall data from the Insular Council of Agua de Tenerife (CIATF), Tenerife's water planning and management organization. The PCA was used to preprocess the rainfall data. The statistical procedure involves converting a set of correlated variables into uncorrelated variables using orthogonal transformation. Exploratory data analysis and predictive models are major outcomes of PCA. This unsupervised statistical technique investigates dependability among variables in the input dataset. A generic factor analysis of the PCA determines the best fitting line, using regression as the basis.

### 8.4.3 Inverse Distance Weighting

Khan et al. [17] sourced data from an Indian meteorological site and applied the IDW model to preprocess the data. The deterministic multivariate interpolation method, IDW, uses a well-known distributed collection of points. A weighted average of the values of the known points is used to assign values for the unknown points. Here, the estimation of the value $z$ at location $\mathbf{x}$ is a weighted mean of the observations which are closer.

$$\widehat{z}(\mathbf{x}) = \frac{\sum_i^n w_i z_i}{\sum_i^n w_i}, \tag{8.1}$$

where

$$w_i = |\mathbf{x} - \mathbf{x}_i|^{-\beta}. \tag{8.2}$$

## 8.5 PERFORMANCE MEASURES IN RAINFALL PREDICTION MODULES

### 8.5.1 Root Mean Square Error (RMSE)

The RMSE is calculated as the square root of the mean of the square of all mistakes. The RMSE error metric is a frequently used error calculation technique for numerical predictions.

$$\text{RMSE} = \sqrt{\frac{1}{n}\sum_{i=1}^n (S_i - O_i)^2}, \tag{8.3}$$

where $O_i$ denotes the number of observations, $S_i$ denotes the projected value of the variable, and $n$ is the total number of observations [12]. Because the RMSE is scale-dependent, it should only be used to compare prediction errors between different models or to configure models with a single variable, and not across variables.

### 8.5.2 Normal Standard Error (NSE)

The NSE method is used to assess the efficacy of a suggested model. The NSE value is calculated by the equation:

$$\text{NSE} = 1 - \frac{\sum_{t=1}^n (Y_t - Y_t')^2}{\sum_{t=1}^n (Y_t - \bar{Y})^2}. \tag{8.4}$$

Khan, in his 2020 [2] paper, mentioned $Y_t$ (mm) and $Y_m'$ (mm) as the observed and expected rainfall at time $t$ (mm) and $Y$ (mm), respectively, as the means of the predicted and observed rainfall, with $n$ as the total number of data points. It is found, from Ref. [13], that the value of the NSE is between $(-\infty, 1)$. A score higher than 0 indicates that the model is more efficient than the said score, which means that the derived values of the model are as expected and in line with the average of the observed values. The model's performance is poor if the NSE is less than zero.

### 8.5.3 THREAT SCORE (TS)

The TS is a widely utilized approach to determine the prediction skill scores that are largely employed in China's meteorological industry. The TS assesses prediction accuracy [3] and calculates short-term rainfall forecast accuracy as well as prediction capabilities, according to the China Meteorological Administration. The TS method was used to preprocess the dataset in Refs. [18,35].

## 8.6 ANALYSIS OF EXISTING WORKS

This section draws a comparison of the papers discussed above, and the findings are depicted in Table 8.1. The authors' names, duration of dataset collection, techniques adopted, evaluation metrics and, finally, the dataset attributes considered for rainfall prediction are taken for the comparison. Manandhar et al. [16] proposed a method to analyze the factors that influence precipitation in the atmosphere. It was discovered that ground-based meteorological aspects, seasonal and diurnal components, and weather variables are crucial for rainfall event prediction, and were therefore incorporated into the model. Of all of the characteristics that are critical to rainfall classification, however, the pulse wave velocity (PWV) feature achieves the highest detection rate, while factors like SR (solar radiation), Date of Year (DoY) (seasonal data), and Hour of Day (HoD) (diurnal data) help reduce false alarm rates. Khan and Maity [17] collected data from the Indian Meteorological Department for the period of 1941–2005, and a multi-step daily rainfall forecast from 1 to 5 days was undertaken. A hybrid Conv-1D MLP model was employed to forecast rainfall, with Conv1D, max pooling, and dropout present in one or more layers. Filters capture features from the input data in the signal format, while kernels influence filter size. Zhang et al. [18] took MICAPS data for surface mapping, altitude, and ground mapping to estimate rainfall 3 hours ahead of time in several Chinese cities in order to enhance short-term forecasting accuracy. Ground-mapping data were used to determine the rainfall forecasting region.

Mapping data from 2015 and 2016 served as the model's training dataset, while data from 2017 served as its testing dataset. The PCA, which is an MLP input, was used to lower dataset dimensions. The MLP structure was determined using the greedy approach. The proposed unique MLP-based Dynamic Regional Combined short-term rainfall forecasting model (DRCF) solution outperformed existing techniques such as the ARIMA, Radial Basis Function Neural Network (RBFNN), and Support Vector Machine (SVM) in terms of the RSME and TS value. Moon et al. [19] sourced regional meteorological data from the South Korean Automatic Weather Station (AWS). A ML - driven test was run in 652 South Korean locations (Moon, 2019) between 2007 and 2012 to study an early warning system for very short-term rainfall. Logistic Regression (LR) was applied to determine the result, which has improved with time. It was observed that the longer the LR model trains the data, the better the accuracy with each passing day. Lazri et al. [20] designed a model to predict precipitation estimates for the given satellite data. Their model incorporated several techniques, including the RF, Weighted k-Nearest Neighbors, SVM, ANN, Naive Bayes (NB), and the K-means algorithm. The findings were divided into six groups, depending on the six different precipitation-level categories of *very high*, *moderate*, *moderate to high*, *light to moderate*, *light*, and *no rain*. In this particular case, the rainfall data acquired and the performance of the constructed model are superior to that of others. Diez-Sierra et al. [21] used rainfall data from the CIATF, Tenerife's water planning and management

**TABLE 8.1**

**Analysis of Existing Works**

| Paper Reference | Methodologies | Performance Metrics |
|---|---|---|
| [16] | SVM, SVR | RMSE |
| [17] | Conv 1D-MLP | RMSE, NSE |
| [18] | MLP | RMSE, TS |
| [19] | Logistic Regression | RMSE |
| [20] | RF, ANN, SVM, NB, weighted K-nearest neighbor (WkNN) | RMSE |
| [21] | SVM, RF, k-NN | RMSE |
| [22] | MLP, ARIMA | NSE |
| [23] | CNN | Mean absolute error (MAE), root mean square error (RMSE), Nash-Sutcliffe efficiency (NSE), Pearson Correlation |
| [24] | LSTM | RMSE |
| [25] | Autoencoder network, MLP | MSE, RMSE |
| [32] | MLR, ANN, Levenberg–Marquardt algorithm with Multilayer Perceptron | Pearson correlation, Statistical error |
| [33] | Adaptive-network-based fuzzy inference system (ANFIS)-Particle Swarm Optimization (PSO), ANFIS- fuzzy c means clustering (FCM), motivation, ability, role perceptions, and situational factors (MARS) and M5Tree, multi-model simple averaging (MM-SA) | NSE, RMSE, Scatter Index and adjusted index of agreement |
| [34] | Weather Research and Forecasting Model (WRF) | Sensitivity analysis of microphysics |
| [35] | Dislocation machine learning method based on a support vector machine | Threat score |
| [36] | MLR, ANN, Levenberg–Marquardt algorithm with Multilayer Perceptron | Pearson correlation, Statistical error |
| [37] | Long Short-Term Memory (LSTM) and Gated 23 Recurrent Unit (GRU), artificial neural network (ANN) models | NSE, Mean Absolute Error (MAE), MSE, EQ |
| [38] | Ensemble numerical weather prediction (NWP)-based ensemble prediction system (TAPEX) | CC, CE, RMSE |
| [39] | Particle Swarm Optimization Neural Network (PSO-NN) Ensemble Prediction (PNNEP) model | Prediction scores (Ps), root mean square errors, mean relative errors |
| [40] | WRF, ANN, SVM, partial mutual information (PMI) | NS, SMER, Mean Absolute Percentage Error (MAPE), Percent of Total Volume Error (PTVE) |

organization, which comprises data for 125 gauges for an average duration of 15 years, from 1979 to 2015. In Tenerife, Spain, the performance of a few statistical models and ML algorithms was tested for long-term daily rainfall, driven by synoptic atmospheric patterns. Statistical and ML approaches were applied, depending on the use of conventional linear regression models and modified linear models. The quality of the occurring rainfall and intensity of rainfall forecasts are examined in this article. The primary distinguishing feature of this study, in comparison to others, is its application of cross-validation utilizing the RF, SVM, and other techniques.

Chen et al. [22] provided a new data fusion system based on ML to improve precipitation based on satellite extraction by merging dual-polarization readings from a ground radar network. In this work, a MLP model was developed by geostationary satellite infrared (IR) data and low earth orbit (LEO) satellite passive microwave (PMW)-based retrievals as inputs to produce rainfall estimates. Rain of exceptional quality outputs from the ground radar network served as target labels for training this MLP model. The Climate Prediction Centre morphing technology (CMORPH) was used to preprocess the satellite data. The findings show that the suggested data fusion architecture might be utilized to provide credible precipitation forecasts and to construct future satellite retrieval methods as a backup tool. Haidar et al. [23] gathered information from five different sources, including climate data from the BoM (also known as the Bureau of Meteorology), the Royal Netherlands Meteorological Institute, Climate of the Twenty-First Century (C20C), Earth System Research Laboratory (ESRL) 30, and Solar Influences Data Analysis Center (SIDC). A deep CNN approach was used to forecast each month's rainfall for a specific eastern Australian region using several features, such as the Southern Oscillation Index, Nino 1.2, Nino 3.0, Nino 3.4, Nino 4.0, and Dipole Mode Index, among others. The created model was compared to a MLP model, and the first iteration of the Australian Community Climate and Earth System Simulator (ACCESS-S1) was executed. The proposed model performed better. Future directions include improving prediction accuracy with the addition of new places and datasets.

Swapna et al. [24] collected data from the Hudhud cyclonic storm that struck the Andhra Pradesh coast to forecast the average rainfall for the following month. The LSTM technique was applied as the training algorithm, and their model was compared with others like it. Although the proposed model outperformed the rest, the results do show occasional deviations, possibly owing to delayed monsoons and cyclonic storm occurrences. Future research will only focus on cyclone-based datasets, with the same model being used to enhance performance. Hernandez et al. [25] forecast the cumulative daily precipitation in Manizales city, Columbia, for the following day. The data used are from 2002 to 2013, and obtained from a meteorological station at Manizales. An autoencoder network technique was applied to preprocess the data, with the MLP for prediction. The results showed that the autoencoder and MLP reduce the RMSE and MSE values, compared to other algorithms. Future directions include improving the architecture used for heavy and light rainfall and testing assorted DL techniques.

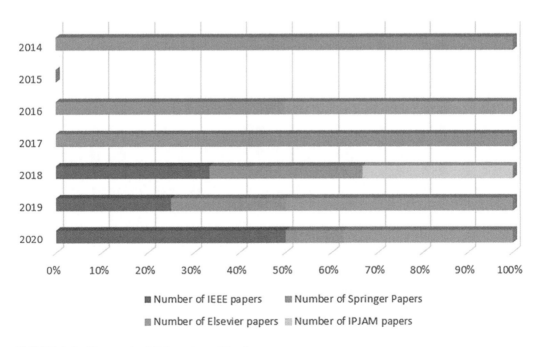

**FIGURE 8.4** Year- and publisher-wise publication.

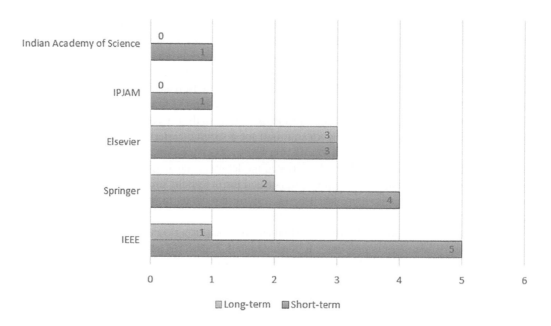

**FIGURE 8.5** Prediction term-wise publications.

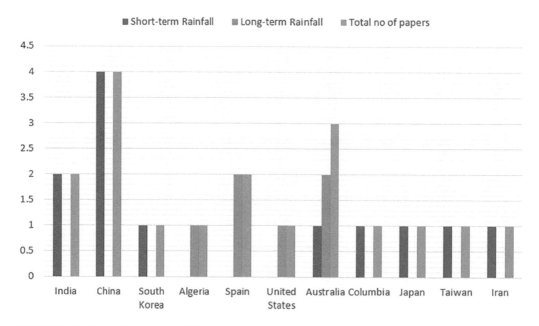

**FIGURE 8.6** Research arenas by different countries.

### 8.6.1 PUBLICATION STATISTICS

Figure 8.4 describes the number of papers collected across different years from different publishers, showing that IEEE accounts for 50% of all papers in 2020, with 40% from Elsevier. Likewise, around 25% of 2019 papers are from IEEE, 50% from Elsevier, and the rest from Springer.

Figure 8.5 depicts the number of long-term and short-term papers on rainfall prediction systems from each publisher.

One paper from IEEE is on long-term prediction and the rest (5) are on short-term prediction. Two papers and four more from Springer are on long-term and short-term prediction, respectively. From Elsevier, three papers are on long-term and three on short-term prediction. Figure 8.6 describes the number of papers worked on in different areas in different countries. The two papers from India are on short-term prediction, and none on long-term prediction. Likewise, of the total of four papers from China, all are on short-term prediction, while the sole Algerian paper is on long-term prediction. Of the total of three papers from Australia, one is on short-term and two are on long-term prediction. One paper each on short-term prediction is from South Korea, Columbia, Japan, and Taiwan, with none on long-term prediction. Figure 8.7 shows that 11 papers are based on ML, nine on DL, and five on a fusion of both.

## 8.6.2 COUNTRY-WISE SYSTEMS

This section briefly discusses the countries in which the work above has been done. It is seen that much of the research has been based in China. The work in Refs. [18,35,37,39] scrutinizes rainfall in different Chinese areas like Fujian Province, Guanxi, and Shenzhen, to name a few.

The reason for the extensive work on rainfall in China is the plentiful availability of datasets and the lack of clarity in rainfall prediction. Two papers [17,24] have studied the rainfall in India, particularly in places like Akola, Amravati, Nagpur, Maharashtra, and Andhra Pradesh. In Maharashtra, the monsoon rainfall is so heavy that several places are flooded. Accurate and timely rainfall prediction is much needed, especially when a heavy downpour is likely to occur.

Three papers [19,34,40] have dealt with the Kinu Watershed in Japan, as well as areas in Korea and Taiwan, respectively. Another three papers [23,32,36] have discussed rainfall prediction systems in different Australian provinces. Much research has been undertaken here, owing to the availability of public datasets and a dire need for rainfall prediction. The papers [2,21,33] have studied prediction systems in Algeria, Spain, and Italy, respectively. The reason for the rather small body of

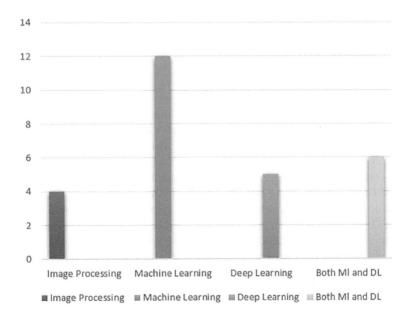

**FIGURE 8.7** Methodology-wise distribution.

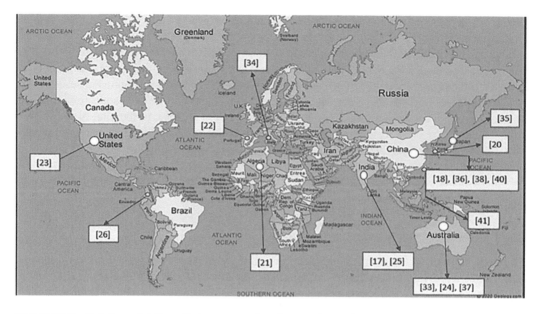

**FIGURE 8.8**   Existing rainfall prediction systems locations on the world map.

work in these countries is that much of the research was undertaken prior to 2015, and the present paper has considered only the latest research. There are, likewise, only two papers [22] and [25] from the United States and Columbia. Despite the plethora of research on this topic in the United States, there are very few readily available papers simply because their weather dataset is not publicly accessible. Figure 8.8 depicts the research done on the world map pictorially.

## 8.7   DATASETS FOR RAINFALL PREDICTION MODEL

The datasets used are classified into two types and described in Table 8.2 as (a) weather numerical data and (b) satellite images, i.e., GIS images.

### 8.7.1   MODULES WITH WEATHER NUMERICAL DATASET

Numerical data are information that is presented in the form of numbers rather than language or a descriptive format. Numerical data, also known as quantitative data, is always gathered in number form and differs from other forms of number data in that it can be statistically and arithmetically computed. The capacity to carry out arithmetic operations with these numbers distinguishes numerical data from other number form data types. Manandhar et al. [16] worked on a dataset from the Wide-Angle High-Resolution Sky Imaging System (WAHRSIS), taking the temperature, relative humidity, and dew point as its attributes. In 2020, Khan [17] used a dataset from the Indian Meteorological Department, with the maximum air temperature, maximum relative humidity, and sea-level pressure as attributes. In 2018, Zhang [18] employed surface mapping data and numerical forecasting results from the China Meteorological Administration as inputs. Figure 8.9 details papers that used the weather numerical dataset and the ones that used a fusion of GIS images and the numerical dataset. Moon et al. [19] sourced data from the AWS, South Korea, with wind, temperature, humidity, and precipitation as attributes. Lazri et al. [20] used the MSG geostationary satellite and MSG multispectral data and chose attributes like radar data and rain gauge data. In 2020, Diez-Sierra [21] took data from the 2018 CIATF and chose attributes like daily values,

**TABLE 8.2**
**Dataset Description**

| Paper Reference | Name of the Country/City | Dataset Types (Global Information System (GIS) Image/ Numerical) | Dataset Detail | Dataset Source | Years of Collection | Methodologies | Performance Metrics |
|---|---|---|---|---|---|---|---|
| [16] | Global | GIS Image, Numerical | Temperature, Relative Humidity, Dew Point, Solar Radiation, PWV along with Seasonal and Diurnal variables | Wide-Angle High-Resolution Sky Imaging System (WAHRSIS) | 4-years (2012–2015) | SVM, SVR | RMSE |
| [17] | Akola, Amravati, Aurangabad, Chandpur, Kolhapur, Latur, Nagpur, Nanded, Nashik, Navi Mumbai, Pune, and Solapur, in Maharashtra, India | Numerical | Maximum air temperature, geopotential height, long-wave radiation, maximum relative humidity, minimum relative humidity, u-wind speed, v-wind speed, sea-level pressure | Indian Meteorological Dept. | 64-years (1941–2005) | Conv 1D-MLP | RMSE, NSE |
| [18] | China | Numerical | Ground surface wind direction, ground surface air pressure, 3 hours pressure change, air pressure, etc. | Surface mapping data, altitude (500 hPa) mapping data and numerical forecasting results released by the China Meteorological Administration | 2 years (2015–2017) | MLP | RMSE, TS |
| [19] | South Korea | Numerical | Wind, temperature, humidity, horizontal wind speed, rain sensor, precipitation | AWS, South Korea | 5 years (2005–2012) | Logistic Regression | RMSE |
| [20] | North-East Algeria | GIS Image, Numerical | Radar data classification, Rain gauge data classification | MSG geostationary satellite, MSG multispectral data | 6 years (2006–2012) | RF, ANN, SVM, NB, WkNN | RMSE |

*(Continued)*

**TABLE 8.2 (*Continued*)**
**Dataset Description**

| Paper Reference | Name of the Country/City | Dataset Types (Global Information System (GIS) Image/Numerical) | Dataset Detail | Dataset Source | Years of Collection | Methodologies | Performance Metrics |
|---|---|---|---|---|---|---|---|
| [21] | Tenerife, Spain | Numerical | Rainfall information of 125 gauges, daily values, sub-daily values, average values | CIATF, 2018 database | 25 years (1979–2015) | SVM, RF, k-NN | RMSE |
| [22] | United States | GIS Image, Numerical | Geostationary satellite infrared (IR) data and low earth orbit satellite passive microwave (PMW)–based retrievals as inputs | Several GEO satellites and number of LEO satellites | 3 years (2013–2015) | MLP, ARIMA | NSE |
| [23] | Innisfail, Eastern Australia. North Queensland | Numerical | Source, minimum, maximum, average, median, and standard deviation of each weather attribute from BoM, Meteorological Institute Climate Explorer (KNMI), Climate of the 20th century (C20C), Earth System Research Laboratory (ESRL), Solar Influences Data Analysis Center (SIDC) | Bureau of Meteorology (BoM) | 104 years (January 1908–December 2012) | CNN | MAE, RMSE, NSE, Pearson Correlation® |
| [24] | Visakhapatnam, Andhra Pradesh, India | Numerical | Temperature Max, Temperature Min, Wind, Pressure Visibility, etc. | Hudhud cyclone data | Hudhud cyclone data (October12, 2014) | LSTM | RMSE |
| [25] | Manizales | Numerical | Temperature, relative humidity, barometric pressure, sun brightness, speed and direction of wind | A meteorological station located in a central area of Manizales | 12 years (2002–2013) | Autoencoder network, MLP | MSE, RMSE |

(*Continued*)

**TABLE 8.2 (Continued)**
**Dataset Description**

| Paper Reference | Name of the Country/City | Dataset Types (Global Information System (GIS) Image/Numerical) | Dataset Detail | Dataset Source | Years of Collection | Methodologies | Performance Metrics |
|---|---|---|---|---|---|---|---|
| [32] | Western Australia. | Numerical | Oceanic climate drivers, El Niño Southern Oscillation (ENSO) and Indian Ocean Dipole (IOD) | Australian Bureau of Meteorology (BoM), The Climate Explorer website | 48 years (1975–2013) | MLR, ANN, Levenberg–Marquardt algorithm with Multilayer Perceptron | Pearson correlation, Statistical error |
| [33] | Samoggia, Italy | Numerical | Hourly rainfall and runoff data, land cover data, soil data | Italian Geographic Military Institute, CORINE database, soil map provided by the local administration | 3 years (2014–2016) | ANFIS-PSO, ANFIS-FCM, MARS and M5Tree, multi-model simple averaging (MM-SA) | NSE, RMSE, Scatter Index and adjusted index of agreement |
| [34] | Kinu Watershed, Japan | Numerical | Radar rainfall data, lower-tropospheric wind, temperature, and water vapor | Japan Institute of Country-ology and Engineering | 2015 | Weather Research and Forecasting Model (WRF) | Ensitivity analysis of microphysics |
| [35] | Shenzhen, southern China | GIS Image, Numerical | Top brightness temperature (TB), cloud top temperature (CTT), gradient of the pixel TB (GT), difference in TB (DT), cloud total amount (CTA), cloud type (CT), and middle and upper tropospheric water vapor content (WVC) | FengYun II geostationary meteorological satellite imagery and ground in situ observational data | – | Dislocation machine learning method based on a support vector machine | Threat score |

*(Continued)*

**TABLE 8.2 (*Continued*)**
**Dataset Description**

| Paper Reference | Name of the Country/City | Dataset Types (Global Information System (GIS) Image/Numerical) | Dataset Detail | Dataset Source | Years of Collection | Methodologies | Performance Metrics |
|---|---|---|---|---|---|---|---|
| [36] | Western Australia. | Numerical | Oceanic climate drivers, El Niño Southern Oscillation (ENSO) and Indian Ocean Dipole (IOD) | Australian Bureau of Meteorology (BoM), The Climate Explorer website | 48 years (1975–2013) | MLR, ANN, Levenberg–Marquardt algorithm with Multilayer Perceptron | Pearson correlation, statistical error |
| [37] | Fujian Province, Southeast China | Numerical | Hourly flow discharge data, hourly precipitation data, flow discharge data | Bureau of 147 Hydrology in Sanming City, Fujian Province | 14 years (2000–2014) | Long Short-Term Memory (LSTM) and Gated 23 Recurrent Unit (GRU), artificial neural network models (ANN) | NSE, MAE, MSE, EQ |
| [38] | West-central Taiwan | Numerical | Hourly rainfall data | Central Weather Bureau, Taiwan | 5 years (2010–2015) | NWP-based ensemble prediction system (TAPEX) | CC, CE, RMSE |
| [39] | Guangxi, China | Numerical | Sea surface temperature, 500 hPa temperature, 200 hPa geopotential height | Guangxi Meteorological Information Center, China | 46 years (1959–2004) | Particle Swarm Optimization Neural Network (PSO-NN) Ensemble Prediction (PNNEP) model | Prediction scores (Ps), root mean square errors, mean relative errors |
| [40] | Karun-4 basin in southwest of Iran | | Type between Rain Gauge and Hydrometry, Latitude, Longitude, Height, Rainfall, Mean Temperature, Relative Humidity, Wind Speed | Karun-4 basin | – | WRF, ANN, SVM, PMI | NS, SMER, MAPE, PTVE |

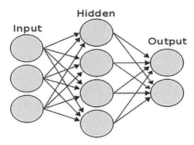

**FIGURE 8.9**   Dataset for rainfall prediction systems.

sub-daily values, average values, and information from 125 gauges. Chen et al. [22] utilized data from the Hudhud cyclone with the maximum temperature, minimum temperature, and wind pressure as attributes. Verma et al. [23] chose a dataset from the BOM, with attributes like the average, median, and standard deviation of each weather attribute. Hernández et al. [25] obtained data from the meteorological station at Manizales, with temperature, barometric pressure, and the brightness of the sun as attributes. Hossain et al. [32] sourced data from the Australian Bureau of Meteorology, while Adnan [33] chose a dataset from the CORINE database of the Italian Geographic Military Institute, with hourly rainfall, runoff data, and land cover data as attributes. Lu et al. [34] utilized a dataset from the Japan Institute of Country-ology and Engineering, with radar rainfall data, temperature, lower-tropospheric wind, and water vapor as attributes. Chen et al. [35] worked on data from the FengYun II geostationary meteorological satellite imagery and ground in situ observational data with brightness, cloud top temperature, and cloud type as attributes. Gao et al. [37] took data from the Bureau of Hydrology and Water Resources in Sanming City, Fujian Province, with hourly flow discharge data, hourly precipitation data, and flow discharge data as attributes. In 2017, Wu [38] used data from the Central Weather Bureau, Taiwan. Jin et al. [39] sourced a dataset from the Guangxi Meteorological Information Center, China, with sea surface temperature, 500 hPa temperature, and 200 hPa geopotential height as attributes. Kisi et al. [40] obtained data from the Karun-4 basin with the rain gauge, hydrometry, latitude, longitude, height, rainfall, mean temperature, relative humidity, and wind speed as attributes. The papers discussed above used weather numerical datasets as inputs.

### 8.7.2   MODULES WITH FUSION OF GIS AND WEATHER NUMERICAL DATASET

Satellite datasets combine a number of images to train and test various predictive scenarios. There are two types of satellite data available, public and private. Public datasets are freely available, of good quality, and have countless use cases. Commercial or private datasets provide even better images for more uses. Image resolution is the first consideration when choosing a dataset. The higher the resolution, the greater the wealth of detail displayed. Manandhar et al. [16] used a fusion of GIS images and numerical data in their choice of data from the WAHRSIS, alongside attributes such as temperature, dew point, relative humidity, PWV with seasonal and diurnal variables, and SR. Chen et al. [22] sourced data from several Geostationary (GEO) and LEO satellites, using attributes like geostationary satellite IR data and low earth orbit satellite PMW-based retrievals. Chen et al. [35] utilized data from the FengYun II geostationary meteorological satellite imagery and ground in situ observational data, which included both GIS images and numerical data.

## 8.8   CONCLUSION

Estimating the quantum of rainfall is critical for water resource management, human survival, and environmental protection. Rainfall estimation, which is heavily influenced by geographical and regional variations and quality, might lead to inaccurate or incomplete results. This study has provided an overview of several rainfall prediction techniques, in addition to which it has raised concerns that are likely to occur when a slew of rainfall forecasting methodologies are employed. Some of the proposed models were only capable of predicting local area forecasts, not global ones. Other models, when applied, turned out to be time-consuming and in need of optimization. Some papers applied different algorithms for different datasets in a process that turned out to be lengthy. Also, certain proposed models were not fitted in extreme situations. The findings show that only a few papers proposed a model for India. Most of the papers worked on medium-sized datasets rather than large ones. Further, they incorporated ensemble techniques with ML and DL to combine model diversities for better prediction.

## BIBLIOGRAPHY

[1] Pankratz, A., 2009. *Forecasting With Univariate Box-Jenkins Models Concepts and Cases*, p. 414. John Wiley Sons, Inc. New York.

[2] Singh, P. and Borah, B., 2013. Indian summer monsoon rainfall prediction using artificial neural network. *Stochastic Environmental Research and Risk Assessment* 27(7), 1585–1599.

[3] Tengeleng, S. and Armand, N., 2014. Performance of using cascade forward back propagation neural networks for estimating rain parameters with rain drop size distribution. *Atmosphere* 5, 454–472. https://doi.org/10.3390/atmos5020454.

[4] Wang, H. and Raj, B., 2015. A survey: Time travel in deep learning space: An introduction to deep learning models and how deep learning models evolved from the initial ideas. *arXiv preprint arXiv:1510.04781.*

[5] Grossi, E. and Buscema, M., 2008. Introduction to artificial neural networks. *European Journal of Gastroenterology & Hepatology* 19, 1046–1054.

[6] Evgeniou, T. and Pontil, M., 2001. Support vector machines: Theory and applications. In *Machine Learning and Its Applications: Advanced Lectures*, pp. 249–257. Springer, Berlin, Heidelberg. https://doi.org/10.1007/3-540-44673-7_12.

[7] Mohamed, H., Negm, A., Zahran, M. and Saavedra, O.C., 2015. Assessment of artificial neural network for bathymetry estimation using high resolution satellite imagery in shallow lakes: Case study El Burullus Lake. In *Proceedings of the Eighteenth International Water Technology Conference, IWTC18, Sharm ElSheikh, Egypt, 12–14 March 2015.*

[8] Jijo, B.T. and Abdulazeez, A.M., 2021. Classification based on decision tree algorithm for machine learning. *Journal of Applied Science and Technology Trends*. 2, 20–28.

[9] Kiranyaz, S., Avci, O., Abdeljaber, O., Ince, T., Gabbouj, M. and Inman, D. J., 2019. 1D convolutional neural networks and applications: A survey. *Mechanical Systems and Signal Processing* 151, 107398.

[10] Harbola, S. and Coors, V., 2019. One dimensional convolutional neural network architectures for wind prediction. *Energy Conversion and Management* 195, 70–75.

[11] Srivastava, N., Hinton, G., Krizhevsky, A., Sutskever, I. and Salakhutdinov, R., 2014. Dropout: A simple way to prevent neural networks from overfitting. *The Journal of Machine Learning Research* 15(1), 1929–1958.

[12] Neill, S. P. and Hashemi, M.R., 2018. Ocean modelling for resource characterization. In *Fundamentals of Ocean Renewable Energy*, pp. 193–235. https://doi.org/10.1016/B978-0-12-810448-4.00008-2

[13] Pichuka, S., Prasad, R., Maity, R. and Kunstmann, H., 2017. Development of a method to identify change in the pattern of extreme streamflow events in future climate: Application on the Bhadra reservoir inflow in India. *Journal of Hydrology: Regional Studies* 9, 236–246.

[14] Rahman, M. M. and Davis, D. N., 2013. Cluster based under-sampling for unbalanced cardiovascular data. In *Proceedings of the World Congress on Engineering*, pp. 3–5.

[15] Basak, D., Pal, S. and Patranabis, D., 2007. Support vector regression. *Neural Information Processing – Letters and Reviews* 11(10), 203–224.

[16] Manandhar, S., Dev, S., Lee, Y. H., Meng, Y. S. and Winkler, S., 2019. A data-driven approach for accurate rainfall prediction. *IEEE Transactions on Geoscience and Remote Sensing* 57(11), 9323–9331. https://doi.org/10.1109/TGRS.2019.2926110.

[17] Khan, M.I. and Maity, R., 2020. Hybrid deep learning approach for multi-step-ahead daily rainfall prediction using GCM simulations. *IEEE Access* 8, 52774–52784. https://doi.org/10.1109/ACCESS.2020.2980977.

[18] Zhang, P., Jia, Y., Gao, J., Song, W. and Leung, H., 2020. Short-term rainfall forecasting using multi-layer perceptron. *IEEE Transactions on Big Data* 6(1), 93–106. https://doi.org/10.1109/TBDATA.2018.2871151.

[19] Moon, S.H., Kim, Y.H., Lee, Y.H. and Moon, B.R., 2019. Application of machine learning to an early warning system for very short-term heavy rainfall. *Journal of Hydrology* 568, 1042–1054. https://doi.org/10.1016/j.jhydrol.2018.11.060.

[20] Lazri, M., Labadi, K., Brucker, J.M. and Ameur, S., 2020. Improving satellite rainfall estimation from MSG data in Northern Algeria by using a multi-classifier model based on machine learning. *Journal of Hydrology* 584, 124705. https://doi.org/10.1016/j.jhydrol.2020.124705.

[21] Diez-Sierra, J. and Del Jesus, M., 2020. Long-term rainfall prediction using atmospheric synoptic patterns in semi-arid climates with statistical and machine learning methods. *Journal of Hydrology* 586, 124789. https://doi.org/10.1016/j.jhydrol.2020.124789.

[22] Chen, H., Chandrasekar, V., Cifelli, R. and Xie, P., 2020, A machine learning system for precipitation estimation using satellite and ground radar network observations. *IEEE Transactions on Geoscience and Remote Sensing* 58(2), 982–994. https://doi.org/10.1109/TGRS.2019.2942280.

[23] Haidar, A. and Verma, B. 2018. Monthly rainfall forecasting using one-dimensional deep convolutional neural network. *IEEE Access* 6, 69053–69063. https://doi.org/10.1109/ACCESS.2018.2880044

[24] Swapna, M. and Sudhakar, N., 2018. A hybrid model for rainfall prediction using both parametrized and time series models. *International Journal of Pure and Applied Mathematics* 119(14), 1549–1556.

[25] Hernández E., Sanchez-Anguix V., Julian V., Palanca J., Duque N., 2016. Rainfall prediction: A deep learning approach. In Martínez-Álvarez F., Troncoso A., Quintián H., Corchado E. (eds) *Hybrid Artificial Intelligent Systems. HAIS 2016. Lecture Notes in Computer Science*, vol. 9648. Springer, Cham. https://doi.org/10.1007/978-3-319-32034-2_13

[26] Albawi, S., Mohammed, T. A. and Al- Zawi, S., 2017. Understanding of a convolutional neural network. In *2017 International Conference on Engineering and Technology (ICET)*, pp. 1–6. https://doi.org/10.1109/ICEngTechnol.2017.8308186.

[27] Royal Netherlands Meteorological Institute Climate Explorer. Accessed: Aug. 4, 2015. [Online]. Available: https://climexp.knmi.nl/start.cgi

[28] Folland, C. K., Shukla, J., Kinter, J. and Rodwell, M. J., 2002. C20C: The Climate of the Twentieth Century Project. [Online]. Available: http://cola.gmu.edu/c20c/

[29] Earth System Research Laboratory. Accessed: Aug. 15, 2015. [Online]. Available: https://www.esrl.noaa.gov/

[30] Saha, B. N. and Senapati, A., 2020. Long short term memory (LSTM) based Deep learning for sentiment analysis of English and Spanish data. In *2020 International Conference on Computational Performance Evaluation (ComPE)*, pp. 442–446. https://doi.org/10.1109/ComPE49325.2020.9200054

[31] Park, Y.-S. and Lek, S. 2016. Chapter 7 – Artificial neural networks: Multilayer perceptron for ecological modeling. In Jørgensen, S. E. (ed) *Developments in Environmental Modelling*, vol. 28, pp. 123–140. Elsevier. https://doi.org/10.1016/B978-0-444-63623-2.00007-4.

[32] Hossain, I., Rasel, H.M., Imteaz, M.A., and Mekanik, F., 2020. Long-term seasonal rainfall forecasting using linear and non-linear modelling approaches: A case study for Western Australia. *Meteorology and Atmospheric Physics* 132, 131–141. https://doi.org/10.1007/s00703-019-00679-4

[33] Adnan, R.M., Petroselli, A., Heddam, S., Santos, C.A.G. and Kisi, O., 2021. Short term rainfall-runoff modelling using several machine learning methods and a conceptual event-based model. *Stochastic Environmental Research and Risk Assessment* 35. https://doi.org/10.1007/s00477-020-01910-0.

[34] Lu, T., Yamada, T. and Yamada, T., 2016. Fundamental study of real-time short-term rainfall prediction system in watershed: Case study of Kinu Watershed in Japan. *Procedia Engineering* 154, 88–93. https://doi.org/10.1016/j.proeng.2016.07.423.

[35] Chen, X., He, G., Chen, Y., Zhang, S., Chen, J., Qian, J. and Yu, H., 2019. Short-term and local rainfall probability prediction based on a dislocation support vector machine model using satellite and in-situ observational data. *IEEE Access*. https://doi.org/10.1109/ACCESS.2019.2913366.

[36] Hossain, I., Rasel, H.M., Imteaz, M.A. and Mekanik, F., 2018. Long-term seasonal rainfall forecasting: Efficiency of linear modelling technique. *Environmental Earth Sciences* 77. https://doi.org/10.1007/s12665-018-7444-0.

[37] Gao, S., Huang, Y., Zhang, S., Han, J., Wang, G., Zhang, M. and Lin, Q., 2020. Short-term runoff prediction with GRU and LSTM networks without requiring time step optimization during sample generation. *Journal of Hydrology* 589, 125188. https://doi.org/10.1016/j.jhydrol.2020.125188.

[38] Wu, M.C. and Lin, G.F., 2017. The very short-term rainfall forecasting for a mountainous watershed by means of an ensemble numerical weather prediction system in Taiwan. *Journal of Hydrology* 546, 60–70. https://doi.org/10.1016/j.jhydrol.2017.01.012.

[39] Jin, L., Zhu, J., Huang, Y., Zhao, H.S., Lin, K.P. and Jin, J., 2015. A nonlinear statistical ensemble model for short-range rainfall prediction. *Theoretical and Applied Climatology* 119(3–4), 791–807. https://doi.org/10.1007/s00704-014-1161-8.

[40] Kisi, O. and Cimen, M., 2012. Precipitation forecasting by using wavelet-support vector machine conjunction model. *Engineering Applications of Artificial Intelligence* 25, 783–792. https://doi.org/10.1016/j.engappai.2011.11.003.

[41] Karimi, B., Safari, M., Mehr, A.D. and Mohammadi, M., 2019. Monthly rainfall prediction using ARIMA and gene expression programming: A case study in Urmia, Iran. *Online Journal of Engineering Sciences and Technologies* 2, 8–14.

# 9 Intelligent Coconut Harvesting System

*B.V.A.N.S.S. Prabhakar Rao*
Vellore Institute of Technology

*Ram Prasad Reddy Sadi*
Anil Neerukonda Institute of Technology and Sciences

## CONTENTS

9.1 Introduction ...................................................................................................98
    9.1.1 Background............................................................................................98
    9.1.2 Motivation.............................................................................................99
    9.1.3 Objectives .............................................................................................99
9.2 Proposed Method...........................................................................................99
9.3 Implementation Modules ..............................................................................99
    9.3.1 Coconut Harvesting Schedule ...........................................................100
9.4 Intelligent Coconut Harvesting System Working Model............................ 101
    9.4.1 Classification of Ground Level before Harvesting Coconuts ................... 102
    9.4.2 The Process........................................................................................ 102
    9.4.3 Measure Economic Feasibility of Harvesting ........................................ 103
    9.4.4 Existing Harvesting Methods ............................................................. 103
    9.4.5 Method of Harvest Task to be Automated ............................................. 103
9.5 Algorithm..................................................................................................... 105
9.6 Pseudo Code ................................................................................................ 107
9.7 Conclusion ................................................................................................... 109
9.8 Future Work ................................................................................................. 109
Bibliography .......................................................................................................... 109

## ABSTRACT

Coconut harvesting needs too many skilled people and a lot of time. Nowadays, finding the right person to harvest is also difficult due to the lack of skilled people who can manage everything on their own. Moreover, the existing models to harvest coconuts involve extensive human intervention. In this scenario, we propose a model that helps farmers to reduce the labor cost and also use it at their convenience. It is easy to use and illiterate farmers with a short training stint can effectively use it. The basic objective is to introduce a convenient model that enables farmers to comfortably harvest coconuts without much human intervention.

## KEYWORDS

AI; Coconut Harvesting; Intelligent System; Machine Learning

DOI: 10.1201/9781003374121-9

## 9.1  INTRODUCTION

Nowadays, harvesting coconuts has become an issue due to several risk factors and the availability of only unskilled labor. Skilled laborers are hardly interested in this work, and they do it only on a part-time basis. This has created a lot of hardship for the farmers' fraternity, and they have almost given up on harvesting coconuts. The dearth of skilled laborers is also a major issue. The current generation desires comfortable jobs, which do not involve much risk. In this context, we need to look for alternate solutions to make farmers' lives more comfortable. This work is an effort toward addressing this issue.

Presently, finding the right person for harvesting coconuts is rather difficult. The farmers are hardly able to locate all the necessary resources including labor to harvest the coconuts. Moreover, this line of work does not provide any insurance coverage for the laborers and their family in case of mishaps. The need of the hour is to come up with an unmanned model to fulfill the needs of the farmer at an affordable price.

### 9.1.1  BACKGROUND

Trees have become a haven for rats, snakes, squirrels, and birds, and coconut trees are no exception. Coconut harvesting is laden with risks and adequate measures have to be taken to mitigate the risk involved. All said and done, it is a risky means of livelihood. In certain cases, the coconut trees are thin and tall but produce a good yield. Laborers have reservations in climbing such trees incurring a loss to the farmer. Also, the harvesting process is exceptionally risky during the rainy season and winter. During these seasons, laborers are hesitant to climb the tress. And, in case they are willing to climb, they charge higher wages in view of the risk involved. Loss of life also could happen due to loose soil.

Considering all of the above risk factors, we propose a model that is unmanned, involves less human intervention, is easy to use, and at the same time supplements most of the needs of the farmers on a daily basis. In this process, the farmer makes a one-time investment and with a little regular maintenance can use the model for a longer duration. Also, there is no training cost involved. This not only provides better margins to the farmer but also makes available the product at an affordable price to the common man. Moreover, the person who uses the model need not be literate.

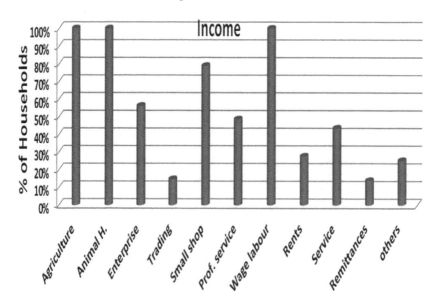

**GRAPH 9.1**   Various income sources for rural areas.

Primary and Secondary Needs have to be fulfilled even under the context of unemployment & poverty. These sector of people have not been able to benefit from the fast economic growth happening in the national economy due to various contextual differences. A major part of the population is engaged in the agriculture sector. They face major impediments like lack of modern technology, dependence on rains, lack of irrigation facilities, and unsuitability of large portions of land for cultivation.

### 9.1.2  MOTIVATION

A system for harvesting the palm fruit using an intelligent palm harvester involves an estimation of the demand for the type of fruit and the capturing of image of at least a tree and at least a parameter from at least an angle; an image-processing module for detecting at least a tree with minimum yield in terms of the number of palm fruits and the condition of the palm fruits to be harvested; a calculation module for calculating the circumference throughout the length of the tree, wherein the diameter of the harvester is adjusted based on the calculated circumference to firmly hold on to the tree; and a plucking module for plucking and dropping the palm fruit on the ground based on the current condition of the fruit, wherein if the fruit is tender or mature, it is dropped from the top of the tree, and if the fruit is young, it is dropped from a safe distance.

### 9.1.3  OBJECTIVES

The objectives of this proposal are as follows:

- The present work relates to developing a field of fruit harvesting systems and a method thereof.
- More particularly, the proposed device relates to a field of harvesting systems that easily harvests palm fruits from taller trees as per the market demand with respect to the season.

## 9.2  PROPOSED METHOD

These days, finding the right persons for harvesting palm fruits is becoming a difficult task. There is a huge demand for disruptive intelligent systems to meet the standards and sustainability [9]. The farmers are finding it difficult to source all the necessary resources including labor to harvest the palm fruits. Also, the lack of insurance coverage for the laborers and their families is a major impediment. Trees have become a resident place for rats, snakes, squirrels and birds. It is widely present in palm fruit trees. People agreeing to do the job of harvesting need to take certain risk to make their livelihood. Necessary care is to be taken to overcome the risk. Moreover, there are certain cases where the palm fruit trees are thin and tall and produce good yield. People usually do not prefer in climbing such tree even if it produces good yield and this results in a loss to the farmer. Harvesting process becomes challenging during rainy and winter seasons. During these seasons, the laborers would not come forward to be part of the process. If they are willing, they tend to ask for more wages than usual as more risk is involved. Human loss might also take place in case of loose soil (Figure 9.1).

## 9.3  IMPLEMENTATION MODULES

1. The system as requested in (1), wherein at least a parameter includes palm fruit inflorescence, dry palm fruit inflorescence, palm fruit leaf, palm fruit stem, palm fruit–type immature palm fruit, young palm fruit, tender palm fruit, and mature palm fruit.
2. The system as demanded in (1), wherein the plucking module plucks the palm fruit from the tree with minimum yield and drops the palm fruit from a few meters' height.

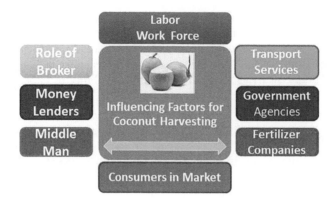

**FIGURE 9.1**    Coconut farming broker roles.

**FIGURE 9.2**    Coconut farming social workers role for better productivity.

3. The system as appealed in (1), wherein the calculation module calculates the actual length of the usual log by subtracting height in meter to the first nut from the average height.
4. The system as demanded in (1), wherein the processing module determines the demand for the marketing of a feature extraction module associated with the pre-processing module for extracting a plurality of features such as color, number of palm fruits on the tree, and texture of the palm fruit; and a classification module associated with the feature extraction module for classifying the plurality of features into a plurality of classes, wherein the plurality of classes include tender, young, and mature palm fruits (Figure 9.2).

### 9.3.1   COCONUT HARVESTING SCHEDULE

A huge amount of fresh coconuts is required as for inclusion in essentials, diets, customary foods. Also, different types of raw coconuts are required for various auspicious occasions. Therefore, coconut production on a large scale can make a major contribution to the Indian economy (Figure 9.3).

| Type of Activity | 1st Year - Process Analysis | | | 2nd Year - Process Synthesis | | |
|---|---|---|---|---|---|---|
| | 1 - 3 m | 4 - 7 m | 8 - 12m | 13 - 15 m | 16 - 20 m | 21 - 24 m |
| Harvesting | ■ | | | | | |
| Cleaning - Preprocessing | ■ | | | | | |
| Collection of Data | | ■ | | | | |
| System Implementation | | | ■ | | | |
| Intelligent Training | | | | ■ | | |
| Model Design | | | | ■ | | |
| Training and Testing | | | | | ■ | |
| Process Verification | | | | | | ■ |
| Design Validation | | | | | | ■ |

**GRAPH 9.2**   Coconut harvesting perspective.

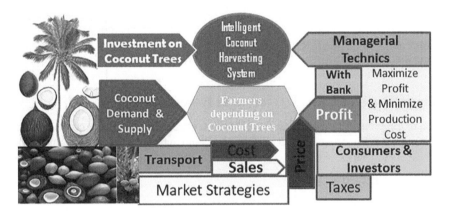

**FIGURE 9.3**   Coconut farmer perspectives.

"India is one of the world's largest producers of coconut, with a turnout of 11,706,343 tons (11,521,459 long tons) in 2018" [8]. But, as per Figure 9.2, we can see how middlemen make huge profits, thereby depriving the farmers of their rightful share (Figure 9.4).

## 9.4   INTELLIGENT COCONUT HARVESTING SYSTEM WORKING MODEL

To build the model, the height of the tree is taken into consideration as shown in Figure 9.4. Besides this, the model should also consider the following aspects: which include

i. The surface where the coconut is being dropped. The surface could be hard, sandy, muddy, or a place where a crop is under cultivation. Taking all these factors into consideration, the coconut is to be dropped in such a manner such that no damage takes place.

ii. If the coconut is being dropped in cultivated land, we need to estimate the damage that is being done to the crop.

**FIGURE 9.4** Coconut tree from the harvesting viewpoint.

iii. We also need to consider the type of coconut that is being dropped. If it is a young coconut, utmost care is to be taken while dropping it. If is a tender coconut, it should be dropped accordingly. However, the process can be a bit relaxed if the coconut is matured.

### 9.4.1 CLASSIFICATION OF GROUND LEVEL BEFORE HARVESTING COCONUTS

The surface of the ground, type of the ground, and objects in and around the ground play an important role in the harvesting process [6]. It might be dry land, hard land, land with rocks or big stones, or wet land; it could also be wet land with crop, sandy soil, or a hut nearby with related things and sensitive objects, and so on.

### 9.4.2 THE PROCESS

An unmanned drone is used to capture the images and send it to the system so as to detect the trees that have the minimum yield in terms of the number of coconuts to be harvested.

The harvesting model is used only for those trees. The harvesting model provides the basic harvesting procedure to cater to the needs of a farmer.

**Phase-1**: The model is fitted to the trunk of the tree. It adjusts automatically based on the circumference of the trunk as the trunk is not uniform from bottom to top.

**Phase-2**: The model on reaching the top captures the image and identifies various objects in the picture which include:
- Coconut inflorescence
- Dry coconut inflorescence
- Coconut leaf
- Coconut stem
- Coconut fruit type (s)
  - Immature
  - Young

**FIGURE 9.5** Coconut inflorescence.

- Tender
- Mature
- Damaged
- Unhealthy

**Phase-3**: The model has to throw down dry coconut inflorescence and dry coconut stem.

**Phase-4**: The goal of the model is to pluck young coconuts or tender coconuts or mature coconuts. If it plucks young coconuts, it has to drop it gently on to the ground. If it plucks tender coconuts or mature coconuts, it can be dropped directly from the top of the tree.

**Phase-5**: The model on completing the task descends gently.

### 9.4.3 MEASURE ECONOMIC FEASIBILITY OF HARVESTING

The cost involved in harvesting the coconut using the model is also an important parameter, as it should not burden the farmer during the process (Figures 9.5–9.9).

### 9.4.4 EXISTING HARVESTING METHODS

- **Auto process**
  Here, the automated system can focus on the climbing cycle or any apparatus.
- **Semi-manual process**
  Many harvesters depend on power tiller operated ladder and such machinery.
- **Fully manual process**
  This is the only conventional approach to climbing coconut trees (Figures 9.10–9.12).

### 9.4.5 METHOD OF HARVEST TASK TO BE AUTOMATED

**FIGURE 9.6**   Immature coconut fruit.

**FIGURE 9.7**   Young coconut.

**FIGURE 9.8**   Tender coconut.

**FIGURE 9.9**   Mature coconut fruit.

**FIGURE 9.10**   Intelligent harvesting implementation.

## 9.5   ALGORITHM

Harvest Coconut
    Input Unmanned Device
    Output Coconuts

1. The device identifies the coconut tree that has the minimum yield.
2. The device is fitted to the coconut trunk manually.
3. The device moves to the top on making necessary adjustments based on the thickness of the trunk of the tree.

**FIGURE 9.11**  Intelligent harvesting system process.

**FIGURE 9.12**  Intelligent harvesting and marketing system.

4. The device identifies the objects and classifies them into coconut inflorescence (dry/live/matured coconut/tender coconut/young coconut/immature coconut), stem (dry/green).
5. If the object is a matured coconut, the device drops them directly to the ground, irrespective of the ground surface.
6. If the object is a tender coconut, the device measures the approximate distance the coconut has to travel and accordingly makes necessary arrangements to drop the coconut so as to avoid any damage or loss based on the surface of the ground (Compute_Height).
7. In the process, the machine also plucks dry coconut inflorescence and dry stems and drops them to the ground

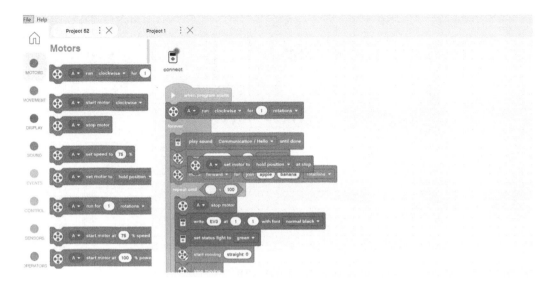

**FIGURE 9.13** Intelligent system module1.

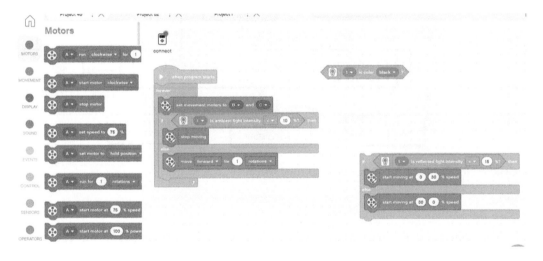

**FIGURE 9.14** Intelligent system main module.

**Compute_Height**: This function records the distance it travels from the spot of fixation to the top of the tree. This measure is used while dropping the tender coconuts to the ground based on the surface with necessary alignments (Figures 9.13–9.15)

**Compute circumference**: This function measures the circumference of the coconut trunk, which is used by the device to align the wheels while moving up or down (Figure 9.16).

## 9.6 PSEUDO CODE

```
// This function returns the height of the object
    Height_of_the_Tree(object) return int
    {
        Compute the approximate aerial distance by using the
unmanned device.
```

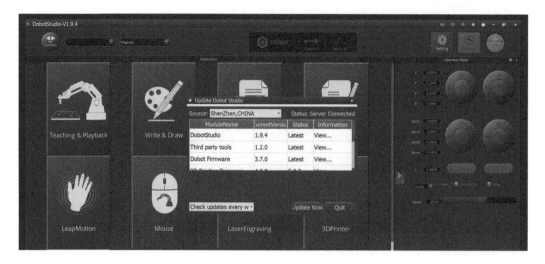

**FIGURE 9.15** Intelligent systems Unmanned Devices (UMD).

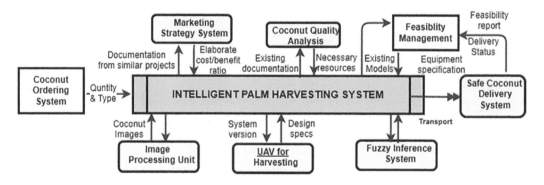

**FIGURE 9.16** Intelligent harvesting and marketing system.

```
            return the height(h);
    }
// This function returns the list of object identifiers based on the type
of theimage
    Identify_the_Object() return object_identifier
    {
        Capture the image
        Classify the object in the image
        return(object_identifier_list);
    }
// This function returns the type of the ground surface based on the type
of the surface
    Test_the_Ground_Surface() return surface_type
    {
        Capture the image of the surface
        Classify the surface
        return (surface_type)
    }
// Driver code
Train the device so as to reach the coconut trees that satisfy the
minimum threshold
```

```
The device is plugged into the tree
     t=Test_the_Ground_Surface();
     if (t==hard_surface || t== muddy_surface)
          load the unmanned device with the rope and the sack
     h = Height_of_the_Tree(object);
     d = 0 // inititalize the distance travelled by the device
     while (d < h)
          d = d+ steps moved by the device
     o = Identify_the_Object();
     if (o in [stem, green in colour])
          pass
     else if (o in [stem, brown in colour])
          pluck and drop
     else if (o in [coconut_inflorescence, o== dry])
          pluck and drop
     else if(o in [live, immature coconut])
          pass
     else if(o in [young, tender, mature])
     {
          t=Test_the_ground_surface()
          if(t==loose_soil)
               drop the object-o
          else
          {
               place the objects in a sack
               drop the sack gently
          }
     }
```

## 9.7 CONCLUSION

A viable technology can support the common man in day-to-day activities in an effective manner to improve productivity. In this context, this chapter illustrates how an intelligent system can support the coconut farmers while harvesting for better marketing. Means of support relates to one's capabilities, access to assets and rights, as well as opportunities for income generation as per the market supply and demand. In agriculture, the self-employed are the farmers who constitute 64% of the total population of agricultural workers. Within the category of farmers, 86% are marginal or small farmers accounting for 45% of the area cultivated. Social workers can play an active role by giving suggestions for ensuring an increase in the awareness of the product in the rural areas, and thereby increase sales and growth prospects for farmers.

## 9.8 FUTURE WORK

Right from the inception, conventional rain fed cultivation has been the foremost basis of revenue generation for deprived people in rural areas. It is supplemented by small handicraft making, livestock rearing, and minimal wages, leading to distress migration. This can be effectively addressed and resolved in future with validation and verification of real datasets if we focus and effectively implement corrective measures in many areas. And, this work can generate many prospects directly or indirectly.

## BIBLIOGRAPHY

1. Wibowo, T.S., Sulistijono, I.A. and A. Risnumawan, End-to-end coconut harvesting. In *Proceedings of 2016 International Electronics Symposium (IES)*, September 2016. DOI: 10.1109/ELECSYM.2016.7861047

2. Abraham, A., Girish, M., Vitala, H. R. and M. P. Praveen, Design of harvesting mechanism for advanced remote-controlled coconut harvesting robot, *Indian Journal of Science and Technology*, 7(10), 1465–1470, 2014. DOI: 10.17485/ijst/2014/v7i10.4

3. Megalingam, R. K., Venumadhav, R., Pavan, A., Mahadevan, A., Kattakayam, T.C. and H. Menon, Kinect based wireless robotic coconut tree climber computer science. In *Proceedings of 3rd International Conference on Advancements in Electronics and Power Engineering (ICAEPE'2013)*, 2013.

4. Megalingam, R.K., Pathmakumar, T., Venugopal, T., Maruthiyodan, G. and A. Philip, DTMF based robotic arm design and control for robotic coconut tree climber. In *2015 International Conference on Computer, Communication and Control (IC4)*, 2015.

5. Bac, C., Hemming, J. and E. V. Henten, Robust pixel-based classification of obstacles for robotic harvesting of sweet-pepper. *Computers and Electronics in Agriculture*, 96, 148–162, 2013.

6. Yamamoto, K., Guo, W., Yoshioka, Y. and S. Ninomiya, On plant detection of intact tomato fruits using image analysis and machine learning methods. *Sensors*, 14(7), 12191–12206, 2014

7. Holland, J., *Adaptation in Natural and Artificial Systems*, University of Michigan Press, 1975.

8. Jaryal, V.B., Singh, D. and N. Gupta, "Graphitic sulphur functionalized carbon sheets as an efficient "turn-off" absorption probe for the optical sensing of mercury ions in aqueous solutions". *New Journal of Chemistry*, 46(12), 5712–5718, 2022.

9. Rao, B.P. and R.K. Singh, Disruptive intelligent system in engineering education for sustainable development. *Procedia Computer Science*, 172, 1059–1065, 2020.

10. Roy, S., Guru, S. and Debnath, S., Design and Performance Analysis of Textile Antenna for Wearable Applications. In *2020 Advanced Communication Technologies and Signal Processing (ACTS)* (pp. 1-4). IEEE, 2020, December.

# 10 IoT-Based Live Ambulance Management and Tracking System

*G. Gayathri, Sarada Manaswini Upadhyayula,*
*O.V. Gnana Swathika, and V. Berlin Hency*
Vellore Institute of Technology

## CONTENTS

10.1 Introduction ........................................................................................................... 111
10.2 Related Works......................................................................................................... 112
10.3 Proposed System..................................................................................................... 114
    10.3.1 Flow Adopted for Designing the System................................................... 115
10.4 Result and Discussion............................................................................................ 118
10.5 Conclusion and Future Scope ............................................................................... 119
References........................................................................................................................ 121

### ABSTRACT

Currently, Internet of Things (IoT) is one of the biggest trends in technology that has altered our lifestyle to a great extent. Each "thing" in the IoT vision has the power of communicating to one another, which brings the thought of the Internet of Everything. A great number of services delivered by IoT can improve our lifestyle and make it more efficient, intelligent, and even reliable. With the help of IoT-based devices, we can design some special services and life-saving systems. In this work, we have demonstrated how IoT can help improve an ambulance tracking and management service for hospital administration. The system is enabled with interactive real-time location tracking on the website coupled with an LCD display highlighting the presence and availability of ambulances at the hospital. This real-time location tracking system installed in the ambulance can help the hospital authorities track the location and convey the necessary details accurately. The system prototype has been designed with NodeMCU, radio frequency identification (RFID), and Global Positioning System (GPS) receiver module linked to the website and ThingSpeak cloud which transmits data to the mobile app through Arduino.

## KEYWORDS

Internet of Things; Ambulance Tracking System; Ambulance Management System; ThingSpeak; GPS; RFID

## 10.1 INTRODUCTION

The idea of inter-connecting basic objects like sensors and actuators and making them equipped with some sort of intelligence was the essence of today's much-talked "Internet of Things (IoT)" system. Despite the concept being coined in the early 1990s, the two important aspects which turned the idea of smart objects into reality were the adoption of IPV6 and wireless sensor networks (WSNs). IoT is a collective term comprising various technologies, applications, and use cases that have certain important strategies or trends like the implementation of Artificial Intelligence;

DOI: 10.1201/9781003374121-10

gathering of data socially, legally, and ethically; data broking; and shift from intelligent edge to intelligent mesh after the WSN development era. Apart from its never-ending list of smart applications, IoT has tremendously transformed the way of implementing vehicular emergency systems. Intelligent transportation systems are one of the example domains of IoT that have witnessed much advancement, especially when vehicle-to-infrastructure communication and real-time tracking systems are being used. A tracking system for vehicles, which is also known as a fleet tracking system, is a system used to track and manage/coordinate vehicles belonging to an enterprise/individual [1].

Nowadays, vehicle tracking is one of the most crucial applications. For instance, maps provide a large role in vehicle tracking and monitoring. The major difficulty in maps is that vehicle owners may not be able to distinguish the vehicle in a place as a result of the overlapping of vehicles, which adversely affects the process of tracking and monitoring/management [2]. It requires some types of systems to identify and detect where objects were.

Since the onset of COVID-19 has had a catastrophic impact on the healthcare system across the globe, throughout the first wave and well into the second wave, finding ambulances was a herculean task for most of the population, and keeping a track of them was difficult for the administrators too, especially in the government-administered hospitals. This situation could have been handled in a more organized manner only if the administrators had the facility to track their ambulances. This would have also helped the administrators to instruct the ambulance drivers about their next patient. With this motivation in mind, we have come up with this work. The system is composed of two parts including the hospital administrator emergency mobile application and the web application that is connected to the things (sensors). The hospital administrator emergency mobile application enables the admin to track the past and present traversed location of an ambulance by showing the time-stamped location of the ambulance and enables them to see the emergency routes and the location of the ambulance(s) anywhere with Internet connectivity. And the web application enables the admin in the hospital to locate the ambulance(s), and also the people in the hospital are aware of the count of ambulances in the hospital.

The remaining sections of the paper are organized as follows. In Section 10.2, we present the most related work to the ambulance tacking/emergency vehicular tracking solutions. In Section 10.3, we give an overview of our smart IoT-based ambulance tracking and management system (ATMS). In Section 10.4, we elaborate on the preliminary assessment/implementation. And finally, in Section 10.5, we conclude the paper and discuss some potential future work.

## 10.2 RELATED WORKS

Being one of the biggest trends in technology, IoT has found its way into all domains including the healthcare industry. A technology like IoT can greatly support the health care industry. The beneficiaries include patients and doctors with devices like smart wearables, families with remote health tracking systems, ambulance and other emergency services and hospitals by enabling efficient management and tracking systems. The authors of Ref. [3] have done a detailed survey and review of such state-of-the-art systems. Their work is mainly focused on the role of IoT and IoT-enabled devices' role in the COVID-19 pandemic. One of the major problems faced during this time was shortage of ambulances and lack of proper management due to overwhelming number of cases.

A major hurdle in coordinating and managing the ambulance service was tracking its location and communicating to the customer. In many underground tunnels, rural and urban areas, Global Positioning System (GPS) signals would be blocked, resulting in ineffective positioning of the ambulance location. To overcome this pitfall, the authors in Ref. [4] has proposed a vehicle location tracking system using radio frequency identification (RFID), GPS and Global System for Mobile communication (GSM). RFID technology is used in areas where the GPS signals are unavailable. The GSM system is used for the wireless transmission of the data which is enabled with a Graphical User Interface (GUI).

A similar system has been proposed in Ref. [5] for child tracking. The child module containing Atmega328 microcontroller, GPS and GSM modules can be installed in the ambulance, whereas the receiver module including a smart phone and database can be administered by the hospital or emergency service provide for the ambulance's location. The system proposed in Ref. [6] uses the commands and sends via a GSM module to mobilize and demobilize the vehicle. The location of the vehicle at the time of request can be extracted in terms of latitude and longitude.

Smart Hospital Emergency System proposed in Ref. [7] is a novel approach which when implemented can come a long way in providing emergency services. Here, the authors have monitored and tracked the health data in the smartphones and processed them securely and efficiently to automatically communicate with the emergency services like ambulance based on GPS and video-calling facility with the operator. The idea proposed in Ref. [8] is similar to this. The authors of Ref. [8] have interfaced various health monitoring sensors to Raspberry Pi 3 along with GPS module which is installed in the ambulance. Any emergency detected is communicated along with the location of emergency to the people added in the system.

Ref. [9] has reviewed IoT-based vehicular emergency system and vehicle tracking system. The emergency systems are enabled with micro-electromechanical systems (MEMS) and GPS sensors which detect any accident and send the location to main server. The main server communicates the location to the ambulance and tracks it with the help of various modern communication and information technologies. The communication system proposed in this work is efficient and quick which is fitting for a high-risk domain such as ambulance services. The authors of Refs. [10] and [11] have also discussed IoT-based tracking systems. The authors of [11] have discussed the implementation of RFID technology for product-based system which can be easily implemented to track ambulances by the hospital or emergency service provider. Ref. [10] talks about using RFID as a wireless object identification technology. The tags are read, the details are sent to the infrastructure where it's compared with the database and categorize the object. This can be a good way to manage ambulances, especially where there are different categories of ambulance services available. Ref. [12] proposed an Autonomous Informative Services for Bus Route Map where GPS is used to obtain the location on Google Maps and RFID technology is used for identification of the bus.

The authors of Ref. [13] have proposed a vehicle monitoring system using GPS/General Packet Radio Services (GPRS) tracking with sensors like collision prevention sensor, door sensor and fuel monitoring sensor on a Raspberry Pi processor. In Ref. [14], the authors have implemented location tracking and emergency communication system using Raspberry Pi 3 and UMTS-HSDPA communication protocol. The system proposed detects the type of emergency, shows a list of nearby rescue centers on the map and also calculates the nearest one from the database. The communication system proposed in Ref. [14] can be coupled with the system proposed in Ref. [13] for a novel ambulance tracking system.

Apart from Ref. [9], the works done in Refs. [8,15,16] have proposed interesting communication systems for ambulances and emergency services. The authors of Ref. [15] have uniquely used RFID technology to track the location. This RFID sends the location to the Directional Antenna which is used to calculate the minimum distance with Received Signal Strength. The signal is sent to the administrator who will track the vehicle. The system proposed in Ref. [16] has two modules – user emergency: ambulance is requested, showing the location of the patient, and paramedic application: locates the patient and appropriate hospital. Once the route is established, it is established as an emergency-route for the ambulance to navigate comfortably. Ref. [17] suggests a similar system with GSM and GPS technologies. The shortest route to the hospital is calculated by monitoring traffic and adopting route optimization technologies and the details are sent to the cloud server.

Refs. [18] and [19] also focus on GPS and GSM-based location tracking and communication. The system in Ref. [18] includes an alcohol and accident detection system which can ensure that the driver is capable to drive. The security and anti-theft system has been included with the help of RFID. The vehicle can be controlled using GSM module and all the details are translated to the ThingSpeak cloud. The authors of Ref. [19] have interfaced SIM8008 and GPS module with Arduino to build a similar system. All the parameters obtained are sent to a web page.

Ref. [20] has taken a unique approach to track ambulances. The proposed system starts out by registering the ambulances as clients with all the required details. They login to the system where each ambulance's location and distances can be calculated with the help of RESTful Application Programming Interfaces (APIs) and web services. These details are returned to the user's system.

## 10.3 PROPOSED SYSTEM

The proposed IoT-based ATMS aims to make use of GPS to track the location of the ambulance with help of the Internet and make use of RFID technology to analyze the count of ambulances in a hospital. The past traversed locations of the ambulance, that is, the latitudinal and longitudinal parameters are displayed in the website and app developed. Thus, the user or the administrator of the hospital can track the past time-stamped location of the ambulance, which is saved and displayed in the ThingSpeak cloud. As shown in Figure 10.1, the system consists of two subsystems, firstly the Ambulance Management system, which is installed in the hospital, and secondly, the Ambulance Tracking system, which is installed in the ambulance. The ambulance management system consists of an RFID Reader and NodeMCU for reading the RFID tags attached to the ambulance along with an LCD to display the count of ambulances available inside the hospital. This, in turn, makes use of the Internet and Arduino IDE to display the count in the website developed. The ambulance tracking system consists of an ESP8266 module along with a GPS sensor and RFID tag embedded in the ambulance. The ESP8266 module makes use of a Wi-Fi hotspot and transmits the GPS parameters to the ThingSpeak cloud, website and mobile app.

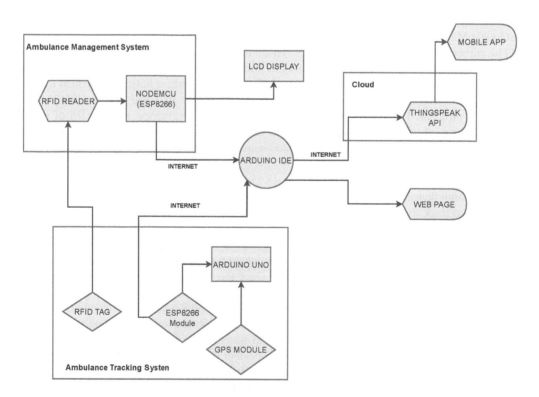

**FIGURE 10.1** IoT-based ATMS block diagram.

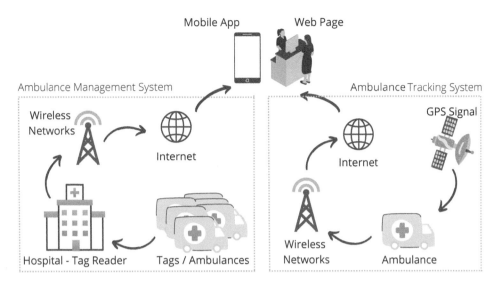

**FIGURE 10.2**   ATMS architecture.

The key focus of the proposed system is to provide an all-in-one setup for the hospital admin-istrators or emergency service providers where they cannot only track their vehicles, but also effi-ciently and effectively plan and manage their ambulance services as shown in Figure 10.2. The user can access all the required information from the website or the mobile app which is connected to the ThingSpeak cloud. The mobile app ensures that the system can be used outside the hospital/company network, making it remote-working friendly. Most of the vehicle tracking systems proposed have adopted GSM technology to communicate location and other details with the server/user, where the ATMS proposed has interfaced Arduino with ESP8266 Wi-Fi module to communicate.

The design of the ambulance management system has been inspired by the FASTag system adopted by the Nationals Highways Authority of India. The RFID tag in the ambulance gets scanned every time it enters or leaves the hospital premises, thereby maintaining a count of the ambulances available for service.

### 10.3.1   Flow Adopted for Designing the System

As shown in Figure 10.3, the algorithm/workflow of the Ambulance Management System is, initially we start with importing the required libraries; in this case, we import #include <LiquidCrystal_I2C. h>, #include<ESP8266WiFi.h>, include<SoftwareSerial.h> and #include<Wire.h> to work with the LCD and NodeMCU Wi-Fi Module and establish a connection using the same to our network through Wi-Fi and for serial communication, respectively. We use NodeMCU as both microcon-troller and Wi-Fi module here. Secondly, we configure the Wi-Fi with NodeMCU by proving SSID and Password to establish the connection and declare a client to display the count in basic HTML webpage code developed for the ambulance management system. As the next step, we develop a simple code to read the RFID tags (embedded in the ambulance) using an RFID Reader and main-tain a count of ambulances that are inside the hospital and display the number of ambulances on the LCD and website developed.

As shown in Figure 10.4, the algorithm/workflow of Ambulance Tracking System follows the initial procedure of importing required libraries; in this case, we import #include<TinyGPS++. h>, #include<ESP8266WiFi.h>, #include<SoftwareSerial.h> and #include<Wire.h> to work with the GPS Module and ESP8266 Wi-Fi Module for establishing a connection between the GPS sen-sor and ThingSpeak cloud through Internet (Wi-Fi Hotspot installed in the ambulance [GPRS]).

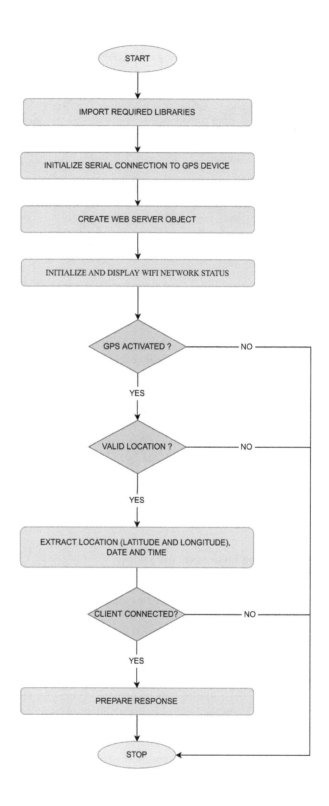

**FIGURE 10.3** Flowchart of ambulance management system.

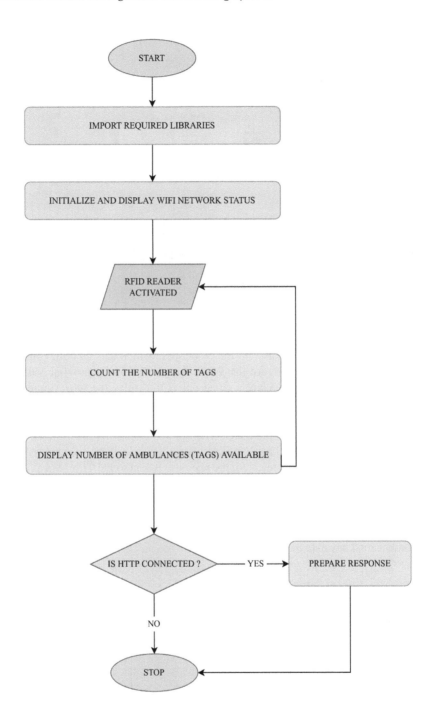

**FIGURE 10.4** Flowchart of ambulance tracking system.

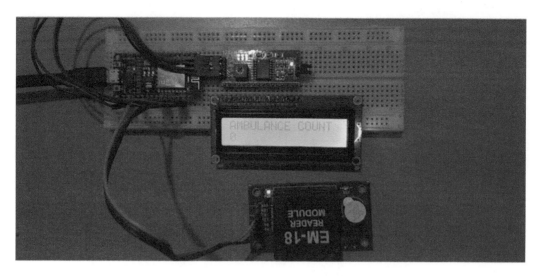

**FIGURE 10.5**   Ambulance management system.

Secondly, we connect the GPS sensor to the NodeMCU which acts as both controller and Wi-Fi module. The next step is to configure the Wi-Fi with ESP8266 by proving SSID and Password to establish the connection and declare a client object to display the sensor readings in the basic HTML webpage developed for the ambulance tracking system, where hospital administration can view the current latitude, longitude, date and time from the GPS sensor inside the ambulance. The next step we perform after connecting the GPS sensor and displaying the network status is to check if the GPS is activated. If the GPS module is activated, it checks if the location so obtained using it is Valid, and if the client is still connected, the response is sent to the ThingSpeak cloud API where the data read from the GPS sensor in the form of latitude and longitude gets recorded in the cloud and this, in turn, can be connected to the app developed using MIT App inventor to remotely monitor the traversed locations by the ambulance. If any of the below steps like GPS activation, Wi-Fi Client Connection establishment and Valid location criteria is violated, the system stops and does not respond or send any data to the cloud.

## 10.4   RESULT AND DISCUSSION

A custom IoT device was devised to test and deploy the idea on real hardware environment.

As shown in Figure 10.5, the Ambulance Management System (AMS) was tested with LCD display, NodeMCU and EM18 RFID Reader module. This setup can be arranged inside the hospital and the RFID tags can be placed near the number plate of all the ambulances. As soon as the RFID tag is detected, the count of the ambulance is displayed in the LCD screen as shown in Figure 10.5.

As shown in Figure 10.6, the Ambulance Tracking System (ATS) was tested with Arduino and GPS module. This setup can be placed inside the ambulances. As soon as the GPS module is powered through the microcontroller, the current location is transmitted to the website developed with all the key parameters like latitude, longitude, date and time as shown in Figure 10.8. Figure 10.7 is the website developed to redirect the web to Figures 10.8 and 10.9, respectively.

As shown in Figure 10.9, the ThingSpeak cloud displays the location obtained from the GPS sensor along with its time-stamp shown in graphical format for better visualization of the ambulance location. Figure 10.10 is the UI developed for App users to visualize the ambulance location from the mobile app which in turn is linked to the ThingSpeak cloud database.

**FIGURE 10.6** Ambulance tracking system.

# Ambulance Monitoring and Tracking System

Location Details through GPS

Click here for GPS tracker Page !

**Past Location Details through Things Speak API/Cloud**

Click here for Seeing Past Locations via Things Speak Cloud!

**FIGURE 10.7** Website to monitor and track the ambulances.

## 10.5 CONCLUSION AND FUTURE SCOPE

In conclusion, the ATMS setup so developed was very reliable in counting the number of ambulances inside the hospital using the RFID tags and thus helps to avoid confusion if any. It also displays the count on the LCD which is very comfortable for everyone to check and act accordingly. In the ambulance tracking system hardware setup, the latitude and longitude obtained from the GPS sensor module are very accurate; hence, this setup is a reliable location tracking system for an emergency service like ambulance, fire services, etc.

The ThingSpeak cloud helps us visualize the past traversed locations of the ambulance which can be very useful for the hospital administration to monitor and keep notes. The app developed

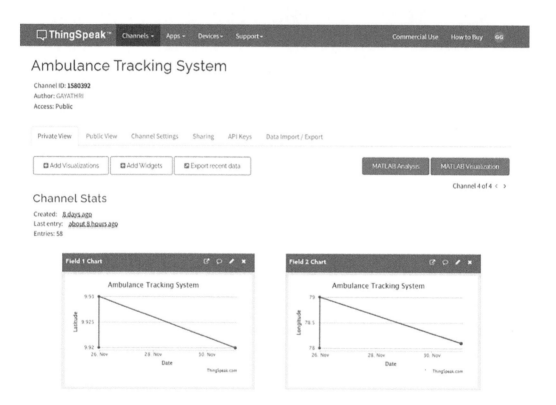

**FIGURE 10.8**　GPS location on website.

**FIGURE 10.9**　GPS location on the ThingSpeak cloud.

using MIT App builder can be helpful for the other workers in the hospital to inform the availability of ambulances during an emergency.

This work can be extended by implementing complex and responsive HTML, CSS and JavaScript code in the webpage development part and integrating that with ThingSpeak API and App developed for interactive, dynamic and colorful visualization remotely.

We can also integrate Google Maps just like the recent delivery/fleet tracking system provide, to the webpage developed for ease of visualizing the live location of ambulances. This system can be implemented in all the emergency services like hospitals, fire stations, etc. especially during the pandemic as it can help communication and work simple, easier and remote. This can provide

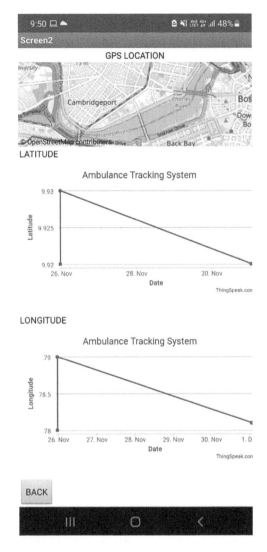

**FIGURE 10.10**   GPS location from ThingSpeak on the mobile app.

a sense of ease for the people as well as the administration to receive and send off ambulances at right time, respectively.

This work can also be integrated with the existing Hospital Management System database with a good front-end-responsive website. The present system can be integrated with all the ambulances in the hospital, and hence, we can calculate the shortest distance between the nearest ambulance and the hospital to quickly send a message or notify people who require ambulance service just like how food and grocery companies implement shortest distance algorithms for finding the least cost past in delivering their goods and services.

## REFERENCES

[1] Priti Jagwani, Manoj Kumar. "IoT Powered Vehicle Tracking System (VTS)", in *Computational Science and Its Applications–ICCSA 2018: 18th International Conference, Melbourne, VIC, Australia, July 2–5, 2018, Proceedings, Part IV 18*. Springer International Publishing, pp. 488–498, 2018, doi: 10.1007/978-3-319-95171-3_38.

[2] Mohammed F. Alrifaie, Norharyati Harum, Mohd Fairuz Iskandar Othman, Irda Roslan, Methaq Abdullah Shyaa, "Vehicle Detection and Tracking System IoT Based: A Review", *International Research Journal of Engineering and Technology (IRJET)*, Vol. 5, No. 8: pp. 1237–1241, 2018.

[3] Mohammad Nasajpour, Seyedamin Pouriyeh, Reza M. Parizi, Mohsen Dorodchi, Maria Valero, Hamid R. Arabnia, "Internet of Things for Current COVID-19 and Future Pandemics: An Exploratory Study", *Journal of Healthcare Informatics Research*, Vol. 4: pp. 325–364, 2020.

[4] Jaco Prinsloo and Reza Malekia, "Accurate Vehicle Location System Using RFID, An Internet of Things Approach", *Sensors*, Vol. 16: p. 825, 2016, doi: 10.3390/s16060825.

[5] Trupti R. Chaudhari, A.J. Patil, R.R. Karhe, "RFID and GPS Based Child Tracking System with Voice Recognition for Security", *International Journal of Scientific Development and Research (IJSDR)*, Vol. 1, No. 12: p. 179, 2016.

[6] Nwukor Frances Nkem, "Implementation of Car Tracking System Using GSM/GPS", *International Journal of Scientific and Research Publications*, Vol. 10, No. 3, pp. 399–403, March 2020.

[7] Mohammed Al-khafajiy, Hoshang Kolivand, Thar Baker, David Tully, Atif Waraich, "Smart Hospital Emergency System Via Mobile-Based Requesting Services", *Multimedia Tools and Applications*, Vol. 78: pp. 20087–20111, 2019.

[8] Vijendra Babu, "Location Tracking System for Vehicular Emergency using IOT", *Elementary Education Online*, Vol. 20, No. 1: pp. 2300–2306, 2021.

[9] H. Patole Gitanjali, A. Shide Jyoti, S. Salve Satish, Vipul Ranjan Kaushik, S.B. Puri, "IOT based Vehicle Tracking & Vehicular Emergency System- A Case Study and Review", *International Journal of Advanced Research in Electrical, Electronics and Instrumentation Engineering*, Vol. 6, No. 10, pp. 8001–8012, October 2017.

[10] Muzhir Al-Ani, "Packages Tracking Using RFID Technology", *International Journal of Business and ICT*, Vol. 1, No. 3–4: pp. 12–20, December 2015.

[11] Ender Aycin, Yilmaz Goksen, Senem kili, "The Product-Based Tracking System by using RFID Technologies in Logistics Applications", *The International Journal of Business & Management*, Vol. 5, No. 11, pp. 284–289, November, 2017.

[12] Anuradha Vishwakarma, Agraja Jaiswal, Ashwini Neware, Shruti Ghime, Antara Marathe, Rashmi Deshmukh, "GPS and RFID Based Intelligent Bus Tracking and Management System", *International Research Journal of Engineering and Technology (IRJET)*, Vol. 3, No. 3: p. 269, March 2016.

[13] Ala Saleh Alluhaidan, Marwan Saleh Alluhaidan, Shakila Basheer, "Internet of Things Based Intelligent Transportation of Food Products during COVID", *Wireless Personal Communications*, Vol. 127: p. 27, 2021.

[14] Subha Koley and Prasun Ghosal, "An IoT Enabled Real-Time Communication and Location Tracking System for Vehicular Emergency", in *2017 IEEE Computer Society Annual Symposium on VLSI (ISVLSI)*, pp. 671–676, 2017, doi: 10.1109/ISVLSI.2017.122.

[15] Akshay Chothani, Jitesh Saindane, Hrudesh Mistari, Nilesh Bhavsar, Rakesh Shirsath, "RFID-Based Location Tracking System Using a RSS and DA", in *2015 International Conference on Energy Systems and Applications*, 04 July 2016.

[16] AbdelGhani Karkar, "Smart Ambulance System for Highlighting Emergency- Routes", in *2019 Third World Conference on Smart Trends in Systems Security and Sustainability (WorldS4)*, p. 255, 21 November 2019.

[17] Aritra Baksi, Mayookh Bhattacharjee, Siddhanta Ghosh, Soham Kanti Bishnu, Arindam Chakraborty, "Internet of Things (IOT) Based Ambulance Tracking System Using GPS and GSM Modules", in *2020 4th International Conference on Electronics, Materials Engineering & Nano-Technology (IEMENTech)*, 30 November 2020.

[18] A. Mounika, Anitha Chepuru, "IOT Based Vehicle Tracking and Monitoring System Using GPS and GSM", *International Journal of Recent Technology and Engineering (IJRTE)*, Vol. 8, No. 2S11, pp. 2399–2403, September 2019.

[19] Mayuresh Desai, Arati Phadke, "Internet of Things Based Vehicle Monitoring System", in *2017 Fourteenth International Conference on Wireless and Optical Communications Networks (WOCN)*, pp. 1–3, 2017, doi: 10.1109/WOCN.2017.8065840.

[20] C S Vikas, Ashok Immanuel, "Ambulance Tracking System Using Restful API", *Oriental Journal of Computer Science & Technology*, Vol. 10, No. 1: pp. 213–218, March 2017.

# 11 Smart Robot Car for Industrial Internet of Things

*Ayushi Chakrabarty, H.R. Deekshetha, S. Reshma,*
*O.V. Gnana Swathika, and V. Berlin Hency*
Vellore Institute of Technology

## CONTENTS

11.1 Introduction ............................................................................................................ 124
11.2 Background .............................................................................................................. 124
    11.2.1 Literature Survey ......................................................................................... 124
    11.2.2 Problem Statement ....................................................................................... 126
    11.2.3 Scope of the Work ....................................................................................... 126
11.3 Design and Implementation of the System .......................................................... 127
    11.3.1 Block Diagram Analysis .............................................................................. 127
    11.3.2 Architecture Analysis .................................................................................. 127
    11.3.3 Algorithm..................................................................................................... 128
11.4 Software Implementation........................................................................................ 129
    11.4.1 Arduino Implementation.............................................................................. 129
    11.4.2 ThingSpeak Implementation ....................................................................... 129
    11.4.3 Google Firebase and Node-RED Implementation....................................... 130
    11.4.4 Android Implementation.............................................................................. 130
11.5 Hardware Implementation ...................................................................................... 130
    11.5.1 Mechanism................................................................................................... 136
11.6 Results and Discussion .......................................................................................... 136
11.7 Conclusion and Future Enhancement ................................................................... 139
References........................................................................................................................ 140

### ABSTRACT

Robots are turning out to be the future with the advancement and growth in the automation sector. It has the potential to play a major role in the forecoming generations. In the recent times, there have been many innovations in the robotics field along with the Internet of Things with special focus on the industrial Internet of Things (IIoT). This work discusses a robot car that can be controlled with the help of an android application remotely to direct the car's movements to capture vital information such as temperature, humidity, and gas concentration values. These robots are targeted to be deployed in the nuclear power plants sector to reduce the vulnerability of a human venturing into dangerous locations. Other alternative applications of these robots encompass war zone areas, spying tools, transportation, and agriculture. The system uses the 802.11 wireless protocol standard (Wi-Fi) in addition to sensors like gas sensor, temperature, and humidity sensor. The robot car uses NodeMCU as the microcontroller board to obtain and process the sensor information. ThingSpeak and Google Firebase are used to send the data to the cloud for further analysis. Node-RED software is used to provide user-friendly Grpahical User Interface (GUI).

DOI: 10.1201/9781003374121-11

**KEYWORDS**

IIoT; NodeMCU; Thingspeak; DHT11; Firebase; Robot; Nuclear Power Plants

## 11.1   INTRODUCTION

Industrial Internet of Things (IIoT) is the application of IoT principles to industrial plants. It also involves monitoring various parameters to enable more efficient, safer, and reliable environment for humans. It is predicted to have enormous impact in the near future. The largest industrial and technology firms are investing in billions for an efficient IIoT platform. Data management also plays a crucial role in deriving conclusive results from the large real-time generated data. Smart Robot car using latest IoT technology is helpful in environments where it is dangerous for the humans to venture. We are building an IoT-based robot car using ESP8266 NodeMCU Module, DHT11 sensor, and gas sensor. This work is controlled by humans using applications and cloud platforms. Data from the different sensors are sent to cloud where it is processed to keep track of the environmental conditions. This system consists of motor driver which drives the motor depending on the input given by the user. The sensor data will be sent to the cloud server in defined interval of time so that it can be monitored from anywhere in the world. IIoT is a rising technology, especially in the context of nuclear power plants. Dangerous accidents can occur through a failing in the cooling system within the nuclear core. It is very difficult and dangerous for a human being to venture in such restricted sections of the plant due to the threat of harmful radiation exposure.

Therefore, keeping these factors in mind, a robot car can be used effectively with prime emphasis on nuclear power plant settings. This robot car can be used to maneuver (using voice commands) into restricted sections of the factory to record critical information like the temperature, humidity, and concentration of gases to enable better control and decision-making. Predictive analytics using Machine Learning algorithms can be incorporated in future to help the nuclear power plant control center to take proactive decision in case of an anomaly or threat, thereby saving human lives.

## 11.2   BACKGROUND

### 11.2.1   LITERATURE SURVEY

Primary investigation in order to understand the existing systems is carried out under this section.

In Ref. [1], the authors have presented the idea of a low-cost autonomous car. This device is controlled by the use of voice commands, and the system is implemented using NodeMCU ESP8266 microcontroller board, Adafruit, and IFTTT. The design is a low-cost solution that helps to increase the productivity in varied different application areas. The primary objective of their work is to help reduce resource wastage.

The authors in Ref. [2] propose a robot car that uses an application for android users in remote mode and is connected to the security system. This is controlled by Raspberry Pi 3. Wireless camera, motion detector, GPS, and gun are the other constituent elements. This system design can be very beneficial in the field of spying-based pertinence, especially in battle zones areas, as it facilitates an easy visualization of war field from the inputs of the wireless camera.

The work in Ref. [3] discusses an IoT-based self-driving car featuring highest and complete automation. There are three major components involved in this setup that consists of the input system including camera and ultrasonic sensor, processing unit such as Raspberry Pi, and RC car control unit. The authors discussed the developed system to be one of the most cost-effective autonomous car models to provide better unrestrictive transport system. Instantaneous (immediate) readings can be accessed using OpenCV by ensuring the driver safety as a result of drunk drivers, leading to lower road accidents.

The authors in Ref. [4] introduced a working model for the self-driving robotic car carrying Raspberry Pi 3 as the working unit and scans for the surrounding Wi-Fi networks to hack into one of the sources to perform penetration testing. The proposed pen tester on wheels can help in dealing with the multitudinous hazards that are belonging to the fraudulents of wireless network.

The work in Ref. [5] uses Arduino as the microcontroller board for their military application-based robot car. Other components include the Bluetooth, NodeMCU ESP8266, DC motors, and camera module. Similar to Ref. [2], it has been utilized for surveillance purpose.

In the survey paper [6], the authors identify solid state-of-the-art robotic applications, important research challenges, and existing technological tools. Many methodologies have been discussed under varied real-life scenario-based applications in addition to effective investigation of the major causes for the IoT-aided robotics applications. Comprehensive study on the literature survey in the prime concept that involved the research fields provides a meticulous overview of the past work. Identifications of research challenges to facilitate in the development of novel solutions have also been included.

The work in Ref. [7] proposes an autonomous vehicle with object detection features using Yolo model. They describe how the live camera input given to the Raspberry Pi board is uploaded to the cloud platform and can be utilized to perform Convolutional Neural Network (CNN). The authors choose Yolo model due to its speed and performance, as it outperforms other algorithms. The system design has the potential to provide good results during real-time scenarios for object detection and is proposed keeping in mind the applications for an autonomous car.

In Ref. [8], an application which is website-based for the purpose of the mobile robot using platform of IoT is proposed by the authors. This developed mobile robot is interfaced with the mobile robot application with the help of microcontroller and NodeMCU board. The authors describe how it can be used to control the robot via the mobile app from anywhere to monitor air quality and position.

The authors in Ref. [9] developed a robot using Raspberry Pi and Google speech reorganization engine for the management of covering a guess estimate distance. An ultrasonic sensor has been used in addition to python speech recognition module configuring 119 languages. The developed system effectively processes speed and performs offline speech recognition.

In Ref. [10], the idea and implementation of an IoT robot for a restaurant has been proposed. The key objective to utilize the system is the waiter robot in this application. The system consists of Arduino nano, Radio Frequency Identification (RFID), and Infrared (IR) sensor to develop the low-cost waiter robot. It avoids obstacles and follows the line. The system can be controlled using multiple users and it has the capability to detect the target table by using the RFID technology.

In Ref. [11], the authors discuss about the health care worker treating a COVID patient as a robot. It aims to give food, water, and medicine, record different readings from patients, and use Bluetooth client-side app via an android phone to send data. In addition to this, server-side python code is used to receive data from phone.

The work in Ref. [12] discusses a mobile/portable/moving surveillance camera using the Raspberry Pi unit and webcam. It is used to capture images in the dark areas by utilizing Secure Shell (SSH) protocol for sending the captured images. It performs real-time surveillance with video using internet and can be used for critical surveillance application purposes. Reference [13] is directed and designed for the domestic areas in isolated supervision using the unit Raspberry Pi through ESP8266 Wi-Fi module, passive infrared (PIR) sensor. The authors retrieve the audio and video as the necessary data from the surroundings. Their work utilizes the web of things concept.

In Ref. [14], there is a robot that is accessed in both manual and automated mode. Blynk application is used to control the movements from anywhere in and around the robot's surroundings. NodeMCU, solar panel, GSM, and IP camera is used to build the mobile (wireless) dusk perception using a camera with real-time streaming for audio and video transmission. The proposed idea will ultimately help reduce the number of lost lives in the battle field and can also be used during natural calamities.

The authors in Ref. [15] have developed a Wi-Fi-controlled electric vehicle. It provides GPS coordinates for the position of the vehicle by exhibiting the necessary pictures and also the real-time streaming of video. Their work also complies with the parameters like acceleration, brake, and steering, and displays current value on the display monitor.

The work in Ref. [16] discusses the working of a vehicle where its movement is controlled by a smartphone using the Bluetooth module. The motor driver helps to calculate different speeds and time of the car. The difference in speed and duration is created due to lower voltage consumption. This process takes about 13 m to send the data and pair.

In Ref. [17], the authors have developed an Android-controlled temperature sensing RoboCar. AirDroid app enables to connect the device to PC through a Wi-Fi controller of wireless network. It can be used to view the location of the car. Similarly, in Ref. [18], a remote-controlled robotic rover maneuvered using a Wi-Fi network is developed. It utilizes Arduino Uno connected with ESP8266 Wi-Fi module as the central controlling board.

In Ref. [19], a cost-effective remote-control car is proposed utilizing Arduino and Bluetooth module. The controls of the vehicle are done using a BotApp.

Extending the application of Ref. [10], in Ref. [20], the authors propose a vehicle controlled by smartphone using the Wi-Fi module. When the vehicle detects an obstacle, a message would be notified or transferred to the smartphone and the robot performs pick and place operation using the robotic arm. Different detectors are used to descry explosion, and buzzer rings with discovery of explosion. A live surveillance is handed, which monitors every movement of the vehicle. Similarly, in Ref. [21], the authors propose a vehicle controlled by smartphone using Bluetooth module where the main objective is to detect obstacles using sensors and hardware components such as the Bluetooth module, PIR sensor, ultrasonic sensor, and buzzers. Software component includes the development and usage of a mobile application.

### 11.2.2  Problem Statement

The IoT-based robot is implemented to tone down the movement of transfer of materials or can be used to spy in the war zones. Generally, the transfer process of accoutrements takes place by using the human strength, and if this process is taking place in repeated mode, it can beget bruises among the drivers. Employing this robot, the driver is not necessary to this fraudulent, thus increasing the efficiency of the work and preventing injuries. Human beings can make mistakes. In the world of industries, they will not able to take any kind of wrongdoings, as all the miscalculation leads to expensive ones, be in terms of interval (duration), currency, and commodity. Therefore, if no effort is put into updating or optimizing the resources, there will be the situation of spending more money for commensurate process. IoT is presumably a result to reduce costs and help loss of resource; this design can be a strong way to attack these kinds of situations.

### 11.2.3  Scope of the Work

On the progress of the technology, these days, there exists the evolvement of the electronics department or field of work. This ultimately impacted many human beings. This is then extended for military or army applications where manual supervision is a challenging task. By merging these with the security aspect, especially in the cyber side, it leads to better conservation in the investments in the near future. Developing a Smart Wireless Sensor in addition to the discussed techniques can aid a farmer in terms of agricultural applications essential in routine life, thereby increasing his/her profits. Replacement of human soldiers with smart robots for the purpose of national security can also pave the way for a smart defense sector. To make the robot automated and self-defensive, it can be implemented with a laser gun. With the help of IP camera, the laser gun will be able to identify the enemy and shoot by utilizing manual or automatic mode. It will be a better option for the surveillance robot to protect the nation from the enemy, as it has the possibility of saving a lot of human lives. This security robot can be used with an additional Artificial Intelligence system that would allow it to fete the world around it and be suitable to descry the interferers and warn the mortal driver. It can also be used in restaurants to give away the orders that is taken and also in other places where transferring occurs.

## 11.3 DESIGN AND IMPLEMENTATION OF THE SYSTEM

### 11.3.1 BLOCK DIAGRAM ANALYSIS

Figure 11.1 represents the presupposed design of IoT robot car where the essential elements are mentioned. The main microcontroller block used is NodeMCU ESP8266. The usage of 9 V Battery is for driving the motors. DHT11 sensor is used for measuring the temperature and humidity value of that particular area. MQ2 gas sensor determines the air quality of that area, and this reading is read through the serial monitor and further sent to ThingSpeak cloud. Motor driver is used to drive the four motors in all the possible eight directions. The software ThingSpeak software platform assists in monitoring the real-time readings of the DHT11 and MQ2 sensor. In addition to this, Node-RED platform is used to develop an interactive dashboard to view the real-time values sent to the cloud by connecting it to the Google Firebase platform.

### 11.3.2 ARCHITECTURE ANALYSIS

The architecture of the IoT robot car in Figure 11.2 represents underlying structure of the proposed model with its associated interdependencies. Laptop, mobile or tablet is used to operate the car in addition to monitoring and analyzing the real-time values received from the sensors. This analysis is done through the internet network. The result of this analysis is processed by software tools like ThingSpeak, Node-RED, and Google Firebase. From the Internet network, the data are sent through the gateway and the data are analyzed by the operator performing/controlling the implementation.

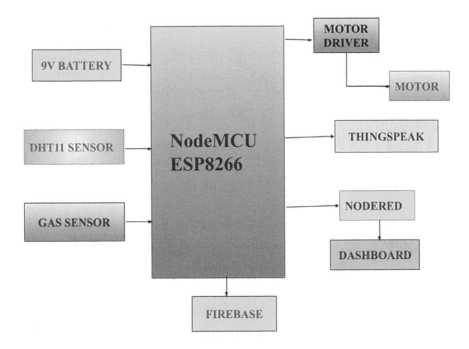

**FIGURE 11.1** Block diagram for the suggested system design constituting the various design components.

### 11.3.3 ALGORITHM

The descriptive steps involved in developing the proposed system are as follows:

- **Step 1:** Configure the pins such as data pin of DHT11 to D2 pin (27th pin), A0 from gas sensor to the A0 pin of NodeMCU, Relay pin to the D0 pin (30th pin), motor to 9 V pin that is the 11th pin.
- **Step 2:** Give the connections to the negative end of the motor to the other end of the relay and also the VDD pin of the DHT11 to the relay which is of 12 V.
- **Step 3:** After connection of the DC motors with the L298N relay module and DHT11, MQ-2 gas sensor with the NodeMCU board verifies the successful working of the components by compiling the module code using Arduino IDE.
- **Step 4:** Attach the car wheels, make the wire connections robust by soldering, and decrease the scope for loose connection failure.
- **Step 5:** Once the hardware configuration setup is over, code the Arduino IDE program to transfer the live sensor values to ThingSpeak account. Create a new ThingSpeak channel and use the private channel application programming interface (API) key.
- **Step 6:** Similarly, program the Firebase code to send the real-time sensor values to Firebase's real-time database which is then connected to Node-RED using Firebase In nodes.
- **Step 7:** Finally, observe the results and derive meaningful inferences.

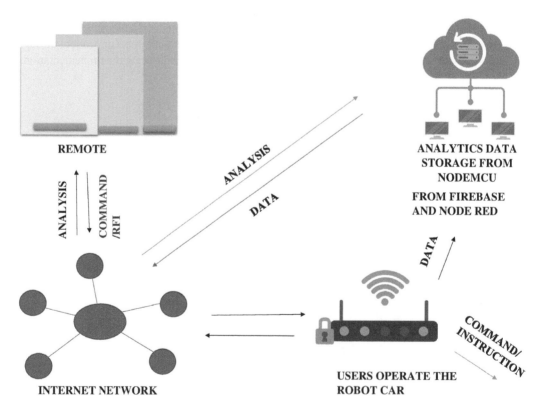

**FIGURE 11.2** System Design Architecture of Smart Robot car highlighting the essential constituting components.

## 11.4 SOFTWARE IMPLEMENTATION

### 11.4.1 ARDUINO IMPLEMENTATION

In Arduino IDE, our code for the proposed model is developed and implemented. The NodeMCU ESP8266 board is connected to the laptop via a USB cable. On successful connection, the port description and board details are specified in the Arduino IDE. The codes for the DHT11 temperature and humidity and MQ2 gas sensors are loaded onto the NodeMCU ESP8266 board after the successful installation of necessary software libraries in the Arduino IDE. Figure 11.3 illustrates the Arduino IDE interface as an example.

### 11.4.2 THINGSPEAK IMPLEMENTATION

ThingSpeak is an IoT analytics platform service that helps to create instant visualizations of the DHT11 (temperature and humidity sensor) and MQ2 gas sensor readings. The analysis can be done readily and instantly. This makes the proposed solution scalable and interoperable. The use of standard APIs ensure interoperability across a variety of different systems and cloud hosting assures that the proposed solution is scalable, making it sustainable and ideal for future use. The ThingSpeak cloud dashboard is illustrated in Figure 11.4.

```
IOT_ROBOT_CAR
#define ENA   14       // Enable/speed motors Right      GPIO14(D5)
#define ENB   12       // Enable/speed motors Left       GPIO12(D6)
#define IN_1  15       // L298N in1 motors Rightx         GPIO15(D8)
#define IN_2  13       // L298N in2 motors Right          GPIO13(D7)
#define IN_3  2        // L298N in3 motors Left           GPIO2(D4)
#define IN_4  0        // L298N in4 motors Left           GPIO0(D3)

#include <ESP8266WiFi.h>
#include <WiFiClient.h>
#include <ESP8266WebServer.h>

String command;              //String to store app command state.
int speedCar = 800;          // 400 - 1023.
int speed_Coeff = 3;

const char* ssid = "Make DIY";
ESP8266WebServer server(80);

void setup() {

  pinMode(ENA, OUTPUT);
  pinMode(ENB, OUTPUT);
  pinMode(IN_1, OUTPUT);
  pinMode(IN_2, OUTPUT);
  pinMode(IN_3, OUTPUT);
  pinMode(IN_4, OUTPUT);

  Serial.begin(115200);
```

FIGURE 11.3 Arduino IDE interface with the code for the proposed system.

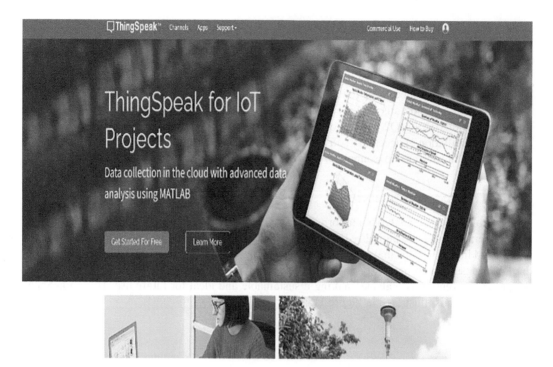

**FIGURE 11.4**  ThingSpeak cloud dashboard for instant monitoring and visualization.

### 11.4.3  Google Firebase and Node-RED Implementation

Node-RED is the browser-based programming tool where a certain range of nodes are deployed to interconnect physical assets to the cloud data. In this work, it helps to create a workflow as shown in Figure 11.5 in correspondence to the proposed hardware setup to obtain the real-time values from the Google Firebase. These values are then observed using the debug monitor and UI dashboard enabling the user with an easy-to-use Grpahical User Interface (GUI).

This is the platform developed by Google for web mobile applications. The Firebase project overview page is shown in Figure 11.6.

### 11.4.4  Android Implementation

For the sake of complying the movement of the robot car, Android application is used. This application helps to maneuver the direction as well as the speed of the car's movement. The user interface (UI) of the car is given in Figure 11.7.

## 11.5  HARDWARE IMPLEMENTATION

The design constitutes several hardware components such as NodeMCU board, DHT11 humidity and temperature sensor, MQ-2 gas sensor, 12 V battery, and L298N motor driver module. These components will work in unison to comply the robot car mode of motion applying the voice commands into restricted sections of the factory to record essential sensor value readings. These recorded values would then be transmitted to the ThingSpeak cloud for easy monitoring from across the globe.

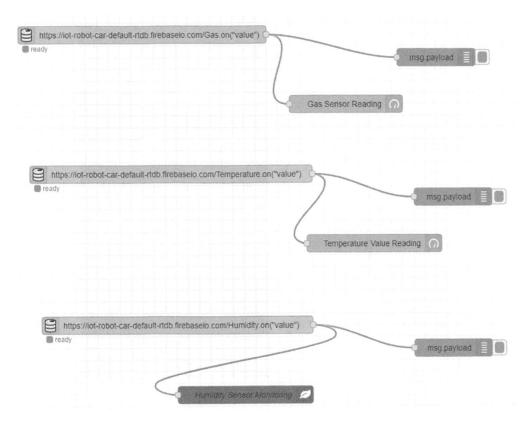

**FIGURE 11.5** Node-RED workflow consisting of Firebase On, Gauge, Debug, and Humid-tree node to obtain and display real-time sensor values.

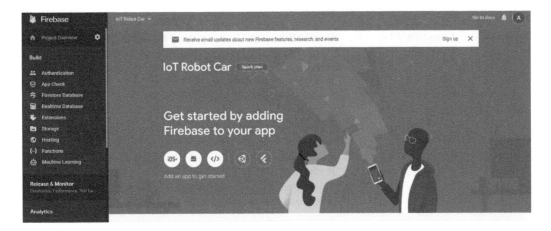

**FIGURE 11.6** Google Firebase project overview window.

### A. NodeMCU ESP8266

NodeMCU is a *free source Lua-based firmware* and development board especially for IoT-based applications. It includes firmware that runs on the ESP8266 Wi-Fi System on Chip (SoC) from *Espressif System* and the hardware which is based on the ESP-12 module.

**FIGURE 11.7**   Android application for the robot car movement control.

**TABLE 11.1**
**Overview of NodeMCU ESP8266 Pin Configuration Details**

| Pin Category | Name | Description |
|---|---|---|
| Power | Micro-USB, 3.3 V, GND, Vin | **Micro-USB:** NodeMCU can be powered through the USB port<br>**3.3 V:** Regulated 3.3 V can be supplied to this pin to power the board<br>**GND:** Ground pins<br>**Vin:** External Power Supply |
| Control Pins | EN, RST | The pin and the button reset the microcontroller |
| Analog Pin | A0 | Used to measure analog voltage in the range of 0–3.3 V |
| GPIO Pins | GPIO1 to GPIO16 | NodeMCU has 16 general purpose input–output pins on its board |
| SPI Pins | SD1, CMD, SD0, CLK | NodeMCU has four pins available for SPI communication |
| UART Pins | TXD0, RXD0, TXD2, RXD2 | NodeMCU has two UART interfaces, UART0 (RXD0 & TXD0) and UART1 (RXD1 & TXD1). UART1 is used to upload the firmware/program |
| I2C Pins | | NodeMCU has I2C functionality support but due to the internal functionality of these pins, you have to find which pin is I2C |

As this board plays one of the major roles on retrieving and transmitting sensor data, a detailed understanding of its pin description, configuration, and system specifications are discussed in Table 11.1, Figure 11.8, and Table 11.2, respectively.

The pin description for the NodeMCU ESP8266 is described in Table 11.1.

The pin configuration for the NodeMCU ESP8266 is shown in Figure 11.8.

NodeMCU ESP8266 specifications and features are mentioned in Table 11.2.

**FIGURE 11.8**    NodeMCU ESP8266 to retrieve and transmit sensor data.

**B. DHT11 temperature and humidity sensor**

DHT11 sensor is a basic, ultra-low-cost, digital temperature and humidity monitoring sensor having a calibrated digital signal output as shown in Figure 11.9. It uses a capacitive humidity sensor and a thermistor to measure the environmental air and outputs a digital signal on the data pin (no analog input pins requirement). This connects to a high-performance 8-bit microcontroller, thereby providing the uses of providing best-quality, fast response and anti-interference ability at an affordable cost. The sensor ensures high reliability and excellent long-term stability. This sensor comes along with a 4.7 or 10 K resistor.

**C. MQ-2 gas sensor**

MQ-2-Robust Metal Oxide Semiconductor is the type of gas sensor otherwise called as chemiresistors as shown in Figure 11.10. It has the potential to sense the concentrations in the atmosphere. The main working principle is based upon the difference in resistance for the perceiving commodity where the gas is conferred around the environment comes in contact with the material. It not only provides a binary indication of the presence of combustible gases but also an analog representation of the concentration in air using a simple voltage-divider network. This analog output voltage changes are proportional to the concentration of the smoke/gas. It has the capacity to detect the concentrations of the above-mentioned gases anywhere from 200 to 10,000 ppm and it works on 5 V DC.

**D. Motor driver**

The L298N motor driver module consists of an L298 Motor Driver IC, 78M05 Voltage Regulator, resistors, capacitor, power LED, and 5 V jumper in an IC. Figure 11.11 shows

**TABLE 11.2**
**NodeMCU ESP8266 Microcontroller Specification Details**

| S. No | Specification |
|---|---|
| 1. | Microcontroller: Tensilica 32-bit RISC CPU Xtensa LX106 |
| 2. | Operating Voltage: 3.3 V |
| 3. | Input Voltage: 7–12 V |
| 4. | Digital I/O Pins (DIO): 16 |
| 5. | Analog Input Pins (ADC): 1 |
| 6. | UARTs: 1 |
| 7. | SPIs: 1 |
| 8. | I2Cs: 1 |
| 10. | Flash Memory: 4 MB |
| 11. | SRAM: 64 KB |
| 12. | Clock Speed: 80 MHz |
| 13. | USB-TTL based on CP2102 is included onboard, Enabling Plug n Play PCB Antenna |
| 14. | Small-sized module to fit smartly inside your IoT projects |

**FIGURE 11.9** DHT11 temperature and humidity sensor image to measure the environmental air for temperature and humidity.

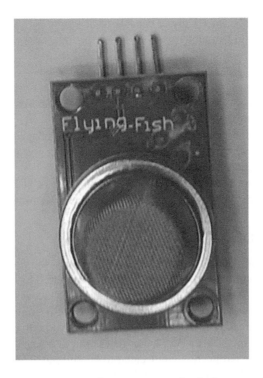

**FIGURE 11.10**  MQ-2 gas sensor to measure the gas concentration in the atmosphere.

**FIGURE 11.11**  L298N motor driver module to control the robot car's DC motors.

a dual-channel H-Bridge motor driver capable of driving a pair of DC motors, i.e., it can individually drive up to two motors.

### 11.5.1 Mechanism

- Two techniques, namely Pulse Width Modulation (PWM) and H-Bridge concepts play an important role to control its speed and rotation direction.
- The speed of a DC motor can be complied by varying its voltage that is given as the input using PWM concept where the mean value of voltage that is given as the input is restricted by transmitting a sequence of on and off pulses.
- In order to control the rotation direction of the DC motor, a H-Bridge technique can be used. It contains four switches with the motor at the center forming an H-like arrangement.

## 11.6 RESULTS AND DISCUSSION

On finalizing the most optimal software and hardware choices, the final integration stage is performed. In this phase, the chosen hardware components were tested module-wise and their respective codes were developed. This was done to ensure that the robot car developed is robust enough to withstand the nuclear power plant settings in terms of connectivity and individual specifications. After integrating all the hardware components with the main NodeMCU ESP8266 microcontroller board, the cloud linkage aspect of the work was investigated. Internet connectivity, latency, and overall system performance parameters are decided to be optimal. The final system prototype for the robot car after the complete integration and development phases is shown in Figure 11.12.

**FIGURE 11.12** Final hardware setup of the developed robot car prototype.

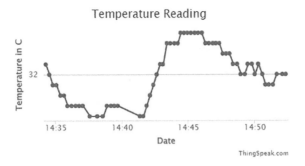

FIGURE 11.13    ThingSpeak real-time values obtained from the DHT11 sensor of the robot car for temperature in Celsius.

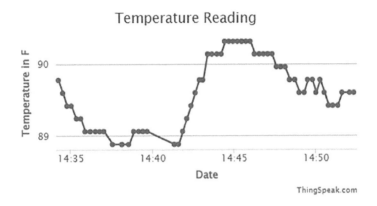

FIGURE 11.14    ThingSpeak real-time values obtained from the DHT11 sensor of the robot car for temperature in Fahrenheit.

The variations in temperature, humidity, heat, and air quality monitoring are observed in real-time using the ThingSpeak cloud platform. In Figure 11.13, the temperature fluctuations over a span of 15 minutes are observed. As the car ventures to different locations in its surroundings, the fluctuations of the temperature values are noticed. The values range based on AC and non-AC conditions.

Similar observations are observed in Fahrenheit scale as shown in Figure 11.14.

In Figure 11.15, the humidity fluctuations over a span of 15 minutes are observed. The values range from as high as 50% to as low as 41%.

In Figure. 11.16, the air quality condition (in ppm) over a span of 15 minutes is observed. The values range from as high as 120 ppm to as low as 55 ppm.

The Node-RED UI Dashboard's illustrative output for humidity is shown in Figure 11.17, and that for gas and temperature sensor is shown in Figures 11.18 and 11.19, respectively.

Similarly, the UI dashboard from the Node-RED software makes the device features much more easily accessible by the user, thereby making it user-friendly. Real-time monitoring of these parameters from the control center can help in detecting anomalies in its initial phase, thereby promoting immediate action in case of an emergency.

Hence, the proposed system can prove to be extremely useful for the IIoT sector, as it acts as the stepping stone in saving human lives working within dangerous nuclear power plant settings. As this novel design uses two NodeMCU ESP8266 boards to achieve the objective of vehicle movement and sensor data retrieval separately, it turns out be an ideal low-latency solution. The design promotes scalability due to real-time cloud monitoring and control facility. The use of standard APIs

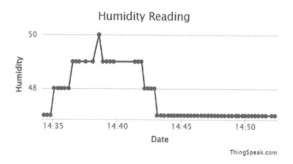

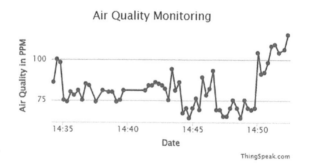

**FIGURE 11.15**  ThingSpeak real-time values obtained from the DHT11 sensor of the robot car for humidity percentage.

**Humidity Sensor Monitoring**

**FIGURE 11.17**  Node-RED UI Dashboard humidity sensor widget.

**FIGURE 11.16**  ThingSpeak real-time values obtained from the DHT11 sensor of the robot car for air quality in ppm.

of the ThingSpeak, Firebase, Node-RED, and Arduino IDE software makes the solution interoperable across different kinds of devices. Also, as NodeMCU ESP8266 microcontroller board has the feature of adjustable clock frequency within the range of 80–160 MHz, the power consumption of the complete system can be controlled effectively, i.e., can be used for low-power requirement conditions. The developed IoT robot car design has the potential to be deployed across various different dangerous settings for the purpose of recording critical information.

Gas Sensor Reading

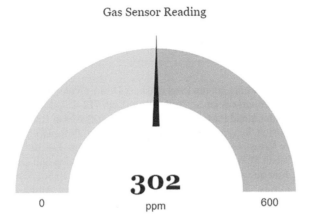

FIGURE 11.18   Node-RED UI Dashboard gas sensor gauge widget.

Temperature Value Reading

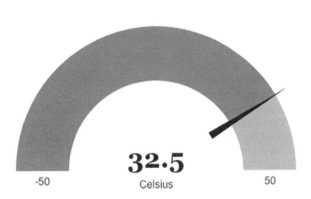

FIGURE 11.19   Node-RED UI Dashboard temperature sensor gauge widget.

## 11.7   CONCLUSION AND FUTURE ENHANCEMENT

In this work, prototype of an IoT-based robot car has been proposed. IIoT is one of the most evolving fields in the present times due to the onset of Industry 4.0. The pandemic has revolutionized the perception of technology, especially in the domain of manufacturing and industrial processes. The novel design proposed aims to reduce the involvement of manpower in dangerous sections of a nuclear power plant by facilitating the control center to record critical information such as temperature, humidity, and air quality index in real-time from any part of the globe. The robot car performs very swift execution of commands taking less processing time due to the usage of an efficient NodeMCU ESP8266 microcontroller having strong onboard processing capabilities. The system is tested under different environmental conditions with satisfactory result of accuracy. This robot car can find its application in various different sectors such as agriculture, medical assistance systems, real-time surveillance purposes, and space exploration applications, to name a few. The design can be extended to have computer vision technology to improve the decision-making process by visualizing the environmental changes even more accurately.

To improve the efficiency and effectiveness of the system, camera-enabled surveillance can be provided. Also, reinforcement learning-based algorithms can be used to train the robot car to perform routine tasks effectively by taking adaptive decisions. NodeMCU ESP8266 boards can be replaced with Field Programmable Gate Arrays (FPGAs) to provide critical data analysis capabilities with improved hardware acceleration for time-sensitive applications.

## REFERENCES

[1] Sumeet Sachdev, Joel Macwan, Chintan Patel, Nishant Doshi, "Voice-Controlled Autonomous Vehicle Using IoT", *Procedia Computer Science*, 2019, vol. 160, pp. 712–717, doi: 10.1016/j.procs.2019.11.022.

[2] Laiqa Binte Imran, Muhammad Farhan, Rana M. Amir Latif, Ahsan Rafiq, "Design of an IoT Based Warfare Car Robot Using Sensor Network Connectivity", in *Proceedings of the 2nd International Conference on Future Networks and Distributed Systems*, 2018, pp. 1–8, doi: 10.1145/3231053.3231121.

[3] Mohammed Maaz, Sabah Mohammed, "IoT Programming to Develop Self Driving Robotics Car Using OpenCV and Other Emerging Technologies", *TechRxiv*, 2020, doi: 10.36227/techrxiv.12055956.v1.

[4] Debajyoti Mukhopadhyay, Sushmita Karmakar, Ankita Meshram, Amita Jadhav, "A Prototype of IoT based Remote Controlled Car for Pentesting Wireless Networks", in *2019 Global Conference for Advancement in Technology (GCAT)*, 2019, pp. 1–7, doi: 10.1109/GCAT47503.2019.8978354.

[5] Dhruv Ketan Kotecha, "Wireless Car Using WIFI – IoT - Bluetooth", *International Research Journal of Engineering and Technology (IRJET)*, 2019, https://www.irjet.net/archives/V6/i6/IRJET-V6I6310.pdf.

[6] Luigi Alfredo Grieco, Alessandro Rizzo, Simona Colucci, Sabrina Sicari, Giuseppe Piro, Donato Di Paola, Gennaro Boggia, "IoT-Aided Robotics Applications: Technological Implications, Target Domains and Open Issues", *Computer Communications*, 2014, pp. 32–47, doi: 10.1016/j.comcom.2014.07.013.

[7] Irfan Ahmad, Karunakar Pothuganti, "Design & Implementation of Real Time Autonomous Car by Using Image Processing & IoT", in *Third International Conference on Smart Systems and Inventive Technology (ICSSIT)*, IEEE Xplore, 2020, pp. 107–113, doi: 10.1109/ICSSIT48917.2020.9214125.

[8] Ravi Kant Jain, Baskar Joyti Saikia, Nitant Pilmo Rai, Partha Pratim Ray, "Development of Web-based Application for Mobile Robot Using IOT Platform", in *2020 11th International Conference on Computing, Communication and Networking Technologies (ICCCNT)*, 2020, pp. 1–6, doi: 10.1109/ICCCNT49239.2020.9225467.

[9] Saad Ahmed Rahat, Ahmed Imteaj, Tanveer Rahman, "An IoT Based Interactive Speech Recognizable Robot with Distance Control Using Raspberry Pi", in *2018 International Conference on Innovations in Science, Engineering and Technology (ICISET)*, 2018, pp. 480–485, doi: 10.1109/ICISET.2018.87456.

[10] Tajim Md Niamat Ullah Akhund, Md Abu Bakkar Siddik, Md Rakib Hossain, Md Mazedur Rahman, Nishat Tasnim Newaz and Mohd Saifuzzaman, "IoT Waiter Bot: A Low Cost IoT Based Multi Functioned Robot for Restaurants", in *2020 8th International Conference on Reliability, Infocom Technologies and Optimization (Trends and Future Directions) (ICRITO)*, 2020, pp. 1174–1178, doi: 10.1109/ICRITO488 77.2020.9197920.56.

[11] Prakash Kanade, Sunay Kanade, "Medical Assistant Robot ARM for COVID-19 Patients Treatment", *International Research Journal of Engineering and Technology (IRJET)*, 2020, vol. 7, no. 10, pp. 105–111.

[12] Bandi Narasimha Rao, Reddy Sudheer, Mohan Aditya Sadhanala, Veerababu Tibirisettti and Sairam Muggulla, "Movable Surveillance Camera Using IoT and Raspberry Pi", *2020 11th International Conference on Computing, Communication and Networking Technologies (ICCCNT)*, 2020, pp. 1–6, doi: 10.1109/ICCCNT49239.2020.9225491.

[13] Mona Kumari, Ajitesh Kumar and Ritu Singhal, "Design and Analysis of IoT-Based Intelligent Robot for Real-Time Monitoring and Control", in *2020 International Conference on Power Electronics & IoT Applications in Renewable Energy and its Control (PARC)*, 2020, pp. 549–552, doi: 10.1109/PARC49193.2020.236673.

[14] Aishwarya K. Telkar and Baswaraj Gadgay, "IoT Based Smart Multi Application Surveillance Robot", in *2020 Second International Conference on Inventive Research in Computing Applications (ICIRCA)*, 2020, pp. 931–935, doi: 10.1109/ICIRCA48905.2020.9183289.

[15] Pooja S. Kulkarni, G. T. Chavan, "Android Application Based Wi-Fi Controlled Robot Car", *International Journal of Science and Research (IJSR)*, 2019, vol. 8, no. 5, pp. 135–140.

[16] Anna Nur Nazilah Chamim, Muhammad Eki Fawzi, I. Iswanto, Rama Okta Wiyagi, Ramadoni Syahputra, "Control of Wheeled Robots with Bluetooth Based Smartphones", *International Journal of Recent Technology and Engineering (IJRTE)*, 2019, vol. 8, no. 2, pp. 6244–6247.

[17] Selvaraj Vijayalakshmi, M. Archana, "Robotic Car Using Arduino with Bluetooth Controller", *International Journal of Intelligence in Science and Engineering*, 2019, vol. 1, no 1, p. 8.

[18] Bharath Kumar, Rashmi Avinash Damle, Sourav Sinha, K. Saravana Kumar, "Wi-Fi Controlled Rover", *International Journal of Trend in Research and Development (IJTRD)*, 2017, vol. 2, pp. 13–16.

[19] Namita Shinde, Shreya Srivastava, Vineet Sharma, Samarth Kumar, "Paper on Android Controlled Arduino Based Robot Car", *International Journal of Industrial Electronics and Electrical Engineering*, 2018, vol. 6, no. 3, pp. 1–3.

[20] Balendu Teterbay, Akshay Bhati, Ayush Srivastava, Abhay A. Deshpande, "Smartphone Controlled Multipurpose Robot Car" *International Journal of Engineering Research & Technology (IJERT)*, 2020, vol. 9, no. 5.

[21] Esra Yılmaz, Sibel Tarıyan Özyer, "Remote and Autonomous Controlled Robotic Car based on Arduino with Real Time Obstacle Detection and Avoidance", *Universal Journal of Engineering Science*, 2019, vol. 7, no. 1, pp. 1–7, pp. 485–488.

# 12 IoT-Based Monitoring System with Machine Learning Analytics of Transformer
## Mini Review

*Aadi Ashutosh Chauhan, Rohan Bhojwani,*
*Rahul Pal, and O.V. Gnana Swathika*
Vellore Institute of Technology

*Aayush Karthikeyan*
University of Calgary

*K.T.M.U. Hemapala*
University of Moratuwa

## CONTENTS

12.1 Introduction ........................................................................................................ 144
12.2 Transformer Health Index Development and Threshold Values Setting with the Help
of Machine Learning ....................................................................................... 144
12.3 IoT-Based Transformer Monitoring .................................................................. 144
12.4 Conclusion ....................................................................................................... 147
References............................................................................................................... 147

### ABSTRACT

Transformer health monitoring is a very important and crucial aspect when it comes to transformers, as it is a major static electro-mechanical device in micro and smart grids, so constant monitoring is of paramount importance. Recent innovations in fields like Internet of Things (IoT), machine learning and artificial intelligence have made the process of monitoring, analysing and protecting transformers more efficient and effective. This has further helped in increasing the lifespan of transformers and made the process of monitoring easier as well as in reducing the errors. These technologies have also made remote monitoring of transformers possible as well. In this chapter, various methods of transformer health monitoring systems along with advanced new-age technologies like IoT and machine learning are reviewed.

## KEYWORDS

Static Electro-Mechanical Device; Smart Grid; Transformer Health Monitoring; Machine Learning; Remote Monitoring

DOI: 10.1201/9781003374121-12

## 12.1  INTRODUCTION

Sensors like temperature sensor, potential sensor and current sensors are used to monitor real-time parameters like temperature, current and potential, and thus quick actions could be taken to avoid any issue. The key aspect of technology used here is the Global System for Mobile communication (GSM)-based system for data transmission which facilitates a better network for data transmission with robust features [1]. The Internet of Things (IoT)-based monitoring system which monitors the parameters like current, voltage, temperature and load ability with the help of various sensors are used for maintaining transformer health. If any of these parameters are above threshold value, then that information is sent through ESP8266 (Wi-Fi) module under HTTP protocol to an IP address that shows real-time data chart form in any web-connected device. The system is also designed in such a way that it could correct some of these abnormalities [2]. IoT helps us to facilitate transformer health monitoring. Parameters like temperature, oil level, and vibration are detected through sensors, and this data is transferred to the control where it checks if any of the threshold values are crossed and if so then a message is sent to the concerned body [3–5]. Sensors are used to measure voltage, current (overvoltage, undervoltage, overcurrent), and other equipment parameters. This information is delivered to the microcontroller, which checks the parameters before sending it to the IoT web server using Adafruit software, ensuring that the operator has the correct information [6–8].

## 12.2  TRANSFORMER HEALTH INDEX DEVELOPMENT AND THRESHOLD VALUES SETTING WITH THE HELP OF MACHINE LEARNING

Defining proper parameters for transformer health is very important when it comes to transformer health. To have greater insight on analytics of transformer health monitoring, we use tools like conditional anomaly detection, correlation, evidence combination, diagnostics, health index estimation and prognostics modules. These tools are used on the data gathered by the sensor and help in early detection of malfunction [3]. Neural networks have been found to be very effective in developing the health index of transformers. A feed-forward artificial neural network (ANN) is used to find health indexes, where parameters like acidity, water content, hydrogen content, break down voltage, methane content, ethylene content, acetylene content, loss factor, furans content, and total solids in oil for each transformers are used to develop the neural network [4,5]. Correlation analysis is used to find the parameters which influence a transformer's health and the factors which come out are dissolved gas analysis (DGA), dielectric loss, breakdown voltage, acid value, micro-water content, furfural content, etc. Using a mix of qualitative analysis and the analytic hierarchy process approach to determine the weight of the index, an index system is created, and a health status evaluation model for the power transformer is created [6]. Transformer health is regularly monitored using an IoT-based edge computing technique. They gathered buzzing sound data from both normal and damaged transformers, and then used a windowing approach to split the buzzing sound into 2-second segments. To extract features from these sounds, Mel frequency cepstral coefficients are employed, and a dataset is formed. Then we train this dataset using the Support Vector Machine (SVM) classifier, and through this, we are able to judge at what values the transformer is healthy and in those it is at fault [7,8].

## 12.3  IoT-BASED TRANSFORMER MONITORING

IoT provides a solution that can help identify internal transformer errors. Sensors are used to detect the malfunction of the transformer, which transmits a certain signal frequency to report abnormalities occurring [9]. A data logger is used to monitor the life of the transformer on loading by regularly checking some of the key parameters that reflect the performance of the transformer. The parameters which are monitored include voltage on load condition, working transformation temperature, transmission into the transformer, the amount of the oil present in the tank and the sensation of burning in the event of a heavy load on the transformer. All these data are sent to the website through sensors where they could be monitored. We use hardware like an Arduino Microcontroller, a DHT11 sensor, Voltage Sensor, ESP8266 Wi-Fi Module, Accelerometer and Ultrasonic Sensor for this purpose [10]. Temperature, humidity and rate of loading are all monitored, and a self-protection mechanism for the transformer is created and implemented. If the transformer is not serviced quickly, it will separate low-importance load and keep the high-importance loads; if the transformer is unable to feed the high-importance loads, it will separate all loads and remain in no-load mode, where the transformer monitors its parameters on its own and automatically returns the loads to service if all parameters return to normal [11]. The importance of real-time monitoring for fault detection cannot be overstated. For this reason, we monitor criterion such as transformer's current and voltage, the oil temperature and the oil level of transformer on a regular basis, as well as whether there is any spark, flame or smoke, and send the appropriate alarm message [12]. We collect a lot of data with IoT, since we use edge computing and an agent-based system design. Without relying on the cloud, autonomous choices can be performed at the edge with locally accessible data, assisting in the management of transformer health [13]. Parameters like temperature, current and oil level are collected with the help of IoT technologies. Then this data is sent to the internet via TCP/IP protocol, and if any parameter is higher than the threshold value, then an alert is sent which could be seen on an android application [14]. For the proper functioning of transformers, some common checks need to be placed to extend its operating time. Some diagnostic checks on a regular basis using IoT sensors facilitate remote monitoring and proper functioning of transformers [15,16]. We monitor the transformer by taking into account several characteristics such as voltage, temperature, current and oil level using various sensors and a microprocessor. Operational data with a unique transformer status is received using IoT server, and this data is stored on a computer server with the help of a specific web address [17]. Detecting hazardous conditions such as cracks which are the most important parameters of current leakage in the substations and gas leaks based on data source file errors from isolated source, and we use this to create an IoT gateway. It contains a NODE-MCU, a red node server and a mosquito server that acts as an intermediary between the NODE-MCU and the red node server and ThingSpeak IoT platform [18]. ThingSpeak IoT platform is used for data processing so that the monitored value can be displayed on a regular or periodic basis. SPSS software package is used to statistically update this analysed result [19].

We use IoT to send real-time data from a transformer to a data centre where the engineer can track the performance and operating life of the transformer. An engineer is needed locally only in case of an emergency; However, the system notifies the engineer using the GSM module via SMS for any restrictions. In addition, automation is introduced into the transformer cooling machine, where the cooler operating capacity is controlled by a microcontroller-supported response system, thus reducing human dependence to a great extent. The results from the experiments and simulations performed are analysed to predict the performance of automatic IoT-based transformers and are compared with the existing model, and real barriers to the same performance are discussed [20]. Real-time monitoring using IoT on the distribution transformer where measurements on parameters like oil level, load current, line voltage and temperature are done along with implementation of MQTT protocol on their IoT divide for optimized high latency in the system [21]. In building a prototype of an IoT system used for monitoring Power Transformers at Nuclear Power Plants, sensors like ultrasonic sensor and gas detecting sensor are used in the prototype to detect the parameters, and these parameters are then displayed on a remote screen. A prototype has been suggested for the real life construction [22]. An IoT-based apparatus for Health Monitoring of the distribution transformer is developed. A prototype is built in which measurements of different parameters using sensors and GSM module are used to communicate the information using which details are shifted to the monitoring mobile phone [23]. An IoT-based apparatus to monitor the basic parameters of prototype is made. The method of IoT communication used is LoRa technology for wireless communication. But the flaw with this apparatus is that LoRa has very restricted range. Therefore, wireless communication over very large distances is not possible. However, a matter of more research is required. Colour sensors and ultrasonic sensors that are used are readily available ones. They have created their own voltage sensing unit [24]. An apparatus is made to monitor a distribution transformer using Bluetooth technology for wireless communication. Prototyping is done on step-down transformers. Measurement of oil level, temperature and voltage values has been monitored to observe any abnormal readings [25–27]. Executing wired apparatus using Arduino Mega is deployed to check different parameters and disconnect the transformer with supply if the values tend to cross the threshold. ThingSpeak is used to display the data [28,29]. Using different sensors which can be directly placed on transformers and then using the data from them enable applying anti-theft measures to protect the transformer from any misuse. IoT has been used as an application. The GSM module has been used for wireless communication [30–32].

A microcontroller based on relay can be utilized to provide the optimum protection for the transformer. The flow sensor, fans, temperature sensor, relay and buzzer are all connected to the protective microcontroller (ESP-32). Real-time data can be supplied to the cloud, and the transformer's protection can be monitored by analysing the data in an online mode [33]. To observe and record properties of a transformer such as temperature, current, oil level, vibration and humidity, embedded systems or GSM/GPRS methods are developed. The transformer site has remote devices that work with an 8-channel analogue-to-digital converter. This system sends an alert message if something unusual happens and or reading exceeds the predefined limits [34].

A relay controls the fan that cools the transformer in the transformer cooling system. Sometimes, transformer temperature fluctuates, causing the air fan to frequently on and off, leading to malfunctioning of the fan. So by using GSM technology we can operate the fan by using IoT remotely [35–37]. It interacts with air due to leakage in oil tanks, creating an oxidation reaction in the transformer oil. The oil comes in the form of iron, copper and various metal alloys. The nature of the oil provided by the transformer will vary as a result of the aforementioned circumstance. As a result, the IoT-based system checks the explosion's vision, clarity and temperature before deciding whether to use the oil or clean and train a person under stress via IoT or GSM [38]. Substation monitoring and control in real time is a critical issue that is typically handled manually or with the help of an expensive Programmable Logic Controller (PLC) and Supervisory Control and Data Acquisition (SCADA) system. Now using IoT has become so much easier and cheap, and also reduces the risk of fault as human intervention will also get reduced [39]. Individual gadgets will be encouraged to use the internet for data sharing through machine-to-machine connectivity. The IoT procedure uses information sent to the microcontroller to evaluate parameter limitations before sending it to the IoT server via IoT modules [40].

The IoT has grown in popularity in recent years. Thanks to the IoT, it can communicate with system devices and use device data for controlling, monitoring and protecting it. Develop an online power consumption monitoring system for loads from a distribution transformer using the IoT. The sensors are pre-processed to meet the controller's analogue input needs. The controller processes the parameters based on the integrated algorithm. The output of the controller is then sent to the IoT server, where it is analysed with the help of a user-friendly Graphical User Interface (GUI) and graphs [41].

The IoT-based monitoring system combines a worldwide system of mobile communication modems with node microcontroller units, sensors and transformers that are processed and recorded in the cloud. Run a simple checkup to check the abnormalities in the system. If found any, a message is sent to the selected mobile at the same time and required action is taken automatically. Basically, the main focus is to reduce human interference and lower the cost and increase the efficiency as far as possible [42].

Transformer health index development and threshold values setting are done with the help of machine learning. Various transformer parameters like voltage imbalances, current volume, transformer oil levels, temperature and vibration are recorded for monitoring real-time error in the transformer. Based on these parameters, the fail state and working state are predicted using the ANN algorithm [16–18]. Two exciting ways of predicting volume have been found. In both processes, the unique usage pattern has been defined using a parametric scale; neural network and backward vector support are being used to determine the total amount of the day-use that is being redistributed according to the day pattern and find the final loading pattern [19].

General Packet Radio Services (GPRS) location technology-based monitoring is used, where the threshold conditions are set and if any of the parameters cross any of these values, the system generates a signal. Distribution transformer parameters like oil temperature, current, etc. are monitored. Database Management System (DBMS) algorithms are used to sort this data [26]. The requirements of smartening of the distribution transformer are very important for its proper functioning. Usage of statistical data about different types of failures is one thing that is found to be useful, as well as finding threshold values. Incorporating new-age technologies facilitate this process [27]. Different faults that can occur in a transformer are studied along with their causes. Statistical data about the probability of occurring of each fault and the possible reasons behind it are researched. A fuzzy network approach is used to recognize the fault [29]. A statistical approach is used to calculate the degradation of transformer by calculating the secondary output of the transformer and using the data in a logistic regression model which is specially designed for this scenario. On this basis, a smart meter is derived, which can be installed on the transformer and can record data and help in proper monitoring of the transformer [31]. Fuzzy logic is applicable on the data collected from the sensor. Sensors used are common, i.e., oil level sensor and temperature sensor, and then, on the basis of the output from the fuzzy logic, prediction is made on the health of the transformer [32].

An ensemble machine learning method and an IoT-based monitoring system are used to detect faults. There are two pieces to the IoT system: a data-measuring subsystem and a data reception subsystem. Transformer vibration signals are measured and sent to a remote server. The EML is made up of deep belief networks and layered denoising autoencoders with distinct activation functions in relevance vector machines. This method has the advantages of being low-power, incredibly trustworthy and long-distance transmission [36]. By monitoring and checking physical and chemical changes in the environment, this technology checks not only the electrical parameters, but also the functional and defect state of the transformers. Different classifiers such as Bayesian networks, multilayer perceptron and random forest were used to classify the operation voltages versus ambient data in order to build a connection between the collected ambient data and transformer operating voltages [37].

## 12.4 CONCLUSION

Monitoring of transformer is one of the key aspects while managing micro and smart grids, and therefore, sensing correct parameters is very important. The methods of IoT-based monitoring of transformer mentioned here in the paper helps to know the real-time condition of a transformer. Analysis of threshold values which can be achieved using machine learning is an advanced approach of monitoring and detection of faults. Further, with the techniques mentioned, a real-time monitoring system can be built.

## REFERENCES

[1] Srivastava, D. and Tripathi, M.M., 2018. Transformer Health Monitoring System Using Internet of Things. In *2018 2nd IEEE International Conference on Power Electronics, Intelligent Control and Energy Systems (ICPEICES)* (pp. 903–908). IEEE.

[2] Bethalsha, C., Jennifer, A., Karthik, M.S. and Sreejavijay, M., 2020. Real-time transformer health monitoring using IOT. *International Journal for Research in Applied Science and Engineering Technology*, 8(9), pp. 521–526.

[3] Aizpurua, J.I., Catterson, V.M., Stewart, B.G., McArthur, S.D., Lambert, B., Ampofo, B., Pereira, G. and Cross, J.G., 2017. Determining appropriate data analytics for transformer health monitoring. In *10th International Topical Meeting on Nuclear Plant Instrumentation, Control and Human Machine Interface Technologies. American Nuclear Society, USA* (pp. 1–11).

[4] Abu-Elanien, A.E., Salama, M.M.A. and Ibrahim, M., 2011. Determination of transformer health condition using artificial neural networks. In *2011 International Symposium on Innovations in Intelligent Systems and Applications* (pp. 1–5). IEEE.

[5] Suja, K. and Yuvaraj, T., 2021. Transformer health monitoring system using android device. In *2021 7th International Conference on Electrical Energy Systems (ICEES)* (pp. 460–462). IEEE.

[6] En-Wen, L. and Bin, S., 2014. Transformer health status evaluation model based on multi-feature factors. In *2014 International Conference on Power System Technology* (pp. 1417–1422). IEEE.

[7] Ahmad, I., Singh, Y. and Ahamad, J., 2020. Machine learning based transformer health monitoring using IoT edge computing. In *2020 5th International Conference on Computing, Communication and Security (ICCCS)* (pp. 1–5). IEEE.

[8] Khandait, A.P., Kadaskar, S. and Thakare, G., 2017. Real time monitoring of transformer using IOT. *International Journal of Engineering Research & Technology (IJERT)*, 6(3), pp. 146–149.

[9] Kamlaesan, B., Kumar, K.A. and David, S.A., 2017. Analysis of transformer faults using IoT. In *2017 IEEE International Conference on Smart Technologies and Management for Computing, Communication, Controls, Energy and Materials (ICSTM)* (pp. 239–241).IEEE.

[10] Zehra, I. and Ansari, M.A., 2021. IOT Based Health Monitoring of Transformer Using Data Logger. In *2021 4th International Conference on Recent Developments in Control, Automation & Power Engineering (RDCAPE)* (pp. 67–72). IEEE.

[11] Hasan, W.K., Alraddad, A., Ashour, A., Ran, Y., Alkelsh, M.A. and Ajele, R.A., 2019. Design and Implementation Smart Transformer based on IoT. In *2019 International Conference on Computing, Electronics & Communications Engineering (iCCECE)* (pp. 16–21). IEEE.

[12] Kumar, R.K., Thilagaraj, M., Vengatesh, P., Rajalakshmi, J. and Babul, M.M., 2021. Remote transformer faults analyzing system using IoT. *International Journal of Modern Agriculture*, 10(2), pp. 2390–2402.

[13] Thangiah, L., Ramanathan, C. and Chodisetty, L.S., 2019. Distribution transformer condition monitoring based on edge intelligence for industrial IoT. In *2019 IEEE 5th World Forum on Internet of Things (WF-IoT)* (pp. 733–736). IEEE.

[14] Blynk, Aralikatti, S., Pradyumna, R., Reddy, R., Sanjay, B.R. and Reddy, S.N.K., 2021. IoT-based distribution transformer health monitoring system using node-MCU & Blynk. In *2021 Third International Conference on Inventive Research in Computing Applications (ICIRCA)* (pp. 1–4). IEEE.

[15] Sharma, R.R., 2021. Design of distribution transformer health management system using IoT sensors. *Journal of Soft Computing Paradigm*, 3(3), pp. 192–204.

[16] Khairnar, V., Kolhe, L., Bhagat, S., Sahu, R., Kumar, A. and Shaikh, S., 2020. Industrial automation of process for transformer monitoring system using IoT analytics. In *Inventive Communication and Computational Technologies* (pp. 1191–1200). Springer, Singapore.

[17] Yamuna, R., Geetha, R., Gowdhamkumar, S. and Jambulingam, S., 2019. Smart distribution transformer monitoring and controlling using IoT. *International Research Journal of Multidisciplinary Technovation*, pp. 111–115.

[18] Mohamad, A.A., Mezaal, Y.S. and Abdulkareem, S.F., 2018. Computerized power transformer monitoring based on the internet of things. *International Journal of Engineering & Technology*, 7(4), pp. 2773–2778.

[19] Singh, M.J., Agarwal, P. and Padmanabh, K., 2016. Load forecasting at distribution transformer using IoT based smart meter data from 6000 Irish homes. In *2016 2nd International Conference on Contemporary Computing and Informatics (IC3I)* (pp. 758–763). IEEE.

[20] Subramanian, P.V., Boddapati, V. and Daniel, S.A., 2022. Automated real-time transformer health monitoring system using the internet of things (IoT). In *Advancement in Materials, Manufacturing and Energy Engineering*, Vol. I (pp. 503–511). Springer, Singapore.

[21] Roy, T.K. and Roy, T.K., 2018. Implementation of IoT: smart maintenance for distribution transformer using MQTT. In *2018 International Conference on Computer, Communication, Chemical, Material and Electronic Engineering (IC4ME2)* (pp. 1–4). IEEE.

[22] Elmashtoly, A.M. and Chang, C.K., 2019. Real-Time Monitoring for Power Transformer Using IoT. In *Autumn Meeting Korean Nucl. Soc.*

[23] Rahman, S., Dey, S.K., Bhawmick, B.K. and Das, N.K., 2017. Design and implementation of real time transformer health monitoring system using GSM technology. In *2017 International Conference on Electrical, Computer and Communication Engineering (ECCE)* (pp. 258–261). IEEE.

[24] Kumar, T.A. and Ajitha, A., 2017. Development of IOT based solution for monitoring and controlling of distribution transformers. In *2017 International Conference on Intelligent Computing, Instrumentation and Control Technologies (ICICICT)* (pp. 1457–1461). IEEE.

[25] Lutimath, J., Lohar, N., Guled, S., Rathod, S. and Mallikarjuna, G.D., 2020. Fault finding system for distribution transformers and power man safety using IoT. *International Journal for Research in Applied Science & Engineering*, 8, pp. 1689–1693

[26] Pawar, R.R., Wagh, P.A. and Deosarkar, S.B., 2017. Distribution transformer monitoring system using Internet of Things (IoT). In 2017 International Conference on Computational Intelligence in Data Science (ICCIDS) (pp. 1–4). IEEE.

[27] Rao, N.M., Narayanan, R., Vasudevamurthy, B.R. and Das, S.K., 2013. Performance requirements of present-day distribution transformers for Smart Grid. In *2013 IEEE Innovative Smart Grid Technologies-Asia (ISGT Asia)* (pp. 1–6). IEEE.

[28] Yaman, O. and Biçen, Y., 2019. An internet of things (IoT) based monitoring system for oil-immersed transformers. *Balkan Journal of Electrical and Computer Engineering*, 7(3), pp. 226–234.

[29] Tran, Q.T., Davies, K., Roose, L., Wiriyakitikun, P., Janjampop, J., Riva Sanseverino, E. and Zizzo, G., 2020. A review of health assessment techniques for distribution transformers in smart distribution grids. *Applied Sciences*, 10(22), p. 8115.

[30] Manu, D., Shorabh, S.G., Swathika, O.G., Umashankar, S. and Tejaswi, P., 2022. Design and realization of smart energy management system for Standalone PV system. In *IOP Conference Series: Earth and Environmental Science* (Vol. 1026, No. 1, p. 012027). IOP Publishing.

[31] Swathika, O.G., Karthikeyan, K., Subramaniam, U., Hemapala, K.U. and Bhaskar, S.M., 2022. Energy Efficient Outdoor Lighting System Design: Case Study of IT Campus. In *IOP Conference Series: Earth and Environmental Science* (Vol. 1026, No. 1, p. 012029). IOP Publishing.

[32] Sujeeth, S. and Swathika, O.G., 2018. IoT based automated protection and control of DC microgrids. In *2018 2nd International Conference on Inventive Systems and Control (ICISC)* (pp. 1422–1426). IEEE.

[33] Patel, A., Swathika, O.V., Subramaniam, U., Babu, T.S., Tripathi, A., Nag, S., Karthick, A. and Muhibbullah, M., 2022. A practical approach for predicting power in a small-scale off-grid photovoltaic system using machine learning algorithms. *International Journal of Photoenergy*, 2022.

[34] Odiyur Vathanam, G.S., Kalyanasundaram, K., Elavarasan, R.M., Hussain Khahro, S., Subramaniam, U., Pugazhendhi, R., Ramesh, M. and Gopalakrishnan, R.M., 2021. A review on effective use of daylight harvesting using intelligent lighting control systems for sustainable office buildings in India. *Sustainability*, 13(9), p. 4973.

[35] Swathika, O.V. and Hemapala, K.T.M.U., 2019. IOT based energy management system for standalone PV systems. *Journal of Electrical Engineering & Technology*, 14(5), pp. 1811–1821.

[36] Swathika, O.V. and Hemapala, K.T.M.U., 2019. IOT-based adaptive protection of microgrid. In *International Conference on Artificial Intelligence, Smart Grid and Smart City Applications* (pp. 123–130). Springer, Cham.

[37] Kumar, G.N. and Swathika, O.G., 2022. 19 AI Applications to. *Smart Buildings Digitalization: IoT and Energy Efficient Smart Buildings Architecture and Applications*, p. 283.

[38] Swathika, O.G., 2022. 5 IoT-Based Smart. *Smart Buildings Digitalization: IoT and Energy Efficient Smart Buildings Architecture and Applications*, p. 57.

[39] Lal, P., Ananthakrishnan, V., Swathika, O.G., Gutha, N.K., Hency, V.B., 2022. 14 IoT-Based Smart Health. *Smart Buildings Digitalization: Case Studies on Data Centers and Automation*, p. 149.

[40] Chowdhury, S., Saha, K.D., Sarkar, C.M. and Swathika, O.G., 2022. IoT-based data collection platform for smart buildings. In *Smart Buildings Digitalization* (pp. 71–79). CRC Press.

[41] Chacko, S.T. and Deshmukh, V., 2019. IoT based Online Power Consumption Monitoring of a Distribution transformer feeding Domestic/Commercial Consumer loads. In 2019 4th International Conference on Information Systems and Computer Networks (ISCON) (pp. 441–445). IEEE.

[42] Balraj, B., Sridevi, A., Amuthameena, S. and Caroline, B.E., 2021. Investigations on the Physical Parameters and Real Time Protection of Distributed Transformers using Internet of Things. In *Journal of Physics: Conference Series* (Vol. 1717, No. 1, p. 012071). IOP Publishing.

# 13 Design of Earthquake Alarm

*K. Lokeswar, D. Subbulekshmi, T. Deepa, and S. Angalaeswari*
Vellore Institute of Technology

## CONTENTS

13.1 Introduction ........................................................................................... 151
    13.1.1 Motivation .................................................................................. 151
13.2 Description ............................................................................................ 152
13.3 Components Information ...................................................................... 152
    13.3.1 Piezoelectric Sensor ................................................................. 152
    13.3.2 IC 555 ......................................................................................... 152
    13.3.3 Transistor (BC547) ................................................................... 152
13.4 Working Principle ................................................................................ 153
13.5 Results ................................................................................................... 153
13.6 Future Scope ........................................................................................ 155
13.7 Conclusion ........................................................................................... 155
Bibliography ................................................................................................... 155

**ABSTRACT**

It is estimated that there are 500,000 detectable earthquakes in the world annually. 100,000 of these are often felt, and 100 of them cause damage. Around ten thousand deaths are reported annually due to earthquakes. The common impacts of earthquake are damage to the buildings, fires, damage to transport facilities, landslides, liquefaction, and tsunami. In older days, the methods used to detect earthquakes included studying animal behavior like fishes in ponds gets agitated, snakes come to the surface, etc. but there was no such a perfect way tt o detect an earthquake. Saving the lives and reducing the damage are the key. Here, we present a piezoelectric sensor-based Early Warning Earthquake Alarm; it detects the SESMIC vibrations and sends the signal to the alarm system to avoid the above-mentioned damages.

## 13.1 INTRODUCTION

Earthquakes cause great dangers to human lives. Earthquakes are caused due to abrupt release of energy in Earth's crust, which produces seismic waves.

This work's objective is to make an earthquake alarm that can sense the seismic vibrations and thereby provide a preparatory time for the people.

### 13.1.1 MOTIVATION

Around 10,000 deaths are reported annually due to earthquakes. The common impacts of earthquake are damage to the buildings, fires, damage to transport facilities, landslides, liquefaction, and tsunami. Our motive is to reduce the loss as much as possible and save people's lives.

DOI: 10.1201/9781003374121-13

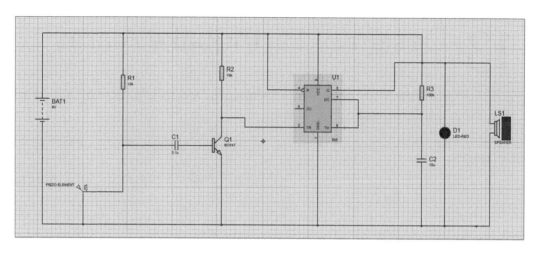

**FIGURE 13.1** Earthquake alarm circuit diagram.

## 13.2 DESCRIPTION

- In our circuit, seismic sensor is used to detect vibrations.
- Due to its height and sensitivity, vibrations induced by motion of animals and other objects might trigger the system to detect vibrations and thereby classify it as an earthquake. This is a major drawback of the alarm system.
- This apparent drawback could be utilized as an extended feature of the system by using it as a security measure which detects undesirable individuals or animals within private property.
- The structure of the circuit comprises readily available electronic parts which lead to a simpler circuit design that is accessible to even an electronics novice.
- To detect pressure variation-induced vibrations, a common piezo sensor is used.
- The piezo element performs like a minuscule capacitor powered by just a few nanofarads of capacitance (Figure 13.1).

## 13.3 COMPONENTS INFORMATION

### 13.3.1 PIEZOELECTRIC SENSOR

Piezoelectric element is made up of lead zirconate crystals, these crystals can store current immediately, and will also release current when the crystals orientation will be disturbed through mechanical vibrations (Figure 13.2).

### 13.3.2 IC 555

In this work, the IC 555 timer is in its monostable mode. In this mode when the voltage on the capacitor will be equal to two-third of the supply voltage, the output pulse ends (Figure 13.3).

### 13.3.3 TRANSISTOR (BC547)

The BC547 transistor is an NPN transistor. BC547 is generally used for current amplification, quick switching, and pulse-width modulation. When power is applied to the base of the transistor, it will

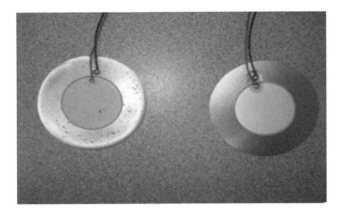

**FIGURE 13.2**    Piezoelectric element.

**FIGURE 13.3**    IC555.

flow from the collector to the emitter. A small current of the base terminal of the transistor controls the large current of emitter and base terminals (Figure 13.4).

## 13.4    WORKING PRINCIPLE

- The piezoelectric sensor converts seismic vibrations from the earth into electric signal.
- The electric signal generated by sensor is given to the transistor BC547 which is used as an amplifier.
- Transistor amplifies this signal and triggers the timer IC555. Here, IC555 is used in monostable mode.
- In normal condition, its output is low. So, the buzzer which is connected to the output remains off.
- When it gets trigger pulse, its output goes high and the buzzer turns on.
- After sometime, its output again goes low and the buzzer turns off.

## 13.5    RESULTS

The pictures of our earthquake alarm are shown in Figure 13.5, when the piezo element detects the seismic vibrations, the loud sound will be produced, thus alerting the nearby people. This is a low-cost

**FIGURE 13.4**   Transistor (BC547).

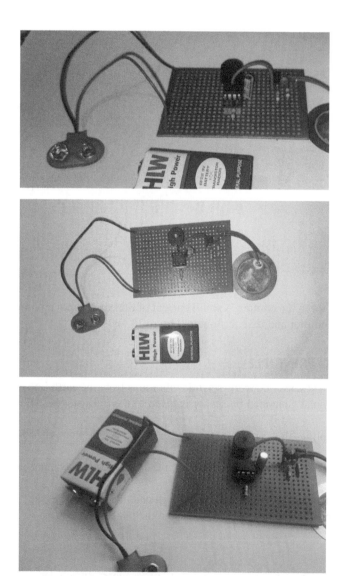

**FIGURE 13.5**   Hardware setup of earthquake alarm: (a) View 1, (b) View 2, and (c) View 3

alarm and made with simple electronic parts, so it can be used in remote villages and can prevent lot of loss of life. The earthquake alarm is sensitive to vibrations; therefore, it detects the earthquake earlier and alerts people and gives people ample time to evacuate the area along with their livestock and their precious belongings, which help them in surviving the after-effects of the earthquake.

## 13.6   FUTURE SCOPE

We can further improvise this activity to help citizen of the city by adding a GSM module (sim card) which will send alert SMS to everyone whenever the alarm detects chances of earthquake.

Also, we are planning to use Artificial Intelligence to increase the accuracy of the device by eliminating inaccurate alerts caused by the animal and vehicle movement.

## 13.7   CONCLUSION

Around 60,089 earthquakes are large enough to be noticed without use of instruments that occur annually over the entire earth. Out of these 50,000 plus earthquakes, 100 have the potential to cause substantial damage if their centers lie near the areas of settlements. Large-scale earthquake occurs about once a year. Till now, they are responsible for many deaths and huge damage to the properties. To help prevent all this damage we have created an earthquake alarm.

Alarm activation works well and the cost and area required for the alarm are greatly reduced because this is a small tool. The structure of the circuit comprises readily available electronic parts which lead to a simpler circuit design which is accessible to even an electronics novice. It works like a normal alarm that produces a loud noise when the earthquake sensor receives a vibration. It is very sensitive and can get vibrated due to the movement of animals or other objects. A standard piezo sensor is used to detect vibrations due to pressure differences. The piezo element behaves like a small capacitor with a capacitance of just a few nanofarads. The alarm must be kept at a distant and isolated place due to its sensitive sensor. This alarm helps in detecting earthquakes and alerting people and giving them enough time to evacuate the area, thereby preventing the loss of lives. This alarm is developed keeping in mind the rural areas where usually most of the deaths occur due to lack of knowledge and money to afford huge technologies. This alarm occupies very less space, and it is low cost, so every village can afford this alarm comfortably and place this alarm in an isolated place in the center of their village so that more people will be alerted and many deaths can be prevented. During earthquakes, the government spends a lot of money to spread knowledge about the earthquake. This money can be saved utilized for a different purpose if every village has their own low-cost earthquake alarm.

## BIBLIOGRAPHY

1. Irshad Khan, Seonhwa Choi, Young-Woo Kwon, "Earthquake Detection in a Static and Dynamic Environment Using Supervised Machine Learning and a Novel Feature Extraction Method", *Sensors* 20(3), pp. 1–21, 2020.
2. Namrata J. Helonde, Punam Suryawanshi, Arun Ankita Bhagwatkar, Arun Wagh, Pradhnya Vetal, "Footstep Power Generation Using Piezoelectric Sensor", *International Journal for Research in Applied Science & Engineering Technology* 9(XII), pp. 1–9, 2021.
3. Nishu Pandey, K. Hari Simha Rao, V. Seshank, Shashank Pokala, Aparna T Sunil, "Earthquake Alert System Based on IoT Technology", *Journal of Emerging Technologies and Innovative Research (JETIR)* 8(5), pp. 1–7, 2021.
4. Nanditha Nandanavanam, "An Imprint of IC555 Timer in the Contemporary World", *International Journal of Engineering and Advanced Technology* 4(6), pp. 1–4, 2015.
5. https://www.avnet.com/wps/portal/abacus/solutions/technologies/sensors/pressure-sensors/core-technologies/piezoelectric/
6. https://www.electroschematics.com/seismic-sensor/

# 14 Energy Demand and Flexibility of Energy Supply
## A Case Study

*R. Rajapriya, Varun Gopalakrishnan, Milind Shrinivas Dangate*
Vellore Institute of Technology

*Nasrin I. Shaikh*
Nowrosjee Wadia College and Abdul Karim Ali Shayad
Faculty of Engineering and Polytechnic, MMANTC

## CONTENTS

14.1 Introduction ........................................................................................................ 158
14.2 Literature Review ............................................................................................... 159
14.3 Materials and Methods ...................................................................................... 161
14.4 Thermotical Expression ..................................................................................... 161
14.5 Considerations in Econometrics ....................................................................... 162
    14.5.1 Data Overview ........................................................................................ 162
    14.5.2 Problems with Using an Aggregate Energy Measure ........................ 163
    14.5.3 Estimation and Outcomes Calculation .............................................. 164
14.6 Conclusion .......................................................................................................... 168
References ...................................................................................................................... 169

**ABSTRACT**

Energy markets are likely to be impacted in several key ways soon. Energy prices are expected to rise in the short and long term, as per the energy information administration's yearly energy outlook 2011 report. The rising consumption of emerging economies like China as well as India is one factor driving up prices. In all likelihood, rising global demand will result in higher energy prices. Increased prices, which result in lower consumption, are likely to slow economic growth. In all likelihood, a higher domestic supply would result in increased energy use. Aside from these factors, climate change is projected to have a considerable impact on energy usage as well as production. In the United States, there is much doubt about the possibility and substance of climate change legislation. Comprehensive climate change legislation might have a big impact on how much energy the country uses. Climate change will have a considerable impact on energy demand, independent of the consequences of climate change legislation. The average temperature in the United States is predicted to rise as a result of climate change. Temperature changes have an impact on the heating and cooling requirements of buildings. This move gives scientists a starting point for assessing the influence of climate change on energy demand. Climate change has resulted in a net increase in energy consumption in the Western region due to major energy requirements for building heating and cooling. Because the number of cooling degree days has risen faster than the number of heating degree days, this is the case. The energy required to generate the additional power rises in tandem with the primary energy necessary to cool buildings.

DOI: 10.1201/9781003374121-14

## 14.1 INTRODUCTION

The goal of this research is to assess energy markets at the state level in the Western United States. The estimation of behavioural parameters is possible with a properly described supply and demand system. Previous elasticity estimates may be out of date due to the overall trend of continuing advances in the energy efficiency of the US economy. Based on the findings, the state-specific expected elasticities will provide for even more precise estimates of behaviour in response to state energy policy. Negative externalities, supply, and other challenges will be addressed as a result of this. Energy strategies are also being developed by regions and governments. The Regional Greenhouse Initiative in the Northeast is one example, as is the multiplicity of state-level renewable portfolio standards (RPS). State-specific elasticity estimations will account for the differences in the economies of Western states. Hamilton has done a substantial study on the macroeconomic effects of energy price shocks. Consumers' first changes to spending pattern is the most important process through which energy price affects economic performance. Due to the relative inelasticity of energy demand's price elasticity, policymakers should be aware that imposing taxes or other price adjustments will almost certainly affect the consumption of other commodities. Whereas the research focuses on total energy use, the findings can be applied to standard economic research in the Western United States. Price rises, according to Metcalf (2008) [1], are effective state-level catalysts for increased energy efficiency. Studies by Keppler et al. (2006) [2] back up Metcalf's conclusions about the relevance of pricing in technological progress. According to Hamilton's research, increased costs will cause technological advancements, which will put downward pressure on economic development. Income fluctuations have a greater impact on energy usage than changes in other demand components, according to the findings presented in this chapter. The contradictory pressures of rising earnings and rising costs emphasize the complexities of many energy concerns. By examining the quality and the quantity of state energy markets, the study adds to the literature on energy-related elasticities. Supply and demand interaction can be controlled with a simultaneous system. In 2003, Gujarati symbolizes the econometric challenges that arise from evaluating a lowered assumption of one side of the economy [3]. The fact that supply and demand, which are both endogenous, influence price and quantity, is one of the most important issues. Lowered estimations of elasticities may not yield accurate results, since they do not enable supply and demand interactions while calculating market prices. Consider the market as in charge of pricing and quantity adjustments without properly for the supply curve. An integrated component of energy usage is a second extension. The majority of research assessed elasticities for a single energy market or source in the literature. Expected elasticity for a given source of power is commonly used as a term for energy elasticity in the literature. The elasticity for domestic power, for example, is referred to it as the price elasticity of energy consumption. Specific energy sources should be considered so as a subset of total energy consumption. Although these calculations are crucial for a thorough understanding, they don't accurately explain total energy advancements since they don't have a good understanding of energy markets. Elasticities generated will be captured via aggregate measures of the impact of shocks on ultimate energy use as energy sources shift. Since energy consumption seems to be a derived demand that is used to offer a service rather than for its intrinsic worth, cumulative energy utilization elasticities will drive policy regardless of whether or not energy sources are replaced. The study's data, except for Hawaii and Alaska, covers states in the Census' Western region. Due to their position beyond the contiguous United States, these states were deleted. The information covers a significant period, from 1970 to 2007. As a result, necessary adjustments in the energy–economic relationship can be detected using the data. Increasing energy efficiency over time (for the majority of the states in the sample), rising real wages and prices, and a slew of energy price spikes are all examples of these dynamics. The investigation proceeds as follows: The relevant literature will be discussed in Section 14.2. Due to the enormous number of research, calculating the analysis would be constrained to a random sample of energy elasticities. The studies that have been addressed will place this study in the context of previous research and justify the data gathering and estimation approach used. The theoretical framework is outlined in Section 14.3. Section 14.4 contains the data. Section 14.5 describes the estimator utilized, as well as the results and estimates of changes in energy consumption caused by a variety of demand determinant shocks. Section 14.6 brings the research to a close.

## 14.2   LITERATURE REVIEW

There is a large body of work dedicated to estimating various elasticities related to energy consumption. Elasticities for several energy sources and regional marketplaces have been calculated. The types of data used in studies span from local to global aggregates. Filippini (2010) [4] is an example at the city level. Data from Swiss cities were used to calculate the price elasticity of power demand at peak and off-peak times. According to a study by He, Wang, and Lai in 2010 [5], oil demand is expected to have a long-run price elasticity of −0.89 globally. This section will include studies that serve as a foundation for the technique. Some outcomes are easily comparable to the conclusions of this study. For decades, energy elasticities have been calculated. More focus will be made on studies from the mid-90s onwards to put the findings of the study in a more current context. Longer time series and updated econometric approaches are now possible in modern investigations. For individuals looking for more information, literature surveys will be highlighted.

A review of recent energy elasticity estimates begins with Taylor's (1975) [6] survey of electricity demand. Taylor summarizes many previous studies on demand elasticity estimation in the residential, commercial, and industrial industries. The large references offered will aid those interested in literature before the 1970s. Most of Taylor's research recommendations already have been implemented. Modelling peak vs. off-peak household usage, seasonal demand, and a closer look at industrial energy consumption are among the recommendations. Many of these extensions have been made possible by data availability and breakthroughs in econometric methodology. Pindyck (1979) [7] is significant research because of the breadth of its examination. In developed countries, Pindyck uses conflicts, to assess if energy as well as other production inputs is replaceable or complements. A panel data technique is used to analyse data from ten industrialized countries from 1959 to 1973. This study is mainly interested in the projected energy use own-price elasticity, which is determined to be −0.8. This is a metric that examines how industrial energy consumption responds to price fluctuations in energy. Before and after the 1974 and 1979 energy price shocks, Bohi and Zimmerman (1984) [8] performed surveys of energy demand. Their studies are focused on certain industries and fuel kinds. Short-run elasticities are frequently lower in exact value than long-run elasticities equivalents in markets where a consensus emerges. The price elasticity of demand for domestic electricity is −0.2 in the short term and −0.7 in the long run, for example. In the gasoline market, similar outcomes are obtained. A frequent motif in the literature is that elasticities becoming more elastic over time. Bohi and Zimmerman (1984) made a few observations that are pertinent to this study. The authors first explain that assessments of exploring elasticities have remained steady across a variety of energy types, including residential electricity, gasoline, and residential natural gas according to the impacts through time and have not been affected by the energy shocks of the 1970s. Because the price shocks they analysed were recent, the authors emphasize that this is a tentative result. Since this study, the US economy has become more energy efficient. As a result, having an updated test of relative elasticity stability is beneficial. The level of aggregation, according to Bohi and Zimmerman (1984) [8], is a significant influence on projected elasticities. Insights around one level of aggregate must not be taken as gospel at a higher level. Using aggregated energy data as discussed in this chapter should allow you to capture shifting energy usage patterns. Maddala, Trost, Li, and Joutz (1997) [9] investigate the use of electricity and natural gas in several states across the United States. The application of Bayesian approaches in their research has made them famous. Except for Hawaii, all of the states' data are from 1970 to 1990. Due to some difficulties with preliminary estimations, Bayesian approaches are applied. The fact that estimates based on only a single state's data produced theoretically conflicting indications is particularly relevant to this investigation. As a result, the authors employ panel-based techniques. The short-run price elasticity of power demand is −0.158, whereas the long-run price elasticity is −0.263, according to Maddala et al. (1997). According to Bohi and Zimmerman (1984), the long-term forecast is less flexible than the consensus forecast. In the short and long run, income elasticities are 0.39 and 0.89, respectively. Natural gas has a price elasticity of −0.099 in the short run as well as −0.28 in the long run. In the short term, income elasticities are 0.28, whereas they are −0.068 in the long run. To provide robust results, the study incorporates results from a variety of methodologies. Using all of the techniques, it is determined that the long-run income elasticity is negative or zero. The authors propose several explanations for natural gas's diminishing income elasticity over time, including households migrating from natural gas to electricity and natural gas usage declining over time. The conclusions of Bohi and Zimmerman on elasticity stability are contradicted by this recent study. Elasticities will not be as constant as Bohi and Zimmerman assumed over a longer time horizon. Olatubi and Zhang (2003) [10] use an approach and data set that is similar to that used in this chapter. From 1977 to 1999, data were collected for 16 southern states. An aggregate energy consumption variable is also used by Olatubi and Zhang. The long-run estimated price elasticity of demand is −0.32, but the income elasticity is 0.4, implying that rising income has a stronger effect on energy use. The income elasticity of demand, according to Olatubi and Zhang, is decreasing over time. The Engel curve predicts this outcome. Energy will represent a smaller budget percentage as earnings rise, resulting in decreasing income elasticity of demand. The short-run and long-run categorization of elasticities is the product of Olatubi and Zhang's (2003) theoretical model and the aforementioned investigations. Because structural characteristics are calculated in this chapter, both temporal dimensions are not directly comparable. The estimate is negative while statistically significant. For derived coefficients, the author doesn't specify significance or critical values thresholds but does say that those without theoretically consistent signals are statistically irrelevant from zero. For source and sector pairings, price elasticities are bigger than income elasticities in absolute value. The authors wish to know if they should estimate pricing elasticities at the national, regional, or state levels, because the majority of the Energy Information Agency's (EIA) research is done at a regional level. The Census' Western region is divided into two sections by the EIA: Pacific and Mountain. California, Oregon, and Washington make up the Pacific area. Regional disparities are larger than state differences within a region, according to Bernstein and Griffin. For state-specific policy challenges, state-specific estimates would be more useful although there is higher variation inter-regionally than intra-regionally. This is a significant finding since studies can benefit from larger time series. Income elasticity declines rather than price elasticity changes. This appears to have influenced demand behaviour. This knowledge could be valuable in policymaking. It suggests that, over time, greater affluence will result in lesser increases in energy use. A pricing change is expected to have multiple effects at the same time. The absence of cooling degree days (CDD) and heating degree days (HDD) symptoms was unexpected. Households are likely to expand their energy use when the number of CDD or HDD days increases. When CDD interacts with a dummy that simulates air cooling and HDD interacts with a dummy that simulates electric heat, the results are as follows: the computed coefficients have the theoretically expected signatures. Their estimations are made using the generalized method of moments. The study uses a unique data collection from 2004 to 2006, the Consumer Expenditure Survey. Arizona, California, Colorado, Oregon, and Washington, according to them, make up the West area. The survey allows for a more in-depth examination of home behaviour than is possible with more aggregated data. The fact that not all states are represented limits their ability to apply their findings to the Western area. Lee and Lee (2010) [11] use data from 25 OECD countries from 1978 to 2004 to conduct a panel data analysis. Both aggregate energy and electrical usage are analysed by the authors. It is evident that the reader will discover an overview of estimated coefficients from similar investigations.

## 14.3 MATERIALS AND METHODS

Several themes from the literature review affected this research. For starters, panel-based techniques are prevalent in the literature. Time-series techniques have some advantages and limitations that panel data approaches do not. The fundamental advantage of combining numerous cross-sections instead of depending on isolated cross-sections is increased diversity. The expectation that computed values are valid for all cross-sections inside the estimation is a limitation of panel approaches. As previously stated, the variability of state economies shows that this premise may not be correct. Estimates will be made on a state-by-state basis for this reason, as well as to guide state-level decisions. To obtain demand-side estimates, the most prevalent technique in energy elasticity research is to employ reduced-form demand curves. The simultaneous calculation of a supply curve is often impossible due to data availability. Estimates in reduced form may not be correct estimates of elasticities because they do not account for the commodity in determining market prices. Then if the demand curve is calculated, quantity and price variations would be assigned to demand factors, rather than supply changes. It's impossible to attribute only one side of the market experiences price and quantity changes without properly developing a supply and demand curve. When using a simultaneous system to estimate a reduced-form specification, in 2003, Gujarati includes a thorough analysis of the econometric issues that can occur. Explanatory variables and error terms are likely to be connected in a supply and demand framework. A quantity-on-price estimation in simplified form, as Gujarati (2003) [3] points out, contradicts "the assumption of no correlation between the explanatory variable(s) and the disturbance term" (p. 719). Supply and demand mutually influence observed price and quantity; both supply and pricing are determined by endogenous factors. Because the price is an explanatory variable in both supply and demand, an error word will be linked to; if the shock is written as the in-demand curve's error term, it becomes an explanatory variable. From estimates that result from test cases, it is seen that a reduced-form demand-only estimation is considered to be effective. Price will not be associated with the error terms if it is presumed (or determined) to be exogenous. This assumption is used by Houthakker, Verleger, and Sheehan (1974) [12] to estimate demand elasticities at the state and regional levels. It would be more accurate to think of measured variables as exogenous when the price is exogenous, "short-run multipliers," as Gujarati (2003) puts it (p. 738). This study concluded that, contrary to what other studies have discovered, energy market elasticities are also not immediately relevant across various energy types or aggregation levels. Collective energy measures appear to be useful when this is combined with the significance of total economic development and energy consumption. Sub-national estimates will be relevant for a range of policy concerns because there is regional or state-specific research in the literature. Many energy policy decisions are made at the state or local level. The findings of this study will be useful to policymakers that represent states in the Western United States.

## 14.4 THERMOTICAL EXPRESSION

The theoretical supply and demand model will be described in this part. The following are some common supply and demand expressions:

$$Q_t^D = f\left(p_t; X_t\right), \tag{14.1}$$

$$Q_t^S = h\left(p_t; Y_t\right), \tag{14.2}$$

where $t$ is a time index, $p$ is the supply-side influence, and $Y$ is the demand-side influence. Each side of the market represents a different state in this research. Weather, price, income, and type of economy are some of the characteristics that the theory proposes should be included in $X$. Price of inputs, climate, and technology are examples of supply-side variables. Which variables are included in estimations will ultimately be determined by the availability of relevant proxies.

When the energy market is in balance,

$$Q_t^D = Q_t^S = Q_t^*, \tag{14.3}$$

where * denotes a point of balance. The equilibrium market price, $p_t^*$, is the energy price that equalizes the quantity requested and supplied in time $t$. The difficulty in accurately attributing energy market shocks to the correct side of the market is inherent in calculating the supply and demand system. Many studies in the literature use Houthakker et al.'s flow-adjustment model (1974) [12]. To indicate desired demand, equation (3.1.1) of Houthakker et al.'s model is modified:

$$Q_{i,t}^{D*} = f(X), \tag{14.4}$$

where * denotes the desired supply. The term "desired" refers to the fact that energy demand is a stock rather than a flow situation. Consider a family's gasoline consumption. A price rise will result in a decrease in the quantity sought due to the law of demand. The flow process of energy consumption is represented by this change in behaviour. A household's first response to a price shift is to alter its consumption. It may, in the long run, upgrade the stock of energy-consuming goods in response to price fluctuations. A price increase in the future may encourage the purchasing of a more fuel-efficient vehicle. When given sufficient time, the capital stock of the choice may change as a result. When the market is only driven by demand, being examined, the flow-adjustment model is used. As a result, the adjustment process is depicted rather than assuming that price and quantity imply market equilibrium. This model obtains structural parameters by estimating a supply curve rather than short- and long-term estimations.

Lin (2011) assumed linear supply and demand functions in this study [13]:

$$Q_t^D = \alpha_1 * p_t + \alpha_i X_t. \tag{14.5}$$

$$Q_t^S = \beta_1^* p_t + \beta_i Y_t. \tag{14.6}$$

The reduced-form system representations of $p_t$ and $Q_t$ could be solved using the premise of market equilibrium. When the estimator is ordinary least squares (OLS) or two-stage least squares (TSLS), this method is commonly used (2SLS).

## 14.5 CONSIDERATIONS IN ECONOMETRICS

The system is estimated using three-stage least squares (3SLS). As Lin (2011) points out, this method is more consistent and efficient than utilizing OLS or 2SLS to estimate equations one by one. The use of instrumental variables, which are not used in OLS calculations, leads to increased efficiency. Due to the simultaneous nature of the estimation, consistency is improved over OLS and 2SLS. 3SLS estimation incorporates a major component of the seemingly unrelated regression estimation (SURE) technique. It controls the relationship between error terms in differential equations. To account for the link between preset variables and error terms, instrument variables are commonly used. 2SLS uses instrument variables as well, but unlike SURE, it doesn't account for the system's simultaneous nature. Kennedy (2003) goes over the various ways in which 3SLS might be viewed as the simultaneous or comprehensive information version of 2SLS in greater detail.

### 14.5.1 DATA OVERVIEW

Energy consumption (E Con), energy price (E Price), gross state product (GSP), manufacturing employment (Mfg), and weather data were used to estimate the demand curve (Clim). The data ranges from 1970 to 2007. States in the Western region of the Census are considered, with the exclusion of Alaska and Hawaii. These states are not included, since they are not part of the United States. The data are shown as natural logs, but the notation is hidden. The nominal figures (GSP and price) were converted to 2000 USD using the Bureau of Labor Statistics' Consumer Price Index (CPI). GSP is the equivalent of gross domestic product (GDP) at the state level, and it is measured in millions of 2,000 dollars. The Mfg time series are impacted by the transition industry classification system, from the Standard Industrial Classification (SIC) to the North American Industrial Classification System (NAICS). A seamless time series is not produced as a result of the reclassification. From 1997 onwards, NAICS metrics of both GSP and Mfg are used. Although this is a flaw in the data, the series still allows for some flexibility in the capturing of major patterns in manufacturing production. The EIA provided energy-related statistics. Two examples are the use of coal by a utility to generate electricity and the use of natural gas by a family to heat. Because the electricity was generated from a different source, it would not be considered primary use. Electricity usage is deemed secondary and is excluded from the calculation to avoid double counting. Data on sales and distribution in each state are used to generate consumption measurements. To reflect the state's entire usage, net power imports are included in consumption. Renewable energy consumption refers to the sum of electricity produced minus any power utilized in its production. The price of energy is calculated using the average price of total energy (E Price).

The study uses a rate of 2,000 dollars per billion British thermal units (BTU) to convert nominal measures to dollars per billion BTU. When possible, taxes are included in the price. Excise and per-gallon taxes are usually included, while municipal sales taxes are not. In-state variances are not reflected because prices are state-level, but aggregate patterns across time are. CDD and HDD are the two climate proxies used. The National Oceanic and Atmospheric Administration publish data for each state. The daily average temperature is displayed as a departure from 65° on CDD and HDD. When temperatures exceed or fall below the standard of 65°, the variable records the cooling or heating requirements for a facility. The CDD for the day would be 35 if the daily average temperature was 30 degrees. Individual weather station observations are weighted according to the population to calculate state-level measures. The estimation will be done separately for CDD and HDD, as well as summed for a particular year, as a robustness check. Urbanization, according to H Kara, O. A. M. A (2014) [14], is a significant predictor of carbon dioxide emissions. In metropolitan regions and, as a result, states with larger urbanized populations, energy usage per capita is lower. The increase in population in the state will affect energy demand. The amount demanded proxy and the weather variable were shown to be highly linked in early results, resulting in erroneous coefficient values. As a result, it was eliminated from contention.

## 14.5.2 PROBLEMS WITH USING AN AGGREGATE ENERGY MEASURE

Although the use of variations in aggregate consumption is not recorded, substitutions between energy fuels are necessary. A large portion of the energy strategy is concerned with specific fuels. Carbon regulation, for example, takes into account the carbon content of fuels. Market (or fuel)-specific elasticities cannot be predicted using aggregate-based techniques. While fuel-specific elasticities are important for policy, overall energy consumption has a big impact on the expansion of the economy. In certain instances, energy cannot be replaced; for example, I need fuel to drive my car. Energy may be a compliment (or substitute) for other components of production in the manufacturing process. Energy, according to Pindyck (1979) [7], is a supplement to labour as well as a replacement for money. The literature on energy and economic growth is summarized by Stern (2003) [15], "it appears that capital and energy work more as substitutes in the long run and more as complements in the short run, and that they may be gross substitutes but net complements." Stern (1997) [16] points out that the delicate nature of conclusions is dependent on the functional form used in the analysis because of the nature of the capital-energy link. Greater manufacturing employment causes more energy consumption if the manufacturing elasticity is positive. Unfortunately, the energy capital link cannot be inferred, since capital is not accounted for in the calculations. For states, there are no readily available capital statistics. The manufacturing elasticity of energy consumption should be regarded with caution because capital is not constant.

### 14.5.3 ESTIMATION AND OUTCOMES CALCULATION

The estimations are based on equations (14.5) and (14.6). Income, output composition, and climate are all represented by proxies in the $X$ vector. GSP, Mfg, CDD, and HDD are the variables that function as these proxies, respectively. The supply vector, $Y$, is made up of input cost and climatic proxies. In estimates, Prim, CDD, and HDD compromise the $Y$ vector. $E$ Price and $E$ Con reflect the equilibrium price and quantity. The system to be estimated is as follows:

$$
E\_Con_t^{D} = \alpha_0 + \alpha_1^{*}E\_Price_t + \alpha_2^{*}GSP_t + \alpha_3^{*}Mfg_t + \alpha_4 * E\_Con_{t-1}
$$
$$
+ \alpha_5 * CDD_t + \alpha_6^{*}HDD + u_{1t}. \tag{14.7}
$$

$$
E\_Con_t^{S} = \beta_0 + \beta_1^{*}E\_Price_t + \beta_2^{*}IC_t + \beta_3^{*}E\_Con_{t-1} + \beta_4^{*}CDD_t + \beta_5^{*}HDD_t + u_{2t}. \tag{14.8}
$$

The random error terms are denoted by $u_{it}$. These are calculated at the same time for each state in the Western United States. To account for potential autocorrelation, a lag in energy use is introduced as an explanatory variable. The instruments for the 3SLS estimations are predetermined variables. The simultaneous nature of the system will be taken into consideration by 3SLS estimation. Tables 14.1 and 14.2 provide the most important findings. The calculated demand elasticities are shown in Table 14.1. Price is behaving as expected (when significant). Demand elasticity is continually inelastic when it comes to price. When significant, it ranges from −0.23 for Idaho to −0.06 for California. Their estimations are −0.98 in absolute value, which is lower than Fell, Li, and Paul's (2010) findings [17]. In their panel research of the Southern United States, Olatubi and Zhang (2003) found a similar range of state-specific elasticities. In the medium term, they forecast short-term price elasticity of demand of −0.08 to −0.11 and the long-term price elasticity of demand of −0.21 to −0.32.

As previously stated, Olatubi and Zhang's (2003) short-run long-run categorization of elasticities is the consequence of their theoretical model. In the short run, except for Idaho and Utah, the price elasticity of demand in most states falls within Olatubi and Zhang's range. Utah's forecast is the most conservative and the only one that falls within its long-term forecast range. The reason why the structural parameters found in this chapter are closer to Olatubi and Zhang's short-run estimates could be due to regional variations, a more recent data set, or structural estimation. The computed coefficients of Olatubi and Zhang (2003) derived reduced-form models based solely on demand short-run multipliers are the most appropriate interpretation of these figures. If the variations in price elasticity estimations revealed in this study vs. Olatubi and Zhang's aren't tied to the data collection, different timeframes, or geographical variances, price elasticity, in the long run, may or may not be a factor acceptable depiction of state energy market behavioural traits. Structural coefficients, on the other hand, appear to be more short-run in character. If this is the case, the tendency towards equilibrium in energy markets should be thought of as a result of short-term shocks rather than a long-term process. The methods presented in this chapter could be used to establish which of the three hypotheses is driving the various results in the study area (Southern United States).

In nine of the 11 states, the income elasticity of demand has the predicted sign. New Mexico and Wyoming are the two states with the lowest per capita income. The short-run range of Olatubi and Zhang (2003) [10] is 0.08–0.1, whereas the long-run range is 0.29–0.44. State-by-state assessments are often higher than Olatubi and Zhang's short-run forecast, but lower than the most conservative long-run forecast. When significant, manufacturing elasticity (output composition) shows mixed results. It is negative in Arizona and Colorado, but positive in Wyoming, Oregon, Washington, and Montana. If state-level capital is taken into account, a positive sign indicates that labour and energy

**TABLE 14.1**
**Demand Elasticities**

| State | Price | Inc | Mfg | CDD | HDD |
|---|---|---|---|---|---|
| AZ | – | 1.3442 | –1.086 | 0.012 | – |
|  | – | –1.0483 | –1.023 | –1.0695 |  |
| CA | –1.0594 | 1.1743 | – | 1.0532 | – |
|  | –1.0303 | –1.038 |  | –1.0305 | – |
| CO | –1.0952 | 1.228 | –1.1344 | – | – |
|  | –1.0448 | –1.0659 | –1.417 | – |  |
| ID | –1.2337 | 1.2743 | – | – | 1.1376 |
|  | –1.576 | –1.647 | – |  | –0.85 |
| MT | – | 1.3494 | 1.3115 | – | 1.2338 |
|  | – | – | – | – |  |
| NM | – | – | – | – | – |
| – | –0.0674 | – | – | – | – |
| – | –0.1046 | – | – | – | – |
| – | – | – | – | – | – |
| – | –0.1128 | – | – | – | – |
| – | – | – | – | – | – |
| NV | –1.0888 | 1.3177 |  | – | – |
|  | 1.0379 | 1.0635 |  | – | – |
| OR | – | 1.2003 | 1.3117 | – | – |
|  | | 1.0308 | 1.642 | – | – |
| UT | –1.1433 | 1.1735 |  | – | – |
|  | 1.0649 | 1.0755 |  | – | – |
| WA | –1.1118 | 1.1713 | 1.3983 | – | – |
|  | 1.0503 | 1.0416 | 1.0804 | – | – |
| WY | – | – | 1.215 | – | – |
|  | – | – | 1.1034 | – | – |

*Note:* The statistically significant coefficients are listed along with their standard errors. Except for ID's HDD, every reported data was accurate.

are complementary; a negative sign implies that they will be substitutes. The author anticipated a complementing partnership from the start. The qualitative results' contradictions make it impossible to come up with a uniform answer. The most straightforward answer is that manufacturing and energy do not have a universal relationship at the state level. The sort of manufacturing that takes place within the state could determine the aggregate link. In California, Arizona, Montana, and Idaho (at the 11% level), climate proxies are important. Climate's importance is varied in other state-level research. The treatment of climate in the literature is not uniform. The findings of estimations using CDD and HDD combined will be briefly reviewed at the end of this section. Climate proxies are negative demand shifters, according to Fell, Li, and Paul (2011) [17], a finding that the authors did not predict. Climate change has regularly been demonstrated to be a positive demand changer in previous studies. Olatubi and Zhang (2003) [10] believe that the climate has no bearing on their findings. Increases in CDD increase energy demand in Arizona and California. This outcome is unsurprising, given the enormous populations and mild temperatures of Los Angeles and Phoenix. HDD is considerable in Montana and Idaho (at the 11% level). Looking at the geography of these states, it's only natural that heating needs would drive up demand. The lack of relevance of climate variables in the majority of states is striking. In eight of the eleven states, the supply price

**TABLE 14.2**
**Supply Elasticities**

| State | Price | IC | CDD | HDD |
|-------|-------|------|------|------|
| AZ | 6.367 | −3.9627 | – | – |
| | 2.3679 | 1.5489 | – | – |
| CA | 3.132 | −2.6838 | 1.0527 | 1.0812 |
| | 1.9406 | 1.388 | 1.308 | 1.0443 |
| CO | | | | |
| ID | 5.9927 | −3.9088 | – | 1.1355 |
| | 2.4366 | 1.5863 | – | 1.0855 |
| MT | 5.3801 | −3.5978 | – | 1.2808 |
| | 2.6567 | 1.6685 | – | 1.1242 |
| NM | 3.6569 | −2.1211 | – | – |
| | 2.3807 | 1.577 | – | – |
| NV | 5.6628 | −3.7155 | – | – |
| | 2.4095 | 1.5638 | – | – |
| OR | 4.9373 | 2.6344 | – | – |
| | 1.9318 | 1.3794 | – | – |
| UT | 4.0886 | 2.2929 | – | – |
| | 2.4234 | 1.5686 | – | – |
| WA | – | – | – | −1.1907 |
| | – | – | – | 1.1196 |
| WY | | | | |

*Note:* The statistically significant coefficients are presented in parenthesis with their standard errors. Except for HDD in ID and WA, everything was reported.

elasticity is in the expected direction. In Colorado, Washington, and Wyoming, it is minor. Supply's estimated price elasticities are substantially higher than demands. The elasticities are calculated to be between 2.43 and 7.67. There are no directly comparable supply-side elasticities because there are no studies of supply and demand at the state or regional level. The price elasticity of supply is inelastic in both Krichene's [18] and Lin's (2011) [13] studies of the global oil market.

This discovery sheds light on the tax consequences of energy-related initiatives. The surplus loss caused by an energy tax will almost probably fall disproportionately on consumers (or policy). Any externalities related to energy consumption or production are omitted in the study's approach. States that have high supply price elasticity also have high input price elasticity. The theoretically predicted negative sign is present in all significant estimations. Although input price elasticity is elastic, its absolute value is always lower than demand price elasticity. For a certain relative price shift in basic energy, the impact on supply will be bigger than the shock (coal, natural gas, and so on). In Arizona, everything is how it should be. Increases in supply act as a brake on regional price declines. One way for increases to occur is through technological advancements in extraction. A second strategy to boost energy supply is to continue to promote renewable resources. The data captures the production and consumption of non-renewable energies, except for energy from hydroelectric dams. Renewable energy sources, such as solar and wind, have zero input costs. A large rise in the percentage of renewable in overall production is required by many RPS in the region. Colorado, for example, has set a target of 30% by 2020. The input cost supply elasticity will most likely be if these percentages are met inaccurate, since it will not account for the significantly different composition of the energy supply. In California, CDD is present, HDD can be found in Montana and Idaho (12% significance level). HDD is negative in Washington at the 1% significance level. These findings may

**TABLE 14.3**

**Increased Energy Demand as a Percentage of Increased Incomes**

| | |
|---|---|
| AZ | 2.5749 |
| CA | 1.5183 |
| CO | 1.8868 |
| ID | 1.8673 |
| MT | 1.7465 |
| NM | |
| NV | 2.715 |
| OR | 1.5992 |
| UT | 1.6987 |
| WA | 1.5593 |
| WY | |

be due to a lack of state-level heterogeneity, as in the case of demand. Although the negative result from Washington was unexpected, negative indicators have been observed in other research (Fell et al., 2011) [17]. This could be due to the state's heavy reliance on hydroelectric power. In 2011, hydroelectric power accounted for 75% of the state's total electrical generation (EIA, 2011). When the snow melts and river flows increase in the spring, the supply of hydroelectric energy increases. HDD is a proxy for heating requirements, which are most often linked to the winter months. The state's energy supply reduces as snow falls, but increases after the snow melts. Elasticities can be used to forecast how energy markets will react to a variety of shocks. Income shocks cause the highest absolute changes in energy use for a given percentage shift in demand. Table 14.3 illustrates the anticipated rise in energy consumption over the data set if state income grows at the same rate as the national average. The biggest increases in energy demand are projected in Arizona and Nevada, while the lowest is expected in California and Oregon.

The reference scenario in the Annual Energy Outlook 2011 forecasts price fluctuations in the Western United States. The forecasts used are the predicted price increases for individual fuels from 2009 to 2035. Estimates of energy prices are not made public on the whole. Motor gasoline is predicted to rise by 3.7%, natural gas by 3.1%, and electricity by 1.8% in the Western area. Table 14.4 depicts the relative changes in quantity demanded and supplied, using the high and low estimations to provide a range of likely price increases. Price fluctuations produce huge difference between the amount of energy given and the amount of energy demanded by a given percentage change.

When you consider the ramifications of this study's findings, you can see how complicated the utilization of energy in the Western United States has a bright future. Increased projected income leads to increased energy use, whereas price increases curb demand. Mostly, on the supply side, the equilibrium quantity has a positive association with market pricing but has a negative relationship with primary energy prices. The total impact of these conflicting pressures will be determined by the number of fundamental changes in energy supply and demand within states. Equations (14.1) and (14.2) were calculated using a single variable that combined CDD and HDD. Throughout the year, both cooling and heating requirements would be captured by this variable. Bernstein and Griffin [19] took a similar strategy (2006). The important findings are presented in Tables 14.5 and 14.6. Because the results are comparable to those of the disaggregated variable, output has been omitted.

The entire output from state estimations with CDD and HDD are separated. Because the results of their combined series are not significantly different from the estimates made with CDD and HDD separately, they are not reported. The estimations perform well when the F-statistics are high. Durbin–Watson statistics are usually always close to 2, indicating that autocorrelation does not affect the outcomes.

**TABLE 14.4**

**Quantity Demanded and Supplied Changes as a Function of Price Changes**

| | Demand | | Supply | |
| --- | --- | --- | --- | --- |
| | Percentage Increase | | Percentage Increase | |
| | 3.70% | 1.80% | 3.70% | 1.80% |
| AZ | – | – | 27.2616 | 13.2625 |
| CA | –1.2194 | –1.1068 | 16.2849 | 8.4359 |
| CO | –1.353 | –1.1713 | – | – |
| ID | –1.8643 | –1.4205 | 26.8726 | 13.5867 |
| MT | – | – | 24.6064 | 12.4843 |
| NM | – | – | 8.8306 | 3.7825 |
| NV | –1.3299 | –1.1598 | 23.6522 | 12.9928 |
| OR | – | – | 13.5676 | 8.0868 |
| UT | –1.5298 | –1.2588 | 12.4275 | 6.5594 |
| WA | –1.4143 | –1.2016 | – | – |
| WY | – | – | – | – |

**TABLE 14.5**

**Significant Elasticities**

| | | Demand | | |
| --- | --- | --- | --- | --- |
| State | Price | Inc | Mfg | Clim |
| AZ | – | 1.3599 | –1.0719 | – |
| CA | –1.066 | 1.1816 | – | – |
| CO | –1.1006 | 1.2255 | –1.1337 | – |
| ID | –1.2319 | 1.2727 | – | – |
| MT | – | 1.3453 | 1.2696 | 1.2279 |
| NM | – | – | – | – |
| NV | –1.0875 | 1.3155 | – | – |
| OR | – | 1.1833 | 1.3157 | – |
| UT | –1.1404 | 1.166 | – | – |
| WA | –1.1117 | 1.1708 | 1.3988 | – |
| WY | – | – | 1.2115 | – |

*Note:* At the 10% level, all reported coefficients are significant.

## 14.6 CONCLUSION

Multiple energy elasticity of demand for Western United States was computed in this study. A further contribution to the literature is more than just a demand curve, the estimation of a supply and demand system. The findings show that short-run shocks in energy markets have a greater impact on demand than long-run shocks. Supply-side estimates show how the quantity delivered a response to different shocks. In most elasticity studies, the supply side of energy markets is ignored. The use of aggregate energy consumption metrics is a second expansion of the study. Renewable energy sources will certainly become increasingly dominant in the future, as will changes in the relative

**TABLE 14.6**
**Significant Elasticities**

| | Supply Elasticities | | |
|---|---|---|---|
| State | Price | IC | Clim |
| AZ | 6.768 | −4.0862 | − |
| CA | 3.3187 | −2.8585 | 1.1127 |
| CO | − | − | − |
| ID | 7.9038 | −3.8722 | − |
| MT | 5.4088 | −3.6085 | 1.2818 |
| NM | 3.4343 | −2.0358 | − |
| NV | 5.6343 | −2.7028 | − |
| OR | 4.4658 | −2.438 | − |
| UT | 3.9607 | −2.2407 | − |
| WA | − | − | −1.2146 |
| WY | − | − | |

*Note:* Coefficients are statistically significant. At the 10% level, all reported coefficients are significant.

pricing of non-renewable. Variables on the demand side have an impact on total energy consumption, and aggregate measures make this easier to see. Estimates of changes in overall consumption are excellent projections of changes in individual energy markets as a supplement because energy is a derived demand and is so crucial in the economic growth process. The study does have certain drawbacks. On both sides of the market, only four of the eleven states investigated had significant and theoretically consistent pricing elasticities. The states in dispute are California, Idaho, Nevada, and Utah. The geographical reach of state-level energy markets may be insufficient. CDD and HDD have a surprising effect on supply and demand.

## REFERENCES

[1] Metcalf, G. E. An Empirical Analysis of Energy Intensity and Its Determinants at the State Level. Risks. *Energy J.* **2008**, *29* (3), 1–26.

[2] Keppler, J. H.; Bourbonnais, R.; Girod, J. The Econometrics of Energy Systems. *Econom. Energy Syst.* **2006**, 1–266. https://doi.org/10.1057/9780230626317.

[3] Brassard, G.; Méthot, A. A. Can Quantum-Mechanical Description of Physical Reality Be Considered Correct? *Found. Phys.* **2010**, *40* (4), 463–468. https://doi.org/10.1007/s10701-010-9411-9.

[4] Filippini, M. Short- and Long-Run Time-of-Use Price Elasticities in Swiss Residential Electricity Demand. *Energy Policy* **2011**, *39* (10), 5811–5817. https://doi.org/10.1016/j.enpol.2011.06.002.

[5] He, Y.; Wang, S.; Lai, K. K. Global Economic Activity and Crude Oil Prices: A Cointegration Analysis. *Energy Econ.* **2010**, *32* (4), 868–876. https://doi.org/10.1016/j.eneco.2009.12.005.

[6] Taylor, L. The Demand for Electricity. *Bell J. Econ.* **1975**, *6* (1), 74–110.

[7] Pindyck, R. S. Interfuel Substitution and the Industrial Demand for Energy: An International Comparison. *Rev. Econ. Stat.* **1977**, *61* (2), 169–179.

[8] Bohi, D. R.; Zimmerman, M. B. Update on Econometric Studies of Energy Demand Behavior. *Annu. Rev. Energy* **1984**, *9*, 105–154. https://doi.org/10.1146/annurev.eg.09.110184.000541.

[9] https://efron.ckirby.su.domains//other/CASI_Chap7_Nov2014.pdf

[10] Olatubi, W. O.; Zhang, Y. A Dynamic Estimation of Total Energy Demand for the Southern States. *Rev. Reg. Stud.* **2003**, *33* (2), 206–228

[11] Lee, C. C.; Lee, J. D. A Panel Data Analysis of the Demand for Total Energy and Electricity in OECD Countries. *Energy J.* **2010**, *31* (1), 1–24. https://doi.org/10.5547/ISSN0195-6574-EJ-Vol31-No1-1.

[12] Houthakker, H. S.; Verleger, P. K.; Sheehan, D. P. Dynamic Demand Analyses for Gasoline and Residential Electricity. *Am. J. Agric. Econ.* **1974**, *56* (2), 412–418. https://doi.org/10.2307/1238776.

[13] Lin, C.-Y. C. Estimating Supply and Demand in the World Oil Market. *J. Energy Dev.* **2008**, *34* (1/2), 1–32.

[14] H Kara, O. A. M. A. No Title No Title. *Pap. Knowl.. Towar. a Media Hist. Doc.* **2014**, *7* (2), 107–115.

[15] Stern, D. I. A Multivariate Cointegration Analysis of the Role of Energy in the US Macroeconomy. *Energy Econ.* **2000**, *22* (2), 267–283. https://doi.org/10.1016/S0140–9883(99)00028–6.

[16] Stern, D. I. *Energy and Economic Growth*; 2019. https://doi.org/10.4324/9781315459653-3.

[17] Van Wilgen, B. W. A New Look at Cheetahs. *S. Afr. J. Sci.* **2017**, *113* (11/12). https://doi.org/10.17159/sajs.2017/a0236.

[18] Krichene, N. World Crude Oil and Natural Gas: A Demand and Supply Model. *Energy Econ.* **2002**, *24* (6), 557–576. https://doi.org/10.1016/S0140-9883(02)00061-0.

[19] Ros, B. A. J.; Romero-Jordán, D.; Pablo del Río, C. P.; Rizal, Y.; Ong, S.; Mckeel, R.; Labeaga, J. M.; López-Otero, X.; Labandeira, X.; Komives, K.; Halpern, J.; Foster, V.; Wodon, Q.; Halpern, J.; Wodon, Q.; Kleven, H. J.; Jooste, M.; Palmer, I.; Guo, J.; Khanna, N. Z.; Zheng, X.; Godoy-Shimizu, D.; Palmer, J.; Terry, N.; General, D.; Internal, F. O. R.; Policy, S.; Galea, J.; Fan, S.; Hyndman, R. J.; EIA; Desroches, G. V.; Benchenna, P.; Zambelli, L.; Dallemagne, A.; David Reeve, S. B.; Braithwait, S.; Hansen, D.; Associates, C.; Consulting, E.; Borenstein, S.; Bernstein, M. A; Griffin, J. ACIL Aleen Consulting. *ACEEE Summer Study Energy Effic. Build.* **2015**, *4* (October), 1–37.

# 15 Internet of Things-Based Toddler Security Monitoring and Management System

*Karmel Arockiasamy, G. Kanimozhi, and Dheep Singh*
Vellore Institute of Technology

## CONTENTS

15.1 Introduction ........................................................................................................... 171
15.2 Literature Survey .................................................................................................. 172
15.3 Proposed System .................................................................................................. 173
    15.3.1 Face Detection Module (FDM) ............................................................. 173
    15.3.2 Face Detection Module: Implementation ............................................. 174
    15.3.3 Face Recognition Module (FRM) ......................................................... 175
    15.3.4 FRM Implementation ............................................................................. 176
    15.3.5 Door Detection Module (DDM) ............................................................ 176
    15.3.6 Mobile Application ................................................................................ 177
15.4 Conclusion ........................................................................................................... 179
References .................................................................................................................... 179

### ABSTRACT

Internet of Things (IoT) alludes to the utilization of resourcefully coherent systems and devices to influence data collected through embedded actuators and sensors in automaton as well as supplementary tangible things. In this present digital era, societal problems like waste management, car accidents, agricultural issues like soil quality, water distribution, child security, etc. are still an open challenge. This chapter mainly concentrates on the child's security. The child security system includes three processes, namely face capture, face recognition and door detection. In the face capturing process, the face of the unknown person entering into the user's house is captured and stored in the cloud server. Utilizing the mobile application, the user can keep track of the persons entering his/her house. The face recognition process recognizes the face of a child, and whenever the child tries to exit from the house, intimation is sent to the parent's mobile. In the door detection module, once a person opens the locked door, a notification is sent on the user's mobile. In this way, he/she becomes acquainted with, if someone opens the door in the absence of the user. An IoT-based prototype is developed and tested to accomplish the child's security at home.

## 15.1 INTRODUCTION

In the kinship, academy and association, offshoot ought to be completely secured so they can endure, develop, peruse and create their entire potential. Many children are not completely secured, and plenty of them need to be secured from viciousness, mistreatment, disrespect, misconduct and rejection along with segregation frequently. Similar infringements are ceiling their chances of abiding, developing, accomplishing and pursuing their dreams. Any child is vulnerable to infringement in a variety of settings, including home [1]. Defending and advancing the child's wellbeing, training, improvement and government assistance can provide care, heading, direction and control in a way

DOI: 10.1201/9781003374121-15

suitable to the child's age and comprehension. The parental responsibility is defined as per the law as a responsibility to decide all parts of childhood and to provide a home, either straightforwardly or by implication, and maintain relations or ordinary contact if not living with the kid, act as the kid's lawful delegate; safeguard and manage any property [2].

Nowadays, the prevailing desire for all parents is by virtue of what to deal with their scion, while the parents are in a far-off spot. Numerous beneficial things are available in the world to all the individuals to lead a quiet life and make an amazing most with their beautiful family. However, just a portion of the individuals makes a mind-blowing most by going through a ton measure of cash for sparing their kin or relatives from issues. The world needs to see extraordinary violations occurring the world over. Numerous crimes are threatened to human life as it were. Although those individuals aren't ready to accept others, they are acceptable in character [3,4]. As a result of every one of these sorts of situation, guardians can't release their kids out without their knowledge. Continuously, it isn't conceivable that parents stay along with their kids all time in a day. Thus, parents need to utilize gadgets from current innovations to follow the kid area or spot and screen the environment where they are currently present [5]. A group of ten international institutions led by the World Health Organization (WHO) produced and endorsed INSPIRE: Seven strategies for stopping aggression of children [6], which is a proof-based specialized package.

## 15.2  LITERATURE SURVEY

For event authority and parents, the security of a kid at a huge unrestricted event is a major worry. In Ref. [7], the authors addressed the importance of child's security in a huge crowd. Also, they proposed an architecture model of the Internet of Things (IoT)-enabled smart child safety tracking digital system. The geographical location of the child in the crowd can be easily tracked using the architecture model that combines the Cloud, Mobile and Global Positioning System (GPS) technology. An IoT paradigm [8] has been proposed with dissimilar localization techniques such as Radio Frequency Identification (RFID) and GPS to provide security for the children on their daily routine from school to home and vice versa. In the real world, the kid's security is a tremendous query in everybody's psyche. Parents consistently expect their kids live in a made sure about the spot where they can invest their energy and psyche with no issue. Normally 50% of them are confronting endless issues.

A child environment monitoring system using IoT is designed in Ref. [9] where sensors are used to remotely monitor whether any unknown person is moving close to the child. The authors also suggested an alcohol sensor, a smoke gas sensor and a sensor to read the blood pressure, to monitor whether the child is in any anomalous condition. These measured input data can be used to make appropriate decisions by the parents to save the children. In Ref. [10], the authors provide a two-level security system to provide unapproved access into the industrial arena that eradicates security threats like spoofing, repudiation, etc., using fingerprint access and a biometric system. Also, an alternate solution is proposed through Virtual Private Network (VPN) and IoT if the authenticated user is not present in the industrial arena.

In Ref. [11], the authors proposed a face recognition system using Local Binary Pattern (LBP) feature extraction. Using wavelet transform, the input face image is divided into several sub-bands of frequencies where each sub-band is decomposed into non-overlapping sub-regions. A single feature histogram is created by concatenating all Local Binary Pattern Histograms (LBPHs) for effective face representation. The similarities of various feature histograms are measured using Euclidean distance and face recognition is done using the nearest-neighbor classifier. A child tracking system using Android mobile applications has been proposed in Ref. [12]. This tracking system helps the parents to track the child's position by getting the exact location of the child.

In Ref. [13], the authors developed an ad hoc network-based children tracking system. Initiation of cluster formation is done when the participating terminals start communicating between themselves using Bluetooth. Since the implementation of this system is high, each child's palm is read using the palm reader before and after the child takes the school bus. The corresponding time gets updated in the cloud server. The school incharge can locate information about the bus and the child. But still, a challenge faced is the accuracy to read the exact values of a child's palm using a palm reader.

An auto door unlock system is an automated technique proposed in Ref. [14] for home automation. The door can be opened using Wi-Fi-enabled mobile phone with a camera that takes the snapshot of the person at the door. The users at home after verifying the person's face at door can proceed further with the door unlock option. This system is proposed specially to help disabled people, to access the gate beyond anyone's help and to lead a normal life. In Ref. [15], the authors used the concept of a smart wearable device for kids. This user-friendly device is a mobile platform-independent and does not depend on the latest smartphones. These devices reduce the burden of parents in tracking the child location. This device also has Save Our Ship (SOS) light and a buzzer which when activated by the parents creates an alarm. The authors aimed at providing the child's security to the parents.

## 15.3  PROPOSED SYSTEM

The proposed security system helps the user to secure home as well as children. The security system is integrated using cameras, IoT devices and some sensors. It enables the user to remotely monitor the persons entering or exiting the home. It also helps to prevent user's kids from going out of the house in their absence. Besides, it also sends intimation to the user when someone opens the door in his/her absence at home. The security system includes their processes, namely face detection, face recognition and door detection. Haar cascade machine learning algorithm is used for detecting the face of the person. Face recognition is done by using LBPH and Eigenfaces algorithms. These algorithms help the camera to recognize the person's face. Door detection is done by using NodeMCU and a magnetic door sensor.

### 15.3.1  Face Detection Module (FDM)

The FDM shown in Figure 15.1 is designed to detect the face of a person using Haar cascade algorithm. Haar cascade includes all the facial features that help to detect the face by converting the data or image into grayscale. Once converted, it identifies the Haar features.

At a given position in a detection window, a Haar-like feature analyzes nearby rectangular parts, sums the pixel intensities in all sectors and computes the variance between these sums. This distinction is then utilized to categorize image subsections. The detection of human faces is an example of this. The skin around the eyes is usually darker than the skin on the cheeks.

A collection of two neighbor rectangular parts above the eye and cheek regions is an example of a Haar-like feature for face detection [15]. Haar cascade converts the data or image into grayscale to detect the face. When the image is converted into grayscale, it identifies the Haar features. The cascade classifier [16] is composed of several phases, each of which displays a group of debilitated learners. The system can recognize the objects in question by sliding a window over the image. All phase of the classifier assigns a positive or negative value to the precise region characterized by the actual location of the window. The presence and absence of an object in the image is indicated by positive and negative values, respectively.

**Harr cascade features:** Facial recognition is everywhere, from the surveillance camera on the front porch to the sensor on the phone. Haar classifiers were used in the first real-time face detector. A Haar cascade classifier is Machine Learning (ML) object recognition software that recognizes objects in images and videos. It is an Object Detection Algorithm [15] that recognizes faces in images and real-time video. The algorithm is given numerous positive images with faces and negative images with no faces to train on.

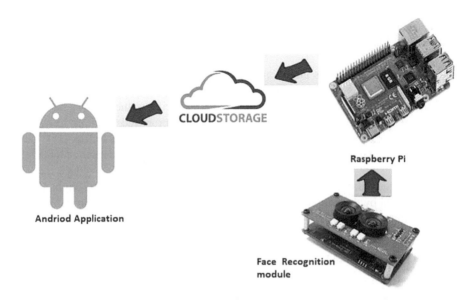

**FIGURE 15.1** Face detection module.

The method can be broken down into four phases, as portrayed in Figure 15.2: (a) Haar Features Calculation, (b) Integral Images creation, (c) Training using Adaboost and (d) Cascading Classifiers implementation. This approach, like other machine learning models, necessitates numerous positive photos of faces and negative photos of non-faces for training the classifier.

The classifier decides whether to mark an item as found (positive) or to move on to the next zone based on this prediction (negative). Stages are set up to reject negative samples as quickly as possible because the majority of the windows contain nothing of interest. Since identifying an item as a non-object would severely impede the object detection system, a low false negative rate is crucial. Haar cascades are one of several object detection methods now in use. To reduce the false negative rate, hyper parameters are set appropriately with Haar cascade during training the model.

### 15.3.2 Face Detection Module: Implementation

The Haar cascade available in the OpenCV library is implemented in Raspberry Pi. This Haar cascade file contains all the relevant features to detect the face. The face image of a person as presented in Figure 15.3 is captured, and it is uploaded on the server using Python's FTP library. To detect the face through a camera, OpenCV library is utilized. In the OpenCV library, cascades are the .xml files that help to detect the face. The face detection requires only a cascade file to detect the face, and the Haar feature algorithm identifies the values of black and white pixels. The white pixels are eliminated from black pixel. When the value is closer to 1, the more likely it finds the Haar feature.

The image is stored on the server using 000webhost with the help of library files. The File Transfer Protocol Library (FTPLIB) in Python helps to connect with the server and uploads the image spontaneously. Duplicate loading of images is not supported on the server FTPLIB. The image captured by the camera is stirred in the local ID. The FTPLIB uses the same ID for uploading to the server. A mobile application is also created by using Android studio. Android studio consists of a web view that helps to convert the website into a mobile app, and with the support of web view, the application is created, assisting the user to see all the people visiting his/her home.

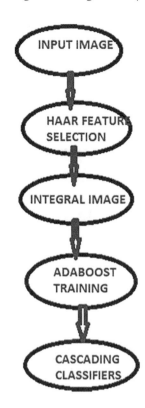

**FIGURE 15.2** Flowchart of the algorithm.

**FIGURE 15.3** Face detection.

### 15.3.3 FACE RECOGNITION MODULE (FRM)

In FRM as shown in Figure 15.4, the captured child's face is recognized using the LBPH face recognizer algorithm, and a SMS is sent to the parent's mobile using way2sms Application Peripheral Interface (API), whether the child is trying to go outside the house. If a parent or guardian is occupied with work at home, this device assists by sending an immediate message to the user's mobile application, instructing them to take the appropriate action if their child leaves the house. The LBPH algorithm divides the window into small windows that have 16×16 dimensions. For each pixel in a small window, the center pixel is compared with the neighbor's pixel. If the numerical quantity of the center pixel is greater than the neighbor's value, mark 0, otherwise, mark 1. It

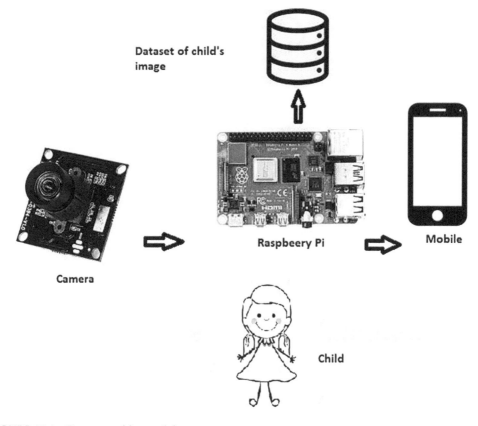

**FIGURE 15.4**   Face recognition module.

creates an eight-digit binary number that is converted to decimal. The decimal number helps to compute the histogram. This histogram combines the histograms of all the small windows into a 256-dimensional feature vector.

### 15.3.4   FRM Implementation

The image dataset is created in the .xml format. The dataset behaves like a two-dimensional array and allots an identity for a person. Whenever a person comes in front of the camera, the Python code detects his/her face and looks in the database for the recognition. The face is recognized as displayed in Figure 15.5. In case, if the camera detects the child's face continuously, the system sends message continuously to the parent's mobile. To avoid this problem, a flag is assigned and set to zero. So, whenever the face is detected and recognized, the flag is set to stops the SMS service.

### 15.3.5   Door Detection Module (DDM)

Figure 15.6 represents the DDM. The purpose of this module is that if unknown person comes to the user's home and opens the door, a notification is sent to the user. In this module, NodeMCU and magnetic door switches are used. NodeMCU is an IoT device that helps to get the input from the magnetic switch. Whenever someone opens the door, the magnetic switch sends its status to the NodeMCU and a notification is displayed on the Blynk app. It comprises firmware that operates on ESP8266 Wi-Fi System on Chip (SoC) and on the ESP-12 device.

**FIGURE 15.5**   Face recognized by camera.

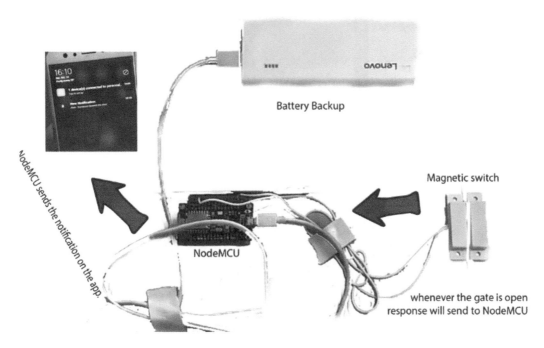

**FIGURE 15.6**   Door detection module.

The magnetic switch is connected to the NodeMCU. The wires from magnetic switch are connected to the D1 and GND (Ground) pins of NodeMCU. When the door is opened, the magnetic switch from the pin D1 sends an input to the NodeMCU. The input must be a zero or one. Arduino IDE microcontroller is used to process the DDM by receiving the signal from magnetic switch and send the notification to the mobile app. The system is receiving the input from the magnetic switch.

### 15.3.6   MOBILE APPLICATION

Mobile application to send a notification to the user's mobile is created as shown in Figure 15.7. Blynk app is used to connect the IoT devices with mobile. A new project is created in the Blynk app and an authentication key is provided for the user. This authentication key needs to be setup in

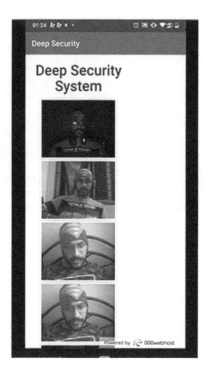

**FIGURE 15.7**    Mobile application.

**FIGURE 15.8**    Notification on Blynk app.

the C code. If unknown person opens the door without the user's knowledge, the user gets notified immediately as portrayed in the Figure 15.8 on the Blynk app.

## 15.4 CONCLUSION

Security is an imperative part of any industry. This work most particularly focuses on criminal identification. The algorithm carried out in this study is a linear binary pattern algorithm. The presented system will be implemented using OpenCV and Raspberry pi. There may be fluctuations in the solution on report of the time interval, camera resolving power and lightning. Advanced processors may be used to reduce the processing time by supplementing more number of recognition servers to reduce the processing time for collection of images.

## REFERENCES

1. Home. https://www.unicef.org/
2. http://iscp.gg/article/118098/Parental-Responsibility/
3. H. L. Burdette; R. C. Whitaker. "Neighborhood Playgrounds, Fast Food Restaurants, and Crime: Relationships to Overweight in Low-Income Preschool Children" *Preventive Medicine* 38.1(2004): 57–63.
4. C. Alison; A. Timperio; D. Crawford. "Playing It Safe: The influence of Neighbourhood Safety on Children's Physical Activity—A Review" *Health & Place* 14.2(2008): 217–227.
5. A., Steve; Y. Rajala. "System and Method for Tracking, Monitoring, Collecting, Reporting and Communicating with the Movement of Individuals" U.S. Patent No. 8,831,627. 9 Sep. 2014.
6. https://www.who.int/news-room/fact-sheets/detail/violence-against-children
7. M. Madhuri; A. Q. Gill; H. U. Khan, "IoT-Enabled Smart Child Safety Digital System Architecture", 2020 IEEE 14th International Conference on Semantic Computing (ICSC), 12 March 2020.
8. L. D'Errico; F. Franchi; F. Graziosi; C. Rinaldi; F. Tarquini, "Design and Implementation of a Children Safety System Based on IoT Technologies," 2017 2nd International Multidisciplinary Conference on Computer and Energy Science (SpliTech), 31 August 2017.
9. R. Kamalraj; M. Sakthivel, "A Hybrid Model on Child Security and Activities Monitoring System Using IoT," 2018 International Conference on Inventive Research in Computing Applications (ICIRCA), 11–12 July 2018.
10. R. J. Prarthana; A. M. Dhanzil; N. I. Mahesh; S. Raghul, "An Automated Garage Door and Security Management System (A dual control system with VPN IoT & Biometric Database)," 2018 Second International Conference on Electronics, Communication and Aerospace Technology (ICECA), 29–31 March 2018.
11. R. D. Rashid; S. A. Jassim; H. Sellahewa, "LBP based on multiwavelet sub-bands feature extraction used for face recognition," IEEE international Workshop on Machine Learning for Signal Processing, Sept. 22–25, 2013, Southampton, UK, 2013.
12. J. Saranya; J. Selvakumar, "Implementation of Children Tracking System on Android Mobile Terminals," International Conference on Communication and Signal Processing, April 3–5, 2013, India.
13. Y. Mori; H. Kojima; E. Kohno; S. Inoue; T. Ohta; Y. Kakuda; A. Ito, "A Self-Configurable New Generation Children Tracking System Based on Mobile Ad Hoc Networks Consisting of Android Mobile Terminals," 2011 10th International Symposium on Autonomous Decentralized Systems (ISADS), pp. 339–342, 23–27 March 2011.
14. M. Muthumari; N. K. Sah; R. Raj; J. Saharia, "Arduino based Auto Door unlock control system by Android mobile through Bluetooth and Wi-Fi", 2018 IEEE International Conference on Computational Intelligence and Computing Research (ICCIC), 13–15 Dec. 2018.
15. P. Viola; M. Jones, "Rapid Object Detection Using a Boosted Cascade of Simple Features," Proceedings of the 2001 IEEE computer society conference on computer vision and pattern recognition. CVPR 2001, vol. 1, pp. I–I. IEEE, 2001.
16. T. M. Inc., "Train a Cascade Object Detector," [Online]. Available: http://www.mathworks.se/help/vision/ug/train-a-cascadeobject-detector.html#btugex8. [Accessed Nov 2014].

# 16 Adaptive Traffic Control

*Yashashwini Dixit, Gyanadipta Mohanty,*
*and Karmel Arockiasamy*
Vellore Institute of Technology

## CONTENTS

16.1   Introduction ................................................................................................... 181
16.2   Literature Survey ........................................................................................... 182
16.3   Proposed Methodology ................................................................................. 184
     16.3.1   System Design ................................................................................. 184
     16.3.2   Implementation ................................................................................ 184
     16.3.3   Results and Discussion .................................................................... 185
16.4   Conclusion ..................................................................................................... 186
Bibliography ............................................................................................................ 186

**ABSTRACT**

Road congestion can be a serious problem as there are many cars on the road. As traffic increases, the length of rows of cars waiting to be processed at intersections increases dramatically, and regular traffic lights cannot schedule them properly. Self-adaptive signal control is an efficient way to reduce traffic congestion in the city. In fact, traffic light junctions employ Computer Vision and Machine Learning (ML) to assess the peculiarity of traffic flow competition. It uses You Only Look Once (YOLO), a Deep Convolutional Neural Networks-based system. Depending on the data collected, such as traffic jams and waiting time per vehicle, the traffic light phase will be optimized to allow as many vehicles as possible to pass safely with as little waiting time as possible. The transfer learning approach can be used to implement YOLO in an embedded controller.

## 16.1   INTRODUCTION

In the process of economic and social expansion, the number of vehicles on the road and the accompanying transportation needs are increasing. Frequent traffic jams on urban roads are detrimental to both the recession and the environment. Due to the defined space reserves of metropolitan areas, effective traffic control measures and improved efficiency of existing transportation facilities have become important research topics for reducing traffic bottleneck. Traffic regulation is one of the most lion's share technology strategies for controlling traffic flow, reducing congestion and even reducing pollution. Advances in data processing, computer technology, and systems science have always been accompanied by their progress and growth. Based on the administrator's control goals (such as minimum delay at intersections) and the influx aspect of traffic outflow at intersections, self-adaptive control systems can change signal timing settings in real-time. Compared to timed and operational controls, self-adaptive control systems can better utilize the traffic capacity of the entire road network and dramatically improve traffic efficiency.

Low speeds, long travel times, and long vehicle waiting times result from uneven and different traffic in different lanes, resulting in inefficient use of the same time slot for each. Using cameras and image processing modules, we aim to design a self-adaptive signal control system that allows a traffic management system to determine the time allocation of a particular lane according to the traffic density of other lanes. It can also include an automatic siren to speed up the car in a

DOI: 10.1201/9781003374121-16

distributed network, dynamic signal times, and interactive views via the dashboard to monitor the above features.

## 16.2  LITERATURE SURVEY

It covers the development of intelligent signal control systems used everywhere the globe, their technical aspects, and therefore the present state of research on self-adaptive control methods and signal control methods for divergent traffic flows from networked autonomous vehicles increase. Finally, signal control supported multi-agent reinforcement learning may be a quite adaptive closed feedback control method that's superior to several counterparts in real-time characteristics, efficiency, and Autodidacticism, and is therefore a crucial study. Within the future, it'll specialize in the "model-free" and "self-learning" characteristics of control methods. Advances in adaptive signal control suggest that they're going to play a crucial role within the destiny advancement of smart cities and therefore the mitigation of traffic jam. Compared to fixed timing techniques, our study demonstrated the adaptability of Reinforcement Learning Traffic Light Control to handle time-varying flows and unpredictable traffic patterns. Our study demonstrates a model that includes three-state depiction from low to high verdict and whose performance in simulation is compared using an asynchronous actuator critical approach using neural network function approximation. The results show that low-resolution state representations (e.g., occupancy and average velocity) are almost exactly superior to high-resolution state representations (e.g., individual vehicle positions and velocities). These results suggest that the implementation of signal control by reinforcement learning can be performed with conventional sensors such as loop detectors, without the need for advanced sensors such as cameras and radars.

It gives a whole evaluation of the connected and autonomous vehicle (CAV) primarily based totally city-site visitors manage system. The evaluate started out with a deterministic and probabilistic technique of estimating site visitors' situations primarily based totally on CAV, that's vital for site visitors manage. In addition, a simple mathematical framework for the improvement of CAV primarily based totally city-site visitors manage turned into proposed. A brief precis of normal effects indicates the strengths and weaknesses of numerous site visitor's situation estimation techniques and site visitors control systems.

Trying to escape the curse of dimensionality in solving the associated Hamilton–Jacobi–Bellman (HJB) optimal control problem, the simulation result obtained from the traffic simulation model of an urban traffic network includes several node-staging variants. The proposed strategy can be adapted to a variety of traffic situations, and the less complex and optimal solution parameterization, linear and bimodal piecewise linear approaches provide the appropriate compromise between computational complexity and network performance. The module is able to identify when a vehicle is being driven recklessly and reports the data to the appropriate authorities. The controller sends a signal to the electronic throttle installed within the car and the vehicle's speed may be regulated and reckless driving can be avoided. Additionally, the status of the car may be communicated to the police, the vehicle's owner, or the appropriate authorities via a Global System for Mobile communications (GSM) module. Creating an automatic speed detection system notices the vehicle speed, and if the speed increases, pull out the vehicle registration number and email it to Toll Plaza to collect the fare. The Doppler Effect phenomenon is used to assess velocity. When an overspeed is detected, the camera immediately records an image of the vehicle and uses a Digital Image Processing algorithm to extract the license plate. The obtained license plate will be sent to Toll Plaza by email.

The goal of creating an intelligent system that can notice and path aggregate automotive objects in a video, while addressing background changes, lighting issues, and environmental conditions, is a difficult suggestion. Vehicle detection is achieved by background removal and is taken from each vehicle that has a scale-invariant feature transformation feature identified. Neural networks and support vector machines (SVMs) are used to classify vehicles. SVM has outperformed artificial neural networks in terms of generalization.

An imaginative and prescient, primarily based on smart machine structure for figuring out site visitors' offenses is described. It utilizes a pc imaginative and prescient machine that aids within-side the detection of site visitors' offenses dedicated at the street intersection. The layout is split into three subsystems: video capture, smart running structure (IOA), and output. The IOA controls many strategies for detecting site visitors' offenses. The test findings display that optical man or woman identity has an accuracy of 86.11% and velocity dimension has an accuracy of 88.45%.

It uses a real-time embedded system, the Number Recognition System (VNPR), to detect the license plate of a car. It provides an option to VNPR by using an open-source library called OpenCV. This system consists of an infrared sensor that recognizes the vehicle. During testing, the microprocessor set recorded the minimum time for the sensor to detect an object. When the timer reaches zero, the camera is activated, the license plate is recorded and the image is saved on the Raspberry Pi (RPi). The RPi processes the captured image and extracts the numbers on the image. Intends to assemble a deep getting to know version that is able to predict the registration of wide variety of automobile at the road. This version change includes the use of the You Only Look Once (YOLO) technique and convolutional neural networks. In this technology, the photo of the car is cut, and most effectively, the photo of the wide variety plate is given as output. The records are in the end surpassed to the Tesseract program, which turns the automobile's registration wide variety to text. The version's performance is measured based on three parameters: recall, precision, and suggest common precision.

It is important to demonstrate a model of vehicle license plate derivation and identification using image processing and wireless transmission over various network. This model utilizes the GSM module for this purpose. The goal of this project is to create a stand-alone vehicle surveillance system that sends fraudulent license plates, speeding license plates, odd license plates, and other information to traffic and other authorities via SMS. The proposed application uses Arduino, while MATLAB® processes the captured photos and provides the data via serial communication.

A speed detection framework is introduced that uses only two speed guns to identify the speed of a vehicle, even if the vehicle is not in the line of sight of the camera. The goal is to determine the average speed of a vehicle moving between two points in a particular area, for example space. It uses an inexpensive RPi module and a regular camera to recognize the license plate of the car's license plate. With a low-power RPi, this hardware is more stable. It can run for hours without crashes or malfunctions.

The purpose is to path the pace of the vehicle, warn of speed fluctuations, and collect data from license plates. Using Artificial Intelligence (AI) with Machine Learning (ML), we visualized clear data captured by sensor cameras placed in strategic locations. Correctly, collected data are identified by ML-enabled Internet of Things (IoT). It is used to section and recognize the character and send it to the server via the Wi-Fi module.

In Ref. [15], traffic video surveillance automatically flags vehicles such as ambulances and trucks. This helped guide the car in an emergency. Here, we used a hybrid SVM with extended Kalman filter to path the car. Then we used the flow-gradient feature histogram to classify the vehicles. Vehicles are calculated based on traffic density and video files, and emergency vehicle signals change dynamically. To detect the ambulance, we set the camera to take traffic records at a distance of more than 500 m from the signal and used a deep learning neural network. As a result, dynamic signal control was quickly implemented. Multinomial logistic regression is also used to reliably predict accidents in real-time livestreaming video.

Reference [16] proposes a method primarily based totally at the dimension of real-site visitors density on the street. Real-time video- and picture-processing strategies are used to reap this. The pictures are recorded and saved at the server, and the density is decided through evaluating them with the real-time pictures captured through the camera. The concept is to modify site visitors

**TABLE 16.1**

**Comparison of Related Works in Terms of Algorithms Used and Metrics Evaluated**

| Method/Algorithm Used | Metric (If Any) | Results/Remarks |
|---|---|---|
| Reinforcement Learning [17] | Queue length, vehicle speed (km/h), waiting time (s) | It is found out that RL-TSC-1 algorithm has least queue length and waiting time. |
| A3C Reinforcement Learning Algorithm [18] | Total throughput (veh/sim), total delay (s/sim), total queue (veh/sim) | When compared to the Actuated algorithm, all reinforcement learning methods outperform it in terms of lowering latency and queue lengths. |
| ANPR Algorithm [19] | Accuracy | MATLAB is used to extract the license plate number using Digital Image Processing methods. |
| For classifying vehicles: [20]: Artificial Neural Network (ANN), SVM | Accuracy | A comparison of the performance of several approaches shows that the support vector machine outperforms the ANN in terms of generalization. |
| Vehicle Detection, Optical Character Recognition [21] | Accuracy | Optical character recognition has an accuracy of 86.11%, while relative speed measurement has an accuracy of 88.45%. |
| YOLOV5 Algorithm, Optical Character Recognition (OCR), and Tesseract [22] | Precision, recall, mean average precision | The YOLOv5 algorithm can identify the existence of the number plate with greater precision, even when the plate is inclined. |
| Hybrid SVM, Multinomial Logistic Regression (MLR) [23] | Accuracy in terms of quality, detection Time | The result validates that the MLR-based system provides a very advanced executing rate in terms of accuracy, precision, recall, sensitivity, specificity, and detection time. |

through figuring out site visitors' densities on each facet of the street and presenting customers with sign manipulate alternatives through a software program (Table 16.1).

## 16.3   PROPOSED METHODOLOGY

### 16.3.1   System Design

The gadget is made of software program and hardware parts. The software programs element handles the stay shooting and time slot distribution for specific avenue lanes. Those time slots are dispatched to visitors lighting for functioning and controlling automobiles at the roads.

This paper works at the software program, a part of this gadget. First, the digital camera feeds snap shots to our gadget at everyday quick intervals. The set of rules then calculate the relative density of the lane when it comes to different lanes with the aid of figuring out the range of vehicles withinside the lane. Then the time allotment module takes this gadget's input (inclusive of visitor's density) and determines the most efficient and green time slot. Finally, the microprocessor sends this record to the suitable traffic lights (Figure 16.1).

### 16.3.2   Implementation

The designed system is first implemented by recording real-time video of every lane at a specific point in time. These videos are processed by the system using the YOLO library. A replacement approach to visual perception is named YOLO. In previous work on object discovery, the classifier was reused to perform the invention. Rather, we consider object detection using spatially isolated

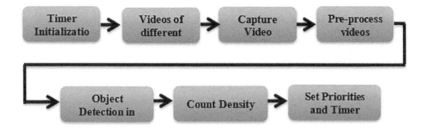

**FIGURE 16.1**  Flowchart of YOLO object detection.

conjugated boxes and connected class quantity as a regression problem. One neural network fore-tells the conjugated box and sophistication quantity immediately from an entire image with one rat-ing. Because the whole discovery pipeline may be a single network, it is often optimized supported end-to-end discovery execution.

Object recognition is the task of recognizing and classifying individual objects in an image. Previous methods, such as region convolutional neural network (RCNN) and its derivatives, used pipelines to perform this process at many stages. This can be time-consuming and difficult to opti-mize, as each component needs to be trained individually. YOLO is a single neural network that can handle all tasks.

This implementation uses the Yolo V3 neural network to count the cars in the video. Detection is done every x-frame, where x is a variable. The dlib library may also be used to path previously detected vehicles. You can also change the confidence detection level, the number of frames to count vehicles detected before removing them from the traceable list, the maximum distance from the center of gravity, and the number of frames to delay detection and skip. It is capable of using the original video size or Yolo V3 size 416×416 as the annotation output. The Yolo V3 model has been downloaded and pre-trained.

The system also determines the number of cars in a lane and calculates the relative density to other lanes. The time allocation module receives input (as traffic density) from this system and determines an optimized and efficient time slot. To do this, specify the base timer and the time range in which the time quota exists. The system provides time slots between time slots, and the total is close to the specified base clock. This value can be sent to each traffic light by the microprocessor. The value is validated by changing the base timer and time range to get the right results for different types of videos from different lanes. The formula used calculates the value according to the base timer and the number of vehicles to get the best and most efficient answer for each lane.

### 16.3.3  RESULTS AND DISCUSSION

The vehicle detection results are accurately detected and various videos are checked. For each video, the vehicles are accurately detected and the total number of vehicles for each video is calcu-lated. The YOLO model calculates the number of cars per lane and their waiting time. The continu-ance of the succeeding phase is determined based on the wait instance and the queue length of each lane to provide the shortest possible wait time. Due to its accuracy and real-time efficiency, YOLO could be used in place of police officers in traffic control optimization. To thoroughly evaluate the solution, you need to complete the hardware implementation. This system of rules is not restricted to cameras. Many sensors are utilized to compensate for poor performance in bad weather to avoid camera errors. The sensor fusion technique is used to integrate the sensors. Using automotive net-work technology, intelligent transportation systems can also collect data and reduce traffic conges-tion (Figure 16.2).

**FIGURE 16.2** Object detection in sample video.

## 16.4 CONCLUSION

The purpose of this study is to improve the traffic system by developing deep learning-based self-adaptive traffic control algorithms. This new method makes it easier for cars to navigate intersections, with benefits such as reduced traffic and reduced $CO_2$ emissions. The need to push the boundaries of object identification, classification, and pathing in real-time applications is demonstrated by a wealth of video data. YOLO achieves very fast inference speeds with low resolution and small objects with little loss of accuracy. Real-time inference is possible, but edge device applications require breakthroughs in the edge device architecture or hardware. Finally, we proposed a new method to reduce latency by optimizing the stage using real-time data from YOLO. Future enhancements include the detection and identification of elevated vehicles using computer vision technology and the generation of automatic fines using a distributed blockchain network. You can also create a dashboard with the above features for monitoring purposes and view real-time traffic on Google Maps

## BIBLIOGRAPHY

1. Wang, Y., Yang, X., Liang, H. and Liu, Y. (2018). A review of the self-adaptive traffic signal control system based on future traffic environment. *Journal of Advanced Transportation*, 2018, 1096123. doi: 10.1155/2018/1096123
2. Mannion P., Duggan J., and Howley E. (2016). An experimental review of reinforcement learning algorithms for adaptive traffic signal control. In: McCluskey T., Kotsialos A., Müller J., Klügl F., Rana O., Schumann R. (eds) *Autonomic Road Transport Support Systems. Autonomic Systems*. Birkhäuser, Cham. doi:10.1007/978-3-319-25808-9_4
3. Genders, W. and Razavi, S. (2018). Evaluating reinforcement learning state representations for adaptive traffic signal control. *Procedia Computer Science*, 130, 26–33. doi:10.1016/j.procs.2018.04.008
4. Guo, Q., Li, L. and Ban, X.J. (2019). Urban traffic signal control with connected and automated vehicles: A survey. *Transportation Research Part C: Emerging Technologies*. doi:10.1016/j.trc.2019.01.026
5. Baldi, S., Michailidis, I., Ntampasi, V., Kosmatopoulos, E., Papamichail, I. and Papageorgiou, M. (2019). A simulation-based traffic signal control for congested urban traffic networks. *Transportation Science*. doi:10.1287/trsc.2017.0754
6. Kumar, V.P., Rajesh, K., Ganesh, M., Kumar, I.R.P. and Dubey, S. (2014). Overspeeding and rash driving vehicle detection system. In *2014 Texas Instruments India Educators' Conference (TIIEC)*, pp. 25–28. doi:10.1109/TIIEC.2014.013
7. Malik, S.M., Iqbal, M.A., Hassan, Z., Tauqeer, T., Hafiz, R. and Nasir, U. (2014). Automated over speeding detection and reporting system. In *2014 16th International Power Electronics and Motion Control Conference and Exposition*. IEEE, pp. 1104–1109. doi:10.1109/EPEPEMC.2014.6980657
8. Ukani, V., Garg, S., Patel, C. and Tank, H. (2017) Efficient vehicle detection and classification for traffic surveillance system. In: Singh, M., Gupta, P., Tyagi, V., Sharma, A., Ören, T., Grosky, W. (eds) *Advances in Computing and Data Sciences. ICACDS 2016. Communications in Computer and Information Science*, vol. 721. Springer, Singapore. doi:10.1007/978-981-10-5427-3_51

9. Billones, R.K.C., Bandala, A.A., Sybingco, E., Lim, L.A.G. and Dadios, E.P. (2016). Intelligent system architecture for a vision-based contactless apprehension of traffic violations. In *2016 IEEE Region 10 Conference (TENCON)*, pp. 1871–1874. doi:10.1109/TENCON.2016.7848346.

10. Shobayo, O., Olajube, A., Ohere, N., Odusami, M. and Okoyeigbo, O. (2020). Development of smart plate number recognition system for fast cars with web application. *Applied Computational Intelligence and Soft Computing*, 2020, 8535861. doi:10.1155/2020/8535861

11. Suresh, M.S. and Suresh, M.S. (2022). Intelligent vehicle license plate recognition by deploying deep learning model for smart cities. *International Journal of Mechanical Engineering*, 7(1).

12. Kohli, S., Sharma, M. and Sehgal, J. (2019). Feature extraction and recognition of vehicle registration for wireless transmission over different network protocols. In *2019 6th International Conference on Signal Processing and Integrated Networks (SPIN)*, pp. 518–521. doi:10.1109/SPIN.2019.8711711

13. Khan, S.U., Alam, N., Jan, S.U. and Koo, I.S. (2022). IoT-enabled vehicle speed monitoring system. *Electronics*, 11(4), 614. doi:10.3390/electronics11040614

14. TK, H.S., JR, D.K. and Priyadharsini, K. (2021). An experiment analysis on pathing and detecting the vehicle speed using machine learning and IoT. In *2021 Smart Technologies, Communication and Robotics (STCR)*, pp. 1–5. doi:10.1109/STCR51658.2021.9587924.

15. Maha Vishnu, V.C., Rajalakshmi, M. and Nedunchezhian, R. (2018). Intelligent traffic video surveillance, and accident detection system with dynamic traffic signal control. *Cluster Computing*, 21, 135–147. doi:10.1007/s10586-017-0974-5

16. Frank, A., Al Aamri, Y.S.K. and Zayegh, A. (2019). IoT based smart traffic density control using image processing. In *2019 4th MEC International Conference on Big Data and Smart City (ICBDSC)*, pp. 1–4. doi:10.1109/ICBDSC.2019.8645568.

17. Mannion, P., Duggan, J., and Howley, E. (2016). An experimental review of reinforcement learning algorithms for adaptive traffic signal control. In: McCluskey, T., Kotsialos, A., Müller, J., Klügl, F., Rana, O., and Schumann, R. (eds) *Autonomic Road Transport Support Systems. Autonomic Systems*. Birkhäuser, Cham. doi:10.1007/978-3-319-25808-9_4

18. Genders, W. and Razavi, S. (2018). Evaluating reinforcement learning state representations for adaptive traffic signal control. *Procedia Computer Science*, 130, 26–33.

19. Malik, S.M., Iqbal, M.A., Hassan, Z., Tauqeer, T., Hafiz, R. and Nasir, U. (2014). Automated over speeding detection and reporting system. In *2014 16th International Power Electronics and Motion Control Conference and Exposition*, pp. 1104–1109. doi:10.1109/EPEPEMC.2014.6980657

20. Ukani, V., Garg, S., Patel, C. and Tank, H. (2017). Efficient vehicle detection and classification for traffic surveillance system. In: Singh, M., Gupta, P., Tyagi, V., Sharma, A., Ören, T., Grosky, W. (eds) *Advances in Computing and Data Sciences. ICACDS 2016. Communications in Computer and Information Science*, vol 721. Springer, Singapore. doi:10.1007/978-981-10-5427-3_51

21. Billones, R.K.C., Bandala, A.A., Sybingco, E., Lim, L.A.G. and Dadios, E.P. (2016). Intelligent system architecture for a vision-based contactless apprehension of traffic violations. In *2016 IEEE Region 10 Conference (TENCON)*, pp. 1871–1874. doi:10.1109/TENCON.2016.7848346

22. Chitra Kiran, N., Suresh, S., Suresh, S. (2022). Intelligent vehicle license plate recognition by deploying deep learning model for smart cities. *International Journal of Mechanical Engineering*, 7(1), 6739–6744.

23. Frank, A., Al Aamri, Y.S.K. and Zayegh, A. (2019). IoT based smart traffic density control using image processing. In *2019 4th MEC International Conference on Big Data and Smart City (ICBDSC)*, pp. 1–4. doi:10.1109/ICBDSC.2019.8645568

# 17 Genuine Investments for Economic Energy Outputs

*Saravanan Chinnusamy*
Chemveda Life Sciences India Pvt. Ltd.

*R. Rajapriya and Milind Shrinivas Dangate*
Vellore Institute of Technology

*Nasrin I. Shaikh*
Nowrosjee Wadia College and Abdul Karim Ali Shayad
Faculty of Engineering and Polytechnic, MMANTC

## CONTENTS

17.1 Introduction ........................................................................................................ 190
17.2 Literature and Background ................................................................................. 190
    17.2.1 Resource Curse ........................................................................................ 190
    17.2.2 Sustainability .......................................................................................... 191
17.3 Model and Methodology .................................................................................... 193
    17.3.1 Theoretical Model.................................................................................... 193
    17.3.2 Empirical Strategy & Empirical Model ................................................. 194
    17.3.3 Data.......................................................................................................... 195
17.4 Relationships to Be Expected ............................................................................ 197
17.5 Data Limitations ................................................................................................ 198
    17.5.1 Estimation and Results ........................................................................... 199
17.6 Potential Weakness ............................................................................................ 201
17.7 Conclusion ......................................................................................................... 201
References........................................................................................................................202

**ABSTRACT**

Savings and investment levels are critical predictors of an economy's production. Changes in manufactured capital are measured by traditional investment measures. "Adjusted net or genuine saving measures a country's true level of saving". This "real level" represents changes in the whole capital stock of an economy or the "productive base," as defined by Arrow. The productive base is defined by Arrow as "society's capital assets and institutions". The studies by Arrow, Hamilton, and Hamilton and Clemens calculated genuine savings (investment) for a variety of countries and regions and found significant disparities around the world. Investment saving measures provides crucial insight into a country's long-term growth potential. In the studies cited above, the importance of organizations and governance on spending time is not taken into account. Instead, they concentrate on determining the specific components of true savings. This includes both the theoretical and practical aspects of the estimation process. This chapter will look at how governments realize how people spend their time in energy-rich countries.

DOI: 10.1201/9781003374121-17

## 17.1  INTRODUCTION

Quantifying and accounting for the variations in management will help researchers better understand how institutional factors affect genuine investment. Several hypotheses will be examined to discover the relationship between governance and true investment. The motivation for hypotheses is a mix of economic theory and econometric issues. Democracy, stability, corruption, efficacy, regulatory quality, and the encouragement of the private sector are the specific governance aspects to be evaluated. Genuine investment measures that do not cause environmental harm will also be evaluated. Real development considers variations in manufactured, human, and natural capital, as well as environmental harm, and considers regional and developmental differences. Due to natural capital depletion, on the other hand, by lowering capital assets, a country's genuine savings are reduced. As a result, other types of capital investments boost genuine savings. Reinvesting a portion of nonrenewable resource depletion economic rents allows for the possibility of offsetting a reduction for one component of spending time with an increase in another. For resource-based economies, the idea of expanding the productive base while diminishing nonrenewable resources has significant ramifications. It indicates that their economic growth will not be constrained by the depletion of nonrenewable resources. Decreases in nonrenewable resources can be offset by investments in these other types of capital, including human, economic, and social capital. The premise is that institution quality is a critical factor in attracting genuine investment and, as a result, long-term growth. In Section 17.4, we'll go over some of the previous research that looked at the relationship between institutions and economic success. Poor governance can stymie long-term resource management in an economy. Resource rents may be used for nonproductive uses as a result of weak institutions, ill-defined property rights, corruption, and other factors. The Hartwick rule serves as the theoretical framework for the analysis's sustainability criterion. This rule is based on Hartwick's and Solow's work (1974 and 1986) [1]. Section 17.4 discusses data and their potential influence on genuine savings. Section 17.5 will look at the results and how they fit into the larger literature. As a robustness check, a variety of empirical specifications will be used. Section 17.6 brings the research to a close.

## 17.2  LITERATURE AND BACKGROUND

### 17.2.1  RESOURCE CURSE

Multiple types of literature devoted to evaluating the relationship between governance, environment, and economy provided the impetus for investigating the impact of governance on genuine investment. This section will offer elements from the resource curse and sustainability literature. Their relative strengths can be leveraged by incorporating approaches and findings from the literature.

A high natural capital endowment would appear to be an overall benefit a priori. According to the literature on resource curses, the opposite is true. Having a lot of natural resources, according to many studies, slows down growth. Sachs and Warner [2] are widely regarded as a seminal works in the field of resource curse research. The writers have revised and expanded their original work several times since then. The majority of resource curse research looks at the link with both resources and growth from the standpoint of income convergence. Two early basic studies in the income converging literature are Baumol [3] and Barro and Sala-i-Martin [4,5]. Income growth is often approximated over some time as a function of a range of institutional and economic variables in income convergence research. Lower-income countries, according to the study, grow more rapidly than greater economies. This conclusion is contingent on some circumstances. Sachs and Warner [6] discuss about literature on resource curses, the income convergence methodology has become mainstream. From 1960 to 1969, a proxy for basic accessibility, the relationship between economic growth, investment, rule of law, trade terms, and economic growth was utilized to regress economic development for some countries from 1970 to 1990 [7]. The abundance of natural resources is a hindrance to income convergence for states in the United States, according to Papyrakis and Gerlagh

[8]. Because of the institutional resemblance between states, this study is noteworthy. States have different government systems; despite this, they are all governed by the same federal system. Many theories have been proposed to explain the resource curse's paradoxical nature. The "Dutch sickness", Rent-seeking behavior, reduced motivations to spend on human capital, declining capital demand, and a lack of economic depths are all factors routes via which resource reliance can stifle progress, according to Gylfason [9]. The Dutch illness is manifested through a country's exchange rate. Foreign investment and commerce in other items could be hampered by volatile currency rates. A financial industry that is reliant on natural resources may be limited in an economy. The channeling of a big portion of a country's investment into the natural resource industry may stifle growth in other industries. As a result, natural resource economic returns are becoming increasingly important for economic growth. The various advantages of having control over resources drive rent-seeking behavior. Natural resource richness, according to Gylfason, reduces the incentives to invest in education since they provide an economic source of non-wage revenue. Individual (or all) of these channels may be exacerbated by insufficient institutions, property rights, and the rule of law. As a result, the domestic economy's natural resource sector may develop to dominate over time. According to Gylfason, an economy that is unduly reliant on a single industry has lower long-term growth potential. Kalyuzhnova, Kutan, and Yigit [10] identify essential methods that can aid in the sustainability of energy-intensive economies. They discovered that lower levels of corruption are associated with less corporate rules and greater democracy. Corruption harms these countries' growth rates and GDP per capita levels. Kaluzhnova et al. data sets are limited in terms of long-term period observation, which is one of their shortcomings (averages from 1989 to 2006). They also employ fixed nonrenewable resource production measures rather than variations in different kinds of natural capital across time. The resource curse is a contentious subject. Norway and the United States, for example, come to mind as counterexamples. The economic progress of countries like these does not appear to be hampered by a lack of resources. According to Brunnschweiler and Bulte [11], analyses that suggest the resource curse are based on model specification. Resource abundance measurements, in particular, are more accurate estimates of resource reliance when utilized in estimations. Differentiating between resource abundance and resource reliance is critical. Dependency indicates a more volatile economy-resource relationship. The economic rewards from the resource become inextricably linked to the economy's growth. Abundance implies that enjoying the economic benefits from a resource has a low opportunity cost. Brunnschweiler and Bulte find no evidential basis and for curse when they control for resource dependence. Rather, they discover that resource scarcity has little impact on growth and that a plentiful supply of resources contributes to strong governance quality and growth. Another denial of the resource curse is Papyrakis and Gerlagh [5]. They discovered that controlling government, for resource extraction availability, conditions of commerce, and education have a favorable impact on growth. Governance is a key control variable in both kinds of research. Neumayer [12] looked into whether resource curses exist when compared to true income metrics. The methodology of the resource curse is linked to Neumayer's research and focuses on the concept of true involvement, which is a measure of sustainability. The presence of the resource curse is confirmed by his findings. The resource curse and environmental Kuznets curve techniques are combined by Costantini and Monni [13], who finish that long-term progress involves human funding and effective capacity reform. It's less obvious whether resources hinder growth or whether the resource stifles growth curse is caused by their impact on institutional development. Important components of the sustainability literature will be used in this study to address this issue.

## 17.2.2 Sustainability

The literature on resource curses is a natural partner to the literature on sustainability. The resource curse literature is inherently retroactive, challenging the impact of resources on economic advancement over time. Sustainability studies, on the other hand, are focused on the future;

what impact will today's decisions have on future growth? Before moving on to the literature on sustainability, it is necessary to define "sustainability" in this chapter. Sustainability is used to address a wide range of concerns across disciplines. Even within the field of economics, the term "sustainability" can refer to a variety of topics. The use of sustainability standards clarifies a phrase that is frequently employed confusingly. The weak and strong sustainability criteria have significantly different implications in the context of resource management. The strong sustainability criterion's essential premise, according to Ayres [14], is that "minimal levels of several various forms of capital (economic, ecological, and social) should be independently maintained". To achieve the weak sustainability criteria, economic profit in an economy's aggregate capital stock must be positive. The question of substitutability between kinds of capital is a significant difference between the two criteria. Capital substitutability is the subject of many studies in *Ecological Economics* (Volume 22, 1997). Among the contributors are Herman Daly, Robert Solow, and Joseph Stiglitz. Acceptable adjustments in capital stocks are subject to stricter conditions under the strong criterion. Because many decisions are irreversible, proponents of the strong criterion support careful ecological management. If a specific resource stock falls, the strong requirement may be broken, whereas the weaker criteria allow for displacement between dropping stocks in exchange for suitable gains in another. Consider an example of a nonrenewable resource like oil. By depleting its oil reserves while investing in education, a country can meet the weak sustainability condition. Because of the strong sustainability, it may be necessary to refrain from extracting a certain amount of oil. The specific limitation's criterion is used in this study. In the coming years, to keep track of resource management, the study could be tweaked or expanded difficulties arising out of sustainable economic criteria. Genuine savings metrics is an important parameter, according to Pearce, Hamilton, and Atkinson [15], the net change in an economy's assets gives vital information, depending on whether the stronger or weaker sustainability criterion is utilized. The Hartwick rule has become a well-known indicator of long-term viability. The formalization of this rule was influenced by Hartwick and Solow [1]. Extensions and clarifications of the Hartwick rule can be found in Dixit, Hammond, Hoel, and Farzin [16]. "A steady level of consumption can be sustained if the value of (net) investment equals the value of extracted resource rents at each point in time." Hamilton outlines the rule and its ramifications. To achieve sustainable growth, according to the Hartwick rule, such decreases in nonrenewable capital should be offset by increases in renewable capital growth in other types of financing. When this happens, there is no need to cut back on consumption. Hamilton, Ruta, and Tajibaeva [17] conduct a hypothetical study to determine the capital stock that would be available in each country if the Hartwick rule were enforced. The Hartwick rule has not been observed in countries where resource rents account for more than 15% of GDP, according to the researchers. These countries are investing far less than is required by law. This demonstrates that resource-rich countries are still not, on average, following Hartwick-defined stable growth paths. Furthermore, the authors demonstrate that consumption is not restricted by a constant positive level of true investment. Section 16.3 will offer a formal statement of the Hartwick rule. To see if a country's investment is satisfying the Hartwick criterion, precise measurements of the various components of its productive base are required. The right valuation of changes in natural capital is likely the most difficult obstacle in accurately evaluating the productive base. Natural capital changes have a less well-defined valuation than produced capital changes. To create appropriate assessments, it is necessary to analyze the economic rents that come from resource extraction. The World Bank's Kirk Hamilton has played a vital role in the creation of natural capital measurements or a net expenditure. These forecasts are used to create real cost-cutting options. The core of meaningful savings initiatives is updating national savings for natural environmental capital. Hamilton summarizes the methodology that has been used in a significant amount of research in this field.

Genuine savings estimates were first published by Hamilton and Clemens in 1999 [18]. In Sub-Saharan Africa, they find that true savings are negative. The theoretical framework for estimating genuine savings measures is also outlined in this study. In Section 16.3, we'll go over this material in further depth. These studies reveal that actual savings harm certain parts of the globe. The estimation of true savings measures is largely an accounting task, once changes in various capital stocks have been produced. These studies don't go into great detail about the factors that contribute to negative true savings. This chapter examines the government's role in further detail. Insufficient governance, according to the resource curse literature, raises the odds of natural resource management failing to achieve economic growth. If the economy's productive foundation is constrained by natural resource availability (or dependency), the economy's growth will be stunted as well.

## 17.3 MODEL AND METHODOLOGY

### 17.3.1 THEORETICAL MODEL

The theoretical foundation for the analysis is the following social welfare function:

$$V(t) = \int_t^\infty e^{-\delta(s-t)} U(C(s)) ds,$$  (17.1)

where $t$ is time, is the discount rate, $U(C)$ is a utility from consumption, $C(t)$ is consumption, and $V(t)$ represents intertemporal social welfare. The consumption of society determines utility and consequently social welfare. According to one interpretation of the sustainability criterion, non-decreasing welfare necessitates:

$$\frac{\partial V(t)}{\partial t} \geq 0;$$  (17.2)

also, the rise in social welfare is not negative with time. The shift in the functional form could be zero in this interpretation and is the case if the true investment is exact enough to satisfy the Hartwick requirement. Arrow, Dasgupta, and Mäler [19] use the Ramsey–Solow model to investigate numerous aspects of economic sustainability in imperfect economies. Their research is predicated on the idea that a country's wealth is derived from its capital assets. The social welfare function $V(\bullet)$ can be written as a function of capital in an economy when social welfare is defined as a function of wealth over time. Non-resource management and capital are the two categories of capital that make up an economy's total capital stock. Make $K(t)$ the vector of non-resource capital and $R(t)$ the vector of resource capital. Examples include manufacturing equipment, buildings, human capital, and other non-resource capital. Let's call a single type of resource capital $Rj$, $j = 1$, $N$, and non-resource capital $Ki$, $I = 1... M$. $V(t)$ becomes $V(t)$ for the social welfare function $(K, R)$. To reflect changes in the productive base of an economy, inequality (3.1.2) will be expanded. The Hartwick rule offers the requirement. When the criterion is met, according to Arrow, "society can continue a continual stream of consumption by investing in reproducible capital goods the competitive rents on its current use of the waste resource". The rule creates a link between consumption, investment, and social well-being. Capital asset changes (the entire capital stock) will be considered first. Appropriate pricing must be developed to measure the economic worth of changes in capital stocks. Let the non-resource and resource capital ($Ki$ and $Rj$, respectively) shadow (or accounting) prices be represented by:

$$\frac{\partial V(K,R)}{\partial Ki} = \lambda i,$$  (17.3)

$$\frac{\partial V(K,R)}{\partial Rj} = \lambda j.$$  (17.4)

As a result of the chain rule (16.2), it becomes:

$$\sum_{i=1}^{M} \lambda i(t) Ki(t) + \sum_{j=1}^{N} \lambda j(t) Rj(t) \geq 0, \tag{17.5}$$

where $\lambda s$ denotes shadow prices and $Kt(t)$ (or $Ry$ $(t)$) denotes the time derivative of the capital stock or investment in question. When a stock's time derivative is negative, it means the stock's physical value is decreasing. Genuine investment is equal to the present value of the development in the economy's capital assets over time, which is represented on the left-hand of equation (17.5). The specific limitations criteria are satisfied, when genuine investment is positive. The wealth management part of the specific limitations criteria and Hartwick rule is captured by inequality equation (17.5). "Investing exhaustible resource rents in generated capital, and its extension, 'zero net investment', is an inherent feature of Solow's [1] original and generalized constant consumption model Hartwick", write Hamilton and Hartwick. The state of net investment can be seen in (16.5). If material capital stocks fall, (16.5) can only be met if non-resource capital stock investments increase. Furthermore, the worth of a non-resource capital investment must be equivalent to the price of the resource capital decline. "Zero net investment" arises when the losses and growth are equal. To indicate the high sustainability requirement, Inequality equation (17.5) could be enlarged or changed. For example, $Ry$ $(t)$ 0 could represent a constraint requiring a non-negative change in a specific form of resource capital over time.

### 17.3.2    EMPIRICAL STRATEGY & EMPIRICAL MODEL

The amendment to Inequality (16.5) creates a reliable test of sustainability:

$$\sum_{i=1}^{M} \lambda i(t) Ki(t) = \left(1 + \alpha(t)\right)\left[-\sum_{j=1}^{N} \lambda j(t) Rj(t)\right], \tag{17.6}$$

where $(t)$ is a parameter which indicates whether an economy's net investment at time $t$ complies with the Hartwick rule. To separate genuine investment, reorganize (3.2.1):

$$\sum_{i=1}^{M} \lambda i(t) Ki(t) + \sum_{j=1}^{N} \lambda j(t) Rj(t) = -\alpha(t)\left[\sum_{j=1}^{N} \lambda j(t) Rj(t)\right], \tag{17.7}$$

However, if the left-hand side is positive, $(t) > 0$, implying that resource capital investment ($Rj$ 0) is not increasing. For some places and countries, genuine savings have been calculated to be negative. However, an economy's productive foundation "consists of society's capital assets and institutions at time $t$", according to Arrow et al. [20]. The shift in capital assets is captured by true savings measures, but institutions are not included. The impact of institutions and other variables on true savings can be discovered by decomposing $(t)$. That is, $(t)$ can be thought of as a function of social features; $(t) = f(Z)$, where $Z$ is a vector with values of $(1, Z1, Z2,\ldots, ZK)$. When you substitute into (18.6) and assume that $f(Z)$ has a linear functional form, you get:

$$\sum_{i=1}^{M} \lambda i(t) Ki(t) = \left(1 + Z'\alpha(t)\right)\left[-\sum_{j=1}^{N} \lambda j(t) Rj(t)\right], \tag{17.8}$$

where $(t)$ is a parameter vector that can be estimated. Equation (16.8) considers more than changes in capital assets to provide a more comprehensive treatment of actual investment. Genuine investment is now determined by social factors as well as changes in the value of resource capital. Let

$VR_t=(j=1)Nj(t)Rj(t)$ and $VK_t=(i=1)Mi(t)Kt(t)$ be the variables. $R_t$ decreases are expressed as a positive value in the data used in the estimations. As a result, (17.8) becomes:

$$VK_t - VR_t = Z'\alpha(t)VR_t. \tag{17.9}$$

Genuine investment (GI$_t$) is represented by $VK_t - VR_t$, and the sign of $Z'(t)$ denotes the sign of genuine investment. Finally, (16.9) may be recast to explicitly state how the various features in $Z$ decompose ($t$):

$$GI_t = \alpha_0 + \alpha_1 Z_{1t}VR_t + \alpha_2 Z_{2t}VR_t + \ldots + \alpha_K Z_{Kt}VR_t. \tag{17.10}$$

With a few tweaks, equation (17.10) can be used to make empirical estimates. First, you must include an error term. This will collect any data with unobservable attributes. In estimations, a panel data technique is used. The particular information that was used was previously discussed. The fixed effects (FE) estimator is used to account for unobservable time-invariant cross-sectional properties. Inside variation, or difference in the data inside this cross-section unit of observation, will be captured by the estimator (nations). To account for occurrences that may have impacted the entire sample, time effects might be applied. The FE estimate for (16.10) is as follows:

$$GI_{it} = y_i + y_t + \alpha_0 + \alpha_1 Z_{1it}R_{it} + \alpha_2 Z_{2it}R_{it} + \ldots + \alpha_K Z_{Kit}R_{it} + \varepsilon_{it}, \tag{17.11}$$

where $I$ represent the country index, $t$ represents the temporal index, GI represents the global index, $R_{it}$ is the return on an asset capital investment, $I$ is the nation FE, $t$ represents the time FE, $Z_{1it}$, $Z_K$ represents the features to be included in the estimation, and it is a non-observable mean-zero random variable. The calculated coefficients will show how the features of a country influence genuine investment (savings). For instance, consider Z1's marginal impact:

$$\frac{\partial GI}{\partial Z1} = \alpha_1 R. \tag{17.12}$$

This means that the impact is dependent on both the $R$-value and the estimated coefficient. $R$ works as a scaling factor in this fashion. According to projections, $Z$ will be harmed by a variety of governance metrics, including the extent of corruption. If Z1 is the corruption proxy in the preceding example, equation (17.12) reflects a percentage increase in determining factors due to a change in corruption.

### 17.3.3 DATA

The data set used in the estimations is made up of countries with a lot of energy. The term "energy-rich" was coined by Kalyuzhnova, Kutan, and Yigit [10]. A country should have at least 0.2% of proven world oil or gas reserves to be termed "energy-rich". The BP Statistical Review of World Energy provides these figures. Energy-rich countries are a fascinating case study because of the wide spectrum of growth of the economy they represent. Nonrenewable energy sources are one of the most well-known elements of data used to assess resource capital changes. The methods needed in evaluating genuine investment and its numerous subcomponents are outlined by Hamilton. Genuine investment is defined as gross national savings minus fixed capital consumption plus education expenditures (Educ) less the cost of resource depletion and environmental harm. There are two techniques to assess genuine investment: either with or without particle emission impacts (denoted GIX as well as GI, respectively).

The present value of growth in non-resource capital, VKit, is calculated using net national savings (NNS) and Educ. Gross national savings minus depreciation equals NNS. It tracks the economy's change in manufactured capital over a year. Education spending can be used as a proxy for human capital investment. Educ covers both running costs and educator pay. Tables 17.1 and 17.2 list quoted definitions of the series. The GDP deflator in the United States was used to convert all data to 2000 US dollars. To reflect the changes in the resource capital stock, several variables are used. Overall rents from the exploitation of crude oil, natural gas, and coal are calculated using energy depletion (Energy). Gold, zinc, nickel, bauxite, lead, copper, iron, phosphate, tin, and mineral depletion (Mineral) is a computation of mineral extraction rents. Discoveries are not included in the development of these metrics.

Fresh finds were not included in the stock of resource capital when these measures are made since the global supply is fixed. As a result, measurements only capture depletion. The term "net forest depletion" (Forest) refers to the rents generated by cutting timber at a rate faster than its natural growth rate. When harvest exceeds growth, the metric is set to zero. It's vital to understand

## TABLE 17.1
### Genuine Investment Component Definitions

| Variable | Definition |
|---|---|
| National Savings Net | Net national savings less fixed capital consumption plus transfers. |
| Expenditure on Education | Current educational running expenses, including capital investments in buildings and equipment, including personnel as well as wages. |
| Depletion of Energy | The total quantity of energy extracted in physical units plus the sum of unit resource rents. It includes, among other things, crude oil, natural gas, and coal. |
| Depletion of Minerals | The total of unit resource rents plus the physical quantity of minerals extracted. Minerals such as bauxite, copper, iron, lead, nickel, phosphate, tin, zinc, gold, and silver are listed. |
| Deforestation on a Net Basis | The surplus of Roundwood harvest above natural growth multiplied by unit resource rents. |
| The Harmful Effects of Carbon Dioxide | The cost of carbon dioxide pollution is estimated to be $20 per tonne (in 1995 dollars) multiplied by the number of tonnes emitted. |
| Damage Caused by Particulate Emissions | Calculated as the amount of money someone would be prepared to pay to avoid death as a result of particle pollution. |

## TABLE 17.2
### Genuine Investment Data Descriptive Statistics

|  | GI | GIX | NNS | EDUC | Energy |
|---|---|---|---|---|---|
| Mean | 51,34,47,253 | 47,05,36,995 | 38,77,00,339 | 18,75,96,589 | 10,51,48,496 |
| Median | 6,25,85,382 | 6,43,25,596 | 9,23,62,584 | 2,13,83,263 | 3,05,34,988 |
| Max | 13,56,79,03,612 | 13,86,02,59,395 | 15,99,05,33,489 | 5,64,43,86,760 | 2,75,89,54,842 |
| Min | −317861420 | −634565675 | −1138452286 | 23,376 | 0 |
| Std. Dev. | 1,44,36,77,879 | 1,36,43,11,979 | 1,00,36,36,379 | 64,00,84,459 | 22,93,04,189 |
| Obs. | 758 | 1,234 | 1,361 | 1,541 | 1,758 |
|  | Forest | CO$_2$ | Part dam | Mineral |  |
| Mean | 22,66,111 | 1,84,48,668 | 1,46,10,723 | 48,61,184 |  |
| Median | 0 | 31,06,275 | 18,76,323 | 52,645 |  |
| Max | 8,63,05,604 | 41,31,15,539 | 34,78,69,013 | 30,26,88,591 |  |
| Min | 0 | 0 | 0 | 0 |  |
| Std. Dev. | 89,51,223 | 5,18,36,264 | 4,42,45,687 | 1,76,74,823 |  |
| Obs. | 1,733 | 1,799 | 948 | 1,799 |  |

*Note:* The table does not include gross national savings, which are used to estimate actual investment.

**TABLE 17.3**

**Governance Measures Descriptive Statistics**

|           | Account | Corrupt | Effect  | Law     | Reg. Qual. | Stable | Polity  |
|-----------|---------|---------|---------|---------|------------|--------|---------|
| Mean      | −1.4638 | −1.1837 | −1.0988 | −1.2182 | −1.1566    | −1.35  | −1.0687 |
| Median    | −1.6839 | −1.3807 | −1.2689 | −1.4503 | −1.2408    | −0.3   | −4      |
| Max       | 2.8368  | 3.4539  | 3.2468  | 2.965   | 2.3455     | 0.455  | 10      |
| Min       | −3.1858 | −2      | −2      | −3.0575 | −3         | −3.09  | −10     |
| Std. Dev. | 2.037   | 2.0424  | 1.9943  | 0.0062  | 0.0308     | 1.989  | 7.7155  |
| Obs.      | 560     | 559     | 559     | 550     | 550        | 551    | 2305    |

*Note:* Polity is published by the Integrated Network for Societal Conflict Research. The World Bank's World Governance Indicators are used for all other governance variables.

that rentals are calculated using market prices. As a result, many resource-specific pricing may be artificially low, since externalities are not taken into consideration. Particulate matter damage (PartDam) is a metric for how much people are ready to pay to prevent death as a result of these kinds of pollution. The prepared-to-pay measurements are based on Pandey et al. [21]. In some computations, just energy, mineral, and forest depletion will be taken into account. The series' environmental damage measures have been deleted. Yemen, Iraq, Libya, Qatar, Turkmenistan, and also Myanmar are among the nations where data for any or all of the resource capital variables are insufficient.

When data is available for a country, it is typically accessible from 1970 to 2008. This is not the case for particle damage measured between 1990 and 2008. The descriptive statistics of the authentic investment data are presented in Table 17.3.

As for governance proxies, some factors are employed. Six governance measures are provided by the World Bank's World Governance Indicators (WGIs). Some of the themes covered are accountability and voice (Account), political stability and the absence of violence/terrorism (Stable), government effectiveness (Effect), regulatory quality (Regqual), rule of law (Law), and corruption control (Control) (Corrupt). For 1996, 1998, 2000, and 2002–2009, each indicator is calculated. These figures are derived from survey data. As a result, they capture how respondents feel about numerous indications. Kaufmann, Kraay, and Mastruzzi provide a more extensive explanation of how the measures are calculated (2010). Estimates will be based on each of these variables separately. Some of the WGI data's shortcomings are discussed in Section 17.3. The Integrated Network for Societal Conflict Research is the second source of governance data. Their Polity2 metric is employed as a proxy for an economy's level of democracy. Yemen, Libya, Myanmar, United Arab Emirates, Nigeria, Qatar, Poland, Australia, Canada, Germany, United Kingdom, and Denmark are excluded from the estimate when Polity is employed as a democracy proxy. The majority of countries' Polity metrics are from 1970 to 2008. Table 17.4 shows descriptive statistics for data on governance.

To see if there are any geographical or economic differences in the data set, dummy variables are used. The regional dummy variable for the Middle East and North Africa (MENA) will be used. There are data constraints for several of these countries, so these should not be included in estimates. The Organization for Economic Co-operation and Development (OECD) dummy will also adjust for high-income countries, except for Mexico and Poland.

## 17.4   RELATIONSHIPS TO BE EXPECTED

Many study findings can be used to form assumptions regarding the relationship between institutional quality and real investment. Stiglitz discusses the relevance of governance in long-term

**TABLE 17.4**

**Correlation Coefficients for Governance Variables in Pairs**

|            | Account | Effect | Law    | Polity | Reg. Qual. | Stable | Corrupt |
|------------|---------|--------|--------|--------|------------|--------|---------|
| Account    | 1       | 1.8044 | 1.7672 | 1.8587 | 1.8223     | 1.584  | 1.7768  |
| Effect     | 1.8044  | 1      | 1.9537 | 1.5158 | 1.9268     | 1.76   | 1.9528  |
| Law        | 1.7672  | 1      | 1      | 1.4286 | 1          | 1.814  | 1       |
| Polity     | 1.8687  | 1.5158 | 1.4286 | 1      | 1.5608     | 1.255  | 1.4503  |
| Reg. Qual. | 1.8334  | 1.9268 | 1.9016 | 1.5608 | 1          | 1.708  | 1.8897  |
| Stable     | 1.5844  | 1.7603 | 1.8134 | 1.2646 | 1.702      | 1      | 1.7556  |
| Corrupt    | 7768    | 1.9628 | 1.9636 | 1.4503 | 0.8989     | 0.757  | 1       |

development and sustainability in a persuasive way. Economic growth has been proven to be harmed by political volatility and instability. While the reasons for political unrest are numerous, economic factors also play a role.

Gylfason explores the phenomenon that countries with more abundant natural resources have higher degrees of corruption. Increases in corruption may boost real investment if they occur within a government framework that encourages it (public works projects). Mo shows, on either hand, that corruption affects private investment, meaning that higher levels of corruption reduce real investment. According to Aidt, Dutta, and Sena [22], a country's institutional context determines the impact of corruption on growth. Another significant institutional component is democracy. Citizens will be able to express their choices for the utilization of revenues from resource depletion vs. public investment as democracy develops [9]. The development and management of sovereign wealth funds, as well as education spending, are examples of how preferences might be communicated. There will be factors to suppose that the level of democracy has a favorable or negative impact on true savings. When the amount of democracy is low, it creates an increase in economic growth. Whereas when the level of democracy is high, a marginal increase causes a drop in growth, according to Barro. Democracy, according to Farzin and Bond [16], allows society to express its environmental preferences. Given the extent to which societal preferences can be expressed, the level of democracy will likely have an impact on actual savings [23].

## 17.5   DATA LIMITATIONS

The association between structural conditions is one of the study's potential flaws. A country's governance scores may trend in the same direction from year to year. These variables are established using surveys, as detailed in Section 17.1. The scores of the governance proxies may be highly connected, if they are evaluating the quality of government. To make matters worse, a few of the proxies evaluate the same qualities of governance. For example, the public's perception of the continuing stance for the private sector is measured by R equal, while policy quality is measured by Effect. The significant correlation among governance factors may produce multicollinearity in the estimations. Gujarati highlights a few of the probable repercussions that can happen when multicollinearity is present. The impact on calculated coefficients is particularly significant to our study. Because variances and co-variances are greater, t ratios are statistically insignificant. The predicted coefficients using Ordinary Least-Squares (OLS) can be sensitive to minor data changes, which is a second potential effect. "Pair-wise correlations may be a sufficient but not required condition for the occurrence of multicollinearity", Gujarati writes. Table 17.5 shows the pair-wise relationships among the institutional factors. Corrupt and Law have the strongest connection (0.95). Account and Stable have the lowest correlation (0.58) among the WGI. It's not unexpected that Account and Polity have such

**TABLE 17.5**

**Estimation with Resource Capital**

| Dep. Var. | GI | | | | | |
|---|---|---|---|---|---|---|
| Variable | Coefficient | Std. Error | Coefficient | Std. Error | Coefficient | Std. Error |
| C | 52597701** | 34970877 | −26932607*** | 16481592 | 12694264*** | 1.60E+06 |
| R | 1.5544*** | 0.1557 | −1.1741** | 0.0685 | −0.026 | |
| GI(-1) | – | | 1.1614*** | 1.0216 | 1.5669*** | 0.0423 |
| GI(-2) | – | | – | | −1.6582*** | 0.0503 |
| $R^2$ | 0.07217 | 0.9578 | | | | |
| adj. $R^2$ | 0.8067 | 0.9647 | 0.9632 | | | |
| F-stat | 50.9514 | 318.5525 | 389.2658 | | | |
| Durbin Watson (DW) stat. | 0.2354 | 1.035506 | 1.8665 | | | |
| N | 758 | 701 | 566 | | | |
| T | 1990–2008 | 1991–2008 | 1992–2008 | | | |
| M | 44 | 44 | 43 | | | |

*Note:* C stands for the entire regression intercept. Fixed-effects estimations with both cross-section and period effects yielded the results. The levels of significance are ***, **, and *, and are symbolized by the letters ***, **, and *, respectively. The letters N, T, and M stand for the number of observations, period, and cross-section units used, respectively.

a strong correlation (0.86), because they're both democracy indicators. Polity isn't as closely linked to the other WGI as Account [22].

More data series will be included in the research in the future, according to the author. This index assesses freedom in many areas, such as business, trade, and labor. The multicollinearity difficulties discussed in Section 17.5 may be eased by increasing organizational and institutional procedures. The use of any data series from various sources should reduce pair-wise correlation, providing a more realistic depiction of the impact of Z's true investment.

### 17.5.1 ESTIMATION AND RESULTS

This section will give estimates that will be used to test the hypotheses stated in the introduction. All governance variables are predicted to have an impact on real investment a priori. Despite the intended uniqueness of WGI, as explained in Section 17.3, there is a strong pair-wise correlation between them. The chance of multicollinearity affecting outcomes will be investigated. The connection between resources required and genuine investment is the first hypothesis to be explored. Given the concept of true investment, a negative association appears to be the most likely outcome. It's feasible that high levels of investment in energy-rich countries will be enough to counterbalance the predicted fall in genuine investment related to the depletion of resource capital before any governance variables are incorporated in the initial estimation. This phenomenon looks like this:

$$GI_{it} = y_i + y_t + \alpha_o + \alpha_1 R_{it} + u_{it}. \tag{17.13}$$

Tables 17.6 and 17.7 contain the results. The $R$ vector is used as the major explanatory variable in these calculations, with no other nation characteristics taken into account. The importance of it is affected by the presence of lags. There is no agreement on the quality of the link between GI and $R$. This could be because other drivers of actual investment aren't taken into account.

**TABLE 17.6**
**Regional and Development Categories**

| | Dep. Var.: GI | |
|---|---|---|
| **Variable** | **Coefficient** | **Std. Error** |
| C | 5948987** | 15006828 |
| R | 1.3221*** | 1.0831 |
| R*MENA | −1.3452** | 1.156 |
| R*OECD | −2.1400*** | 1.1308 |
| GI(-1) | 1.4828*** | 0.0344 |
| GI(-2) | −1.5356*** | 0.0388 |
| $R^2$ | 1.9677 | |
| adj. $R^2$ | 1.9677 | |
| F-stat | 433.8533 | |
| DW stat. | 1.9855 | |
| N | 566 | |
| T | 1992–2008 | |
| M | 43 | |

*Note:* C stands for the entire regression intercept. Fixed-effects calculations of both cross-section and period effects yielded the results. The levels of significance are ***, **, and *, and are symbolized by the letters ***, **, and *, respectively. The letters N, T, and M stand for the number of observations, period, and cross-section units used, respectively.

**TABLE 17.7**
**Estimation with Individual Governance Indicators**

| | Account | | Polity | | Corrupt | | Effect | |
|---|---|---|---|---|---|---|---|---|
| **Variable** | **Coeff** | **Std. Error** | **Coeff** | **Std. Error** | **Coeff** | **Std. Error** | **Coeff** | **Std. Error** |
| C | 138750015 | 28474341 | 79025808*** | 15219333 | 43821272 | 25637948 | −406914 | 2.50E+06 |
| R | −1.5210*** | 1.0981 | 1.0843 | 1.0612 | −1.2143*** | 1.0804 | 1.001 | 1.0765 |
| R*Govi | −1.5286*** | 1.0539 | −1.0653*** | 1.0055 | −1.4504*** | 1.0733 | −1.4616*** | 1.0859 |
| GI(-1) | 1.4681*** | 1.0518 | 1.5086*** | 1.0322 | 2.4832* | 1.0651 | 2.6517*** | 1.0649 |
| GI(-2) | −1.4746*** | 1.0517 | −1.5921*** | 1.0364 | −1.4062* | 1.0553 | −1.4644*** | 1.055 |
| $R^2$ | 0.7993 | 0.9717 | | 1.9757 | 1.9755 | | | |
| adj. $R^2$ | 0.967 | 0.9769 | 1.874 | 0.873 | | | | |
| F-stat | 400 | 542 | 257 | 256 | | | | |
| DW stat. | 2.0758 | 1.9483 | 1.8867 | 2.016 | | | | |
| N | 42 | 42 | 42 | 42 | | | | |
| T | 1996–2008 | 1992–2008 | 1996–2008 | 1996–2008 | | | | |
| M | 397 | 651 | 397 | 397 | | | | |

*Note:* The letter C denotes the whole regression intercept. The results of fixed-effects calculations integrating both cross-section and period effects are presented. The individual governance variable used in the estimation is called Govi. The letters N, T, and M denote the number of observations, period, and cross-section units used, respectively.

The positive correlation could be influencing the results, as indicated by the low Durbin–Watson value of 0.26 for the initial estimation. To account for the possibility of autocorrelation, lag GI terms have been incorporated. As a result, the remainder of the estimates will contain GI lagged twice. The $R^2$ measure is quite large due to the high explanatory power of these lagged factors. High $R^2$ values can be seen in all of the estimations in this section. The effect of environment and growth on GI is the second hypothesis to be examined. In this estimation, two types of dummy variables are used. For countries in the MENA region, a regional dummy is constructed. Tunisia, Iran, Oman, Saudi Arabia, Bahrain, Kuwait, and Egypt are all parts of the MENA area in the data set. Due to a lack of data, other countries, the areas identified as energy-rich in the region, are excluded. The OECD membership dummy is the second dummy used. The OECD dummy is likewise a high-income country control except for Mexico and Poland. The estimate is in the following format:

$$GI_{it} = y_i + y_t + \alpha_o + \alpha_1 R_{it} + \alpha_2 R * MENA_{it} + \alpha_3 R * OECD_{it} + \alpha_4 GI_{it-1} + \alpha_5 GI_{it-2} + u_{it}. \quad (17.14)$$

All of the coefficients of relevance are statically important in this specification. If a country is not part of the OECD or the MENA region, an increase in $R$ leads to an increase in GI. GI has a negative association with both the MENA and OECD dummies. This shows that differences in regional or developmental status among countries have an impact on the connection with GI. The MENA's and OECD's dismal indicators are not surprising. As per the model, OECD membership results in a lower GI than MENA membership. According to Hamilton (2006), many industrialized countries' GI is positive, and their production bases are growing with time, but several countries in the MENA have negative GI levels.

The following series of calculations attempt to represent the impact of governance on GI. This is in contradiction to the prior estimates' findings. A rise in resource depletion causes a decrease in GI when governance is adjusted for.

## 17.6   POTENTIAL WEAKNESS

The data used in this study come with a lot of limitations regarding how reliable the results are. It has been discovered that governance variables have a significant pair-wise connection. A longer time series would be excellent for research like this, which analyzes long-term development and growth difficulties. A longer time series would most likely catch more data variance. It would also capture a larger portion of the economic growth process. The availability of governance proxies is now constraining the time series. From 1996 onwards, the WGI is only available. A longer time frame would almost certainly provide a more realistic picture of how governance influences actual investment. In the theoretical model, the productive base accounts for all types of capital. Several key forms of capital must be removed due to a lack of data. Social capital, for example, is not taken into account. Educ are investments in human capital. Education spending is a common proxy for human capital investments, but the underlying benefit of human capital investment is an improvement in labor skills. Education spending does not account for changes in skills.

Changes in energy, minerals, and forests are measured by resource capital. Fisheries are an obvious exclusion. Environmental damages include the effects of particulate emissions and the expected consequences of climate change. Climate change impact measurements are fraught with uncertainty. The data on environmental damages do not take into account the quality of water and soil. Although the data used covers a large portion of an economy's productive base, it does not fully capture its changes.

## 17.7   CONCLUSION

The impact of governance on genuine investment has been investigated in different ways. Data was used to capture a variety of characteristics of governance, but it was ineffective. The WGIs assesses six major dimensions of governance: private and sector development, corruption, stability, quality, effectiveness, and democracy. Considering this, the indicators' high pair-wise correlation shows that they only measure one facet of governance. Individual WGI can be used to gauge democracy due to the consistency of the democracy results. This conclusion holds even when environmental effects are excluded from genuine investment and resource capital depletion measurements. In terms of the underlying link between resource depletion and true investment, these estimates are inconsistent. Because Polity has a larger time series and more observations, this inconsistency could be explained. In energy-rich OECD countries, estimates that take into account both the location of the region and the level of development demonstrate a negative link between resource depletion and real investment. For members of the MENA region, the conclusion holds, but the relationship's absolute value is lower. Other researchers, such as Arrow and Hamilton, have confirmed the findings relating to the MENA region (2006). The author intends to increase the governance features addressed in the study in the future by including additional data sets. The Heritage Foundation's Index of Economic Freedom and the Wall Street Journal's Index of Economic Freedom, for example, will be used. This data set could help to identify which characteristics of democracy are to blame for the skewed association between democracy as well as actual development. Analyzing the separate components of genuine investment is a second possible development. Measurements of investments, environmental effects, and resource depletion are all part of the genuine investment. To gain an insight into the mechanisms via which governance affects real investment, the effect of governance on such individual variables could be investigated.

## REFERENCES

[1] Solow, M. On the Intergenerational Allocation of Natural Resources. *The Scandinavian Journal of Economics* **2013**, *88* (1), 141–149.
[2] Sachs, J. D.; Warner, A.; Åslund, A.; Fischer, S. Economic Reform and the Process of Global Integration. *Brookings Papers On Economic Activity* **1995**, *1995* (1), 1–118.
[3] Sachs, J. D.; Warner, A. M. Sources of Slow Growth in African Economies. *Journal of African Economies* **1997**, *6* (3), 335–376.
[4] Sachs, J. D.; Warner, A. M. The Big Push, Natural Resource Booms and Growth. *Journal of Development Economics* **1998**, *59* (1), 43–76.
[5] Sachs, J. D.; Warner, A. M. The Curse of Natural Resources. *European Economic Review* **2001**, *45* (4–6), 827–838.
[6] Baumol, W.J. Productivity Growth, Convergence, and Welfare : What the Long-Run Data Show. *The American Economic Review* **1986**, *76* (5), 1072–1085.
[7] Sala, X. Nber Working Paper Series Public Finance in Models of Economic Growth. **1990**.
[8] Paper, W.; Eni, F.; Mattei, E. www.econstor.eu. **2004**.
[9] Gylfason, T.; Zoega, G. Natural Resources and Economic Growth: The Role of Investment (No. 2001–02). EPRU Working Paper Series. **2001**.
[10] Kalyuzhnova, Y.; Kutan, A. M.; Yigit, T. Corruption and Economic Development in Energy-Rich Economies. *Comparative Economic Studies* **2009**, *51*, 165–180. https://doi.org/10.1057/ces.2008.46.
[11] Christa, N.; Erwin, H. www.econstor.eu. **2006**.
[12] Resources, N.; Development, E. Natural Resources, Education. No. 2594. https://cepr.org/publications/dp2594
[13] Paper, W.; Growth, E.; Enrico, F. E. www.econstor.eu. **2006**.
[14] Ayres, R. U.; Ayres, R. U.; Ayres, R. U.; Benjamin, W. Eco-Thermodynamics: Economics and the Second Law. *Ecological Economics* **1998**, *26* (2), 189–209.
[15] Pearce, D.; Hamilton, K.; Atkinson, G.; Pearce, D.; Hamilton, K.; Atkinson, G. Measuring Sustainable Development: Progress on Indicators. *Environment and Development Economics*. **2008**, *1*, 85–101. https://doi.org/10.1017/S1355770X00000395.
[16] Farzin, Y. H.; Bond, C. A. Democracy and Environmental Quality. *Journal of Development Economics* **2006**, *81*, 213–235. https://doi.org/10.1016/j.jdeveco.2005.04.003.
[17] Hamilton, K.; Ruta, G.; Tajibaeva, L. Capital Accumulation and Resource Depletion: A Hartwick Rule Counterfactual. *Environmental and Resource Economics* **2006**, *34*, 517–533. https://doi.org/10.1007/s10640-006-0011-2.

[18] Hamilton, K.; Clemens, M. Genuine Savings Rates in Developing Countries. *The World Bank Economic Review* **1999**, *13*, 333–356.

[19] Arrow, K. J.; Dasgupta, P. Evaluating Projects and Assessing Sustainable Development in Imperfect Economies. *Environmental and Resource Economics* 2003, *26*, 647–685.

[20] Arrow, K.; Dasgupta, P.; Goulder, L.; Daily, G.; Ehrlich, P.; Heal, G.; Levin, S.; Schneider, S.; Starrett, D.; Walker, B. Are We Consuming Too Much ? *Journal of Economic Perspectives* **2004**, *18* (3), 147–172.

[21] Hamilton, K.; Bank, W. Investing Exhaustible Resource Rents and the Path of Consumption. *Canadian Journal of Economics/Revue canadienne d'économique* **2005**, *38* (2), 615–621.

[22] Dutta, J.; Dodlova, M. Governance Regimes, Corruption and Growth: Theory and Evidence Related Papers. *Journal of Comparative Economics* **2008**. https://doi.org/10.1016/j.jce.2007.11.004.

[23] Barro, R. I. © 1994 by Robert J. Barro. All Rights Reserved. Short Sections of Text, Not to Exceed Two. 1994, No. 4909.

# 18 IoT-Based Power Theft Detection
## *Mini Review*

*Ameesh Singh, Harsh Gupta, and O.V. Gnana Swathika*
Vellore Institute of Technology

*Aayush Karthikeyan*
University of Calgary

*Akhtar Kalam*
Victoria University

## CONTENTS

18.1 A Comparative Study of Multiple Machine Learning Techniques for Detecting Electricity Theft................................................................................206
18.2 Power Theft Identification and Alert System Using IoT.....................................207
18.3 Energy Monitoring and Theft System Using IoT ...............................................207
18.4 Hardware Implementation of Power Theft Detection System and Disconnection Using Smart Meter...............................................................................208
18.5 Theft Detection Using Other Technologies ........................................................210
18.6 Conclusion ..........................................................................................................210
References...............................................................................................................211

### ABSTRACT

Electrical power theft is a significant issue in the global power system, and it is illegal and should be avoided at all costs. Power theft is defined as the usage of electricity without permission from the provider. A big problem is that we can never monitor how a theft occurred because the ways in which thefts might occur are limitless, and this issue should have been realized as early as feasible under the circumstances. To eliminate control theft, the area of the theft's intensity must be identified so that appropriate actions against legitimate offenders can be taken. The client's bills are increased as a result of the theft of electricity. We reduce human participation in electrical energy support in this context. This framework will be used to detect power theft. In this chapter, we will look at and plan an IoT-based energy meter theft alert system.

## KEYWORDS

Internet of Thing (IoT); Electric Power; Power Theft; Theft Alert

DOI: 10.1201/9781003374121-18

## 18.1 A COMPARATIVE STUDY OF MULTIPLE MACHINE LEARNING TECHNIQUES FOR DETECTING ELECTRICITY THEFT

Smart grids that integrate information and energy flows can help to reduce power theft. The smart grid is used in providing a massive amount of data and thus helps in detecting the abnormal pattern in the electricity consumption. Based on the Wide and Deep Convolutional Neural Networks (CNNs) model, a novel electricity theft detection (ETD) approach is developed [1].

Electricity theft results in irregular power consumption and causes harm to the power grid. In order to address this issue, this research presents a CNN-random forest (CNN-RF) model for automatic detection of power theft. A CNN uses convolution and downsampling to learn the differences in features across various times of the day [2].

A detector based on a deep neural network is suggested, that can detect electricity theft cyber-attacks efficiently. Sequential grid analysis is used for better performance in the theft detection process. Better results are seen using this detection process compared to other detectors [3].

ETD is proposed here using Supervised Learning Techniques and real electricity data. Adasyn algorithm is used, since the distribution data of electricity consumption is imbalanced. It is found that the proposed method is good in handling large datasets and detecting electricity thieves [4].

Electricity theft is a key source of concern for utilities, since it has a significant impact on the grid. To work on it in the best way possible, this paper describes combinations of various machine learning algorithms and data handling techniques. ROC is a probability curve and AUC represents the degree or measure of separability. Different machine learning algorithms, for example Logistic Regression (LR) and Random Forest, have been implemented and then results are computed, and the Area Under ROC Curve (AUC) and F1 score are compared to check which combination provides the best results [5].

Through data analysis techniques, smart grid adaption can considerably reduce this loss. The smart grid system creates a large amount of data, including individual user power use. With the help of this data, the ML and DL techniques may rightly identify electricity thieves. This study proposes an electrical theft detection system based on CNN and long short-term memory (LSTM) architecture. CNN is frequently used in technologies that automate power like feature extraction and classification. This study also used a novel data pre-processing strategy to find and evaluate missing chunks in the dataset [6].

The mechanism employed by businesses and academia to identify electricity theft is called ETD. The ETD, on the other hand, cannot be utilized effectively due to reasons such as unbalanced data, processing of high-dimensional data, and overfitting difficulties. As a result, this study presents a remedy to the aforementioned restrictions. To detect aberrant patterns in electricity consumption data, a LSTM algorithm is used along with the random under-sampling boosting (RUSBoost) algorithm for optimizing parameters. Support Vector Machine (SVM), Convolutional Neural Network (CNN), and LR are examples of state-of-the-art approaches that outperform the proposed method [7].

This study proposes a deep residual neural network-based model for power theft, which is in turn based on the timing and periodicity trait of smart meter data to mine the enormous amount of data offered by smart meters in a better way and improve the accuracy of power theft detection models. The proposed method has a high detection rate. Different studies on real datasets validate that the proposed model has a higher detection rate as compared to others as well as low false-positive rate [8].

A brief analysis of machine learning research in power theft detection using the data of smart meters is offered in this study. It then summarizes and compares different algorithms utilized in terms of indicators, simulation and analytic environment, and sets of data used in these studies. At the end, it emphasizes the difficulties in detecting power theft. It also acknowledges that these issues have not yet been sufficiently paid attention to or explored in prior studies and that they must be solved to enhance power theft detection [9].

In this research, a gradient boosting theft detector is created for detecting electrical theft using the three most recent gradient boosting classifiers in extreme gradient boosting, categorical boosting, and mild gradient boosting approaches. The XGBoost algorithm is an ML algorithm that provides great accuracy in a short amount of time. The smart meter data are pre-processed and then feature selection is performed in this algorithm [10].

## 18.2 POWER THEFT IDENTIFICATION AND ALERT SYSTEM USING IoT

One of the most serious problems that power companies face is electricity theft. The construction of a cost-effective power theft detection system using the Internet of Things (IoT) is presented in this paper. The IoT is a type of technology. The microcontroller board Arduino MKR1000 is used as a bi-functional component to coordinate operations. A website has been made to keep a visual check on the status of the meter and detect theft if any [11].

The distribution network undergoes a lot of challenges including power theft being the major one, since it results in unbalanced loads and thus impacting the utilities. An IoT-based solution has been developed and described here to deal with the power theft issue. It will sense the power theft in a line by comparing the sending and receiving end current values and thus alert the authority to take the actions if the theft is detected. It also keeps a check on the energy consumption and alerts the authority if it exceeds [12].

This paper describes a strategy for locating power theft and taking appropriate action both under normal circumstances and when power theft is identified. When the controller detects a theft, it sends a signal to the circuit breaker to turn off the power and re-check for the theft. If the theft is not detected after four attempts, the system is reset; otherwise, an alert message with the stolen location is issued [13].

Smart grid technologies and smart meters combined with Information Communication Technology (ICT) give a reliable way to identify power theft. The use of IoT in power theft detection and real-time smart meter monitoring is discussed in this study. The Android application is used to track consumer consumption and inform authorities in the event of power theft [14].

In this paper, it is planned to create a web-based prepaid energy meter with theft prevention. Anyone can replenish their electrical needs according to their needs using this approach. This system will provide the user with real-time access to peak load, power theft, etc. IoT is used to provide the user interface with the real-time information. Location of the theft can also be located. This solution provides a convenient solution to access energy meter [15].

Power theft can take various forms, one of which is a registered customer circumventing the meter (i.e., tampering with the meter to make it read less or no consumption) or connecting around the meter to a live cable on the company side of the meter. To eradicate electricity theft, it is vital to first detect it [16].

The purpose of the proposed project is to create a system that can track the amount of electricity used per load, as well as trace and eradicate electricity theft in the current line and meter. This endeavor also comprises informing Electricity Board (EB) officials about the theft that occurred as a result of IoT. The transmission of real-time data through the Internet is aided by a network of connected devices, such as sensors. The Arduino Uno is used to detect energy theft and send the data to the Global System for Mobile communication (GSM) module, which subsequently sends the data to the EB [17].

Detecting and eradicating power theft in third-world countries like India are incredibly difficult due to infrastructure and human restrictions. This paper describes an IoT solution for monitoring and eliminating power theft, identifying distribution system faults, and upgrading existing grids to smart grids that can be added to existing power grid and discom infrastructure [18].

## 18.3 ENERGY MONITORING AND THEFT SYSTEM USING IoT

In today's linked world, the development of intelligent gadgets has ushered in a new era of innovation. The development has opened new gateways of machine-to-machine communication with

Internet linked to it. This paper describes the design and implementation of energy meter utilizing IoT concept. Controls can be maintained from a distance and any theft if detected will result in automatically switching off of power [19].

IoT plays an important and significant role in smart grid. This paper throws light on the deepening of IoT technology and its application in the power system. An IoT-based status monitoring and early warning system for the power distribution network is proposed. The system can be used in order to display the status of all equipment in the distribution network, and on-line theft monitoring can be done [20].

Electric power is the lane to digitalization. With the increasing demand, many people are undergoing illegal practices of stealing power. This results in massive amount of losses. This paper describes how these losses can be worked upon if IoT technology is incorporated [21].

The paper describes the digitalization of the load energy usage and power theft using various hardware and software tools. Using this smart energy monitoring, the human intervention will be reduced. Using a channel ID the user can monitor the energy consumption using a webpage. The power theft detection is monitored using an infrared (IR) sensor and is fed to the Raspberry Pi which then relay it to the webpage [22].

In this paper, the current values are measured at the shaft and at the meter using current curl and loop. If the current at both points is found to be equivalent, then there is no robbery; else power robbery is detected. The difference in units is sent to the predefined site page using the Wi-Fi module. At the theft point a SMS is sent to the concerned specialist with the house number [23].

The goal of this research is to use IoT to measure power usage in a family unit and automatically generate a bill. The information of electricity usage can be updated automatically, reducing the amount of effort required by humans. The microcontroller is used to enable workouts with the sophisticated metering system and for connecting the system to the Internet and the server. A latent IR sensor is integrated into the framework to detect any unauthorized changes in the metering system. The proposed framework is capable of accurately detecting the amount of energy utilized and improves smart home security [24].

This study focuses on to develop a system for monitoring load consumption and to detect and eliminate power theft in transmission lines and energy meters. This initiative also focuses on using IoT to communicate theft information to the EB. A network of connected devices, such as sensors, has the potential to communicate real-time information via the Internet. Power theft in this system is detected using Raspberry Pi which then sends commands to the GSM module and the GSM module sends the theft information message to the EB [25].

With the transformation in the electric industry, the dependability of power supply and the quality of energy are receiving more attention. Electricity providers and users are both concerned about reliable electricity, regardless of whether the focus is on intrusions, aggravations, or protracted power outages. The primary hurdle to delivering quality control to valued subscribers is power theft [26].

The author of this work has proposed a way for conveniently monitoring power theft. The goal of the project is to build a load-based power monitoring system. This method is also used to send information about theft to the electrical board using an IoT application called Nexmo. The Raspberry Pi is at the heart of the entire network. Liquid Crystal Display (LCD), current sensor, and Analog to Digital Converter (ADC) are interfaced to the Raspberry Pi. The Raspberry Pi is used in power theft detection and to relay the information to Nexmo, which delivers that information to the EB internally [27].

## 18.4 HARDWARE IMPLEMENTATION OF POWER THEFT DETECTION SYSTEM AND DISCONNECTION USING SMART METER

Earlier times, when analog meters were used, permanent magnets were used to slow down its rotation. Later, in digital meters, power was looted through bypassing the meter. Smart meters with communication network and database management system (DBMS) as discussed here can help to locate and thus minimize the power theft [28].

Arduino and GSM-based smart energy meters are proposed in this paper. These smart energy meters can read and send billing and metering data via a wireless network utilizing GSM technology and a GSM modem. Smart energy meters stores the data over the cloud and thus the user can access it using the mobile app. These smart meters will be cost-friendly and at the same time with the advanced technology can also monitor the consumption and power theft [29].

A smart meter is a costly and reliable technology upgrade of an existing meter. Smart meter consists of various monitoring features and communication features inside the meter, but alongside these are in a threat of cyber-attacks which can disrupt the distribution in the area. The paper proposes the smart energy meter with Advanced Encryption Standard (AES) algorithm. The data generated by the meter are stored using AES algorithm in the cloud, and thus the data are protected from cyber-attacks [30].

Smart meters provide a benefit of detecting power theft based on the consumption patterns. The energy theft detector detects theft by making use of prediction on customers' regular and malicious consumption patterns. Distribution transformer meters are used to locate areas with a high potential of energy theft, and suspicious clients are discovered by monitoring abnormalities in usage patterns [31].

With the increasing usage of smart meters, the frequency at which the household data are collected has increased, and thus this opens doors for advanced data analysis. This work describes a temperature-based model for detecting electricity theft in a given area using data from smart meters and distribution transformers [32].

For a long time, illegal electricity use has been a serious worry in the power system industry. An imbalanced demand–supply gap could occur from fraudulent large-scale electricity consumption. With the increasing usage of smart meters and improved metering infrastructure, this study developed a data-driven electricity theft detector based on random matrix theory. The suggested method's important step is the use of an augmented matrix as the data source, which indicates the relationship between power consumption and system operating states under abnormal conditions of electricity use [33].

Due to the power losses constantly at different spots, in this paper a framework is proposed in which one can ceaselessly screen the power theft utilizing sensors. The input voltage and the yield voltage drawn are also checked. The sensor values are verified on a regular basis. If a power burglary is detected, the customer receives an alarm message through the GSM Module. The alarm message includes the timing of the break-in as well as the current sensor values [34].

Theft of electricity has been identified as a major danger to meter infrastructure. Because this issue poses a significant danger to the whole grid architecture, there has been a lot of research done on it. This study explores the most common thefts and dangers in such a setting, as well as the critical safety requirements that a smart meter must meet. To determine important hazards, a risk model is created utilizing attack characteristics. Various forms of power theft detecting strategies are investigated and analyzed in order to precisely identify and take into account the attack vulnerability in smart meters [35].

This study shows how to build a low-cost method for detecting and preventing electricity theft based on IoT technologies. To unlock the feature in the distribution network and to the supply cutoff from the customer, the required data requirements for smart meters and distribution substations are defined. It becomes possible to generate bill for power utilized and also to identify those who have not paid. A load control meter is an intelligent energy meter that has a novel invention from load control [36].

Advances in energy meter research have resulted in the development of a number of novel and cost-effective technologies for the construction of energy meters, which assist to improve both the billing and payment systems at the same time. In this work, a new energy meter concept is introduced using embedded technology, a power factor meter, and a instrument transformer. The proposed energy meter is intelligent enough to function as a billing and payment system as well [37].

## 18.5   THEFT DETECTION USING OTHER TECHNOLOGIES

Energy theft is an increasing problem with the growing consumers. This theft results in a lot of revenue losses untraced. In this work, a new process based on the Microcontroller Atmega328P is used to identify and control energy meter power theft and to fix the problem by remotely disconnecting and reconnecting a specific consumer's service (line). A to-and-fro message will be communicated between the utility central server and the microcontroller, resulting is disconnecting the unauthorized supply [38].

Theft detection can be done at a cost-friendly way using Zigbee module. The irregular consumption pattern signal and the regional/individual theft detection algorithm are meant to identify regions with a high chance of theft and pick clients suspected of fraud. An alarm will be raised to the actual user if the theft is detected [39].

A power theft detection system is proposed here which detects the theft by periodically measuring the current. The GSM/General Packet Radio Services (GPRS) module is used to send data from the distributor box to the server database. And similarly the electric meter current reading is also posted in the server database using the same module. These values are then compared and if there is a marginal difference then theft is detected [40].

In a smart grid setting, the authors propose a framework based on Universal Anomaly Detection and the Lempel–Ziv universal compression method. The statistics on energy usage, the rate of change in energy consumption, the date stamp, and the time signatures are all monitored by this method. This method can give alerts in case any difference in pattern from the normal data is observed [41].

The design and implementation of an energy meter using the PIC18F46k22 microcontroller and the IoT concept are presented in this paper. The suggested system architecture eliminates human involvement in electricity maintenance. When an energy meter is tampered, the theft detection unit linked to it will alert the company, delivering theft detection information through Programmable Logic Controller (PLC) modem and displaying the theft on the firm's terminal window. The IoT operation is carried out by the Wi-Fi unit, which sends energy meter data to a webpage that can be viewed through IP address [42].

This research offers a detection method for energy theft attempts in advanced metering infrastructure based on principal component analysis approximation (AMI). The AMI data are reconstructed using principal components, which are then utilized to determine relative entropy. The proposed method compares the similarity of two probability distributions created from a reconstructed consumption dataset using relative entropy [43–46].

The present power grid has evolved into the smart grid. Increased use of contemporary technology, combined with advances in high-speed connectivity and low-cost sensors, gives utilities more information to operate the grid. It consists of a two-way communication system in which the consumer and utility exchange electricity and information in order to maximize efficiency. With the use of smart transformers and smart energy meters, the control center ensures that the smart grid optimizes circuit Volt-Amps Reactive (VAR) flow and voltages, allowing power theft to be tracked [46–50].

## 18.6   CONCLUSION

In today's world, each and every machine takes a shot at electricity from the oscillating brush to tremendous engines. An existence without electricity is impossible today. With the headway of innovation tremendous measure of electricity is created today. This electricity can be transmitted to far away remote spots to be utilized by the general population. The intense interest of electricity has made it a multi-billion-dollar industry. It is the biggest business in the whole world with an interconnection of machines like none other.

Proposed IoT-based smart energy meter surveillance concerning IoT, IoT as an emerging topic, and IoT-based gadgets have ushered in a revolution in electronics and information technology. The primary goal of this initiative is to raise knowledge about energy usage and how to utilize home appliances more efficiently to save energy. The current electricity billing system has significant flaws due to manual labor. Using IoT, the suggested system will provide information on meter readings as well as power cuts when power use exceeds the specified limit. With the help of the GSM module, the Arduino esp8266 microcontroller is programmed to achieve the goals. It is intended to eliminate all of the flaws in the current energy meter. All of the information is delivered to the consumer's phone via IoT and GSM, and it is also shown on the LCD. It helps to avoid human intervention when using IoT and thus saves time.

## REFERENCES

[1] Zheng, Z., Yang, Y., Niu, X., Dai, H.N. and Zhou, Y., 2017. Wide and deep convolutional neural networks for electricity-theft detection to secure smart grids. *IEEE Transactions on Industrial Informatics*, *14*(4), pp. 1606–1615.

[2] Li, S., Han, Y., Yao, X., Yingchen, S., Wang, J. and Zhao, Q., 2019. Electricity theft detection in power grids with deep learning and random forests. *Journal of Electrical and Computer Engineering, 2019*, p. 4136874.

[3] Ismail, M., Shahin, M., Shaaban, M.F., Serpedin, E. and Qaraqe, K., 2018, April. Efficient detection of electricity theft cyber attacks in AMI networks. In *2018 IEEE Wireless Communications and Networking Conference (WCNC)* (pp. 1–6). IEEE.

[4] Khan, Z.A., Adil, M., Javaid, N., Saqib, M.N., Shafiq, M. and Choi, J.G., 2020. Electricity theft detection using supervised learning techniques on smart meter data. *Sustainability, 12*(19), p. 8023.

[5] Pereira, J. and Saraiva, F., 2020, July. A comparative analysis of unbalanced data handling techniques for machine learning algorithms to electricity theft detection. In *2020 IEEE Congress on Evolutionary Computation (CEC)* (pp. 1–8). IEEE.

[6] Hasan, M., Toma, R.N., Nahid, A.A., Islam, M. and Kim, J.M., 2019. Electricity theft detection in smart grid systems: A CNN-LSTM based approach. *Energies, 12*(17), p. 3310.

[7] Adil, M., Javaid, N., Qasim, U., Ullah, I., Shafiq, M. and Choi, J.G., 2020. LSTM and bat-based RUSBoost approach for electricity theft detection. *Applied Sciences, 10*(12), p. 4378.

[8] Chen, Y., Hua, G., Feng, D., Zang, H., Wei, Z. and Sun, G., 2020, September. Electricity theft detection model for smart meter based on residual neural network. In *2020 12th IEEE PES Asia-Pacific Power and Energy Engineering Conference (APPEEC)* (pp. 1–5). IEEE.

[9] Maamar, A. and Benahmed, K., 2018, January. Machine learning techniques for energy theft detection in AMI. In *Proceedings of the 2018 International Conference on Software Engineering and Information Management* (pp. 57–62).

[10] Bamane, R., Vinod, M., Shah, J., Ahuja, S. and Sahariya, A., 2020. Smart meter for power theft detection using machine learning. *International Journal of Scientific Research and Engineering Development, 3*(1), pp. 526–528.

[11] Ogu, R.E. and Chukwudebe, G.A., 2017, November. Development of a cost-effective electricity theft detection and prevention system based on IoT technology. In *2017 IEEE 3rd International Conference on Electro-Technology for National Development (NIGERCON)* (pp. 756–760). IEEE.

[12] Rajagiri, A.K., Ajitha, A. and Thalluri, A.K., 2021, January. Development of an IoT based solution for Smart Distribution Systems. In *2021 International Conference on Sustainable Energy and Future Electric Transportation (SEFET)* (pp. 1–6). IEEE.

[13] Uvais, M., 2020, February. Controller based power theft location detection system. In *2020 International Conference on Electrical and Electronics Engineering (ICE3)* (pp. 111–114). IEEE.

[14] Jeffin, M.J., Madhu, G.M., Rao, A., Singh, G. and Vyjayanthi, C., 2020, July. Internet of things enabled power theft detection and smart meter monitoring system. In *2020 International Conference on Communication and Signal Processing (ICCSP)* (pp. 0262–0267). IEEE.

[15] Ananth, S., Parthasarathy, S., Kala, R. and Banumathi, S., 2021. Web based prepaid energy meter with theft control. *European Journal of Molecular & Clinical Medicine, 7*(11), pp. 7693–7697.

[16] Leninpugalhanthi, P., Janani, R., Nidheesh, S., Mamtha, R.V., Keerthana, I. and Kumar, R.S., 2019, March. Power theft identification system using IoT. In *2019 5th International Conference on Advanced Computing & Communication Systems (ICACCS)* (pp. 825–830). IEEE.

[17] Kumaran, K., 2021. Power theft detection and alert system using IOT. *Turkish Journal of Computer and Mathematics Education (TURCOMAT), 12*(10), pp. 1135–1139.

[18] Nalinaksh, K., Pathak, L. and Rishiwal, V., 2018, February. An internet of things solution for real-time identification of electricity theft and power outrages caused by fault in distribution systems (converting existing electrical infrastructure of third world countries to Smart Grids). In *2018 3rd International Conference On Internet of Things: Smart Innovation and Usages (IoT-SIU)* (pp. 1–8). IEEE.

[19] Uddanti, S., Joseph, C. and Kishoreraja, P.C., 2017. IoT based energy metering and theft detection. *International Journal of Pure and Applied Mathematics, 117*(9), pp. 47–51.

[20] Qing-Hai, O., Zheng, W., Yan, Z., Xiang-Zhen, L. and Si, Z., 2013, October. Status monitoring and early warning system for power distribution network based on IoT technology. In *Proceedings of 2013 3rd International Conference on Computer Science and Network Technology* (pp. 641–645). IEEE.

[21] Sirisha, B.L., 2020. Minimizing electricity theft using IOT. *SAMRIDDHI: A Journal of Physical Sciences, Engineering and Technology, 12*(SUP 3), pp. 49–52.

[22] Saritha, K. and Murthy, A.S.N., 2020. Energy monitoring and theft system using IoT. *International Journal of Research, 8*(6), pp. 1705–1711.

[23] Shravani, K. and Reddy, G.S., 2018. Implementation of embedded web server based power theft detection and smart monitoring system. *International Journal of Advanced Technology and Innovative Research, 10*(1), pp. 0068–0070.

[24] Vineeth, V.V., Ambrish, V., Haricharann, D.V., Harshini, V. and Abilash, C., 2021, May. Power theft recognition and data security in smart meter reading of a smart grid. In *Journal of Physics: Conference Series* (Vol. 1916, No. 1, p. 012216). IOP Publishing.

[25] Meenal, R., Kuruvilla, K.M., Denny, A., Jose, R.V. and Roy, R., 2019, November. Power monitoring and theft detection system using IoT. In *Journal of Physics: Conference Series* (Vol. 1362, No. 1, p. 012027). IOP Publishing.

[26] Rajashekar, M., Govardan Kumar, R.R. and Rashmi, K., 2021. Monitoring the power theft using IoT. *Journal of Advances in Computing and Information Technology, 1*(1), pp.46–53.

[27] Reddy, B.S., Sanjana, C.M., Reddy, C.P. and Suraj, D., Power monitoring and theft detection system using Raspberry Pi and IoT application, pp. 1–7.

[28] Sathyapriya, R. and Jeyalakshmi, V., 2020. Hardware implementation of IOT based energy management theft detection and disconnection using smart meter. *Malaya Journal of Matematik, 5*(2), pp. 4177–4180.

[29] Mir, S.H., Ashruf, S., Bhat, Y. and Beigh, N., 2019. Review on smart electric metering system based on GSM/IOT. *Asian Journal of Electrical Sciences, 8*(1), pp. 1–6.

[30] Manu, D., Shorabh, S.G., Swathika, O.G., Umashankar, S. and Tejaswi, P., 2022, May. Design and realization of smart energy management system for Standalone PV system. In *IOP Conference Series: Earth and Environmental Science* (Vol. 1026, No. 1, p. 012027). IOP Publishing.

[31] Swathika, O.G., Karthikeyan, K., Subramaniam, U., Hemapala, K.U. and Bhaskar, S.M., 2022, May. Energy Efficient Outdoor Lighting System Design: Case Study of IT Campus. In *IOP Conference Series: Earth and Environmental Science* (Vol. 1026, No. 1, p. 012029). IOP Publishing.

[32] Sujeeth, S. and Swathika, O.G., 2018, January. IoT based automated protection and control of DC microgrids. In *2018 2nd International Conference on Inventive Systems and Control (ICISC)* (pp. 1422–1426). IEEE.

[33] Patel, A., Swathika, O.V., Subramaniam, U., Babu, T.S., Tripathi, A., Nag, S., Karthick, A. and Muhibbullah, M., 2022. A practical approach for predicting power in a small-scale off-grid photovoltaic system using machine learning algorithms. *International Journal of Photoenergy, 2022*, 9194537.

[34] Odiyur Vathanam, G.S., Kalyanasundaram, K., Elavarasan, R.M., Hussain Khahro, S., Subramaniam, U., Pugazhendhi, R., Ramesh, M. and Gopalakrishnan, R.M., 2021. A review on effective use of daylight harvesting using intelligent lighting control systems for sustainable office buildings in India. *Sustainability, 13*(9), p. 4973.

[35] Swathika, O.V. and Hemapala, K.T.M.U., 2019. IOT based energy management system for standalone PV systems. *Journal of Electrical Engineering & Technology, 14*(5), pp. 1811–1821.

[36] Swathika, O.V. and Hemapala, K.T.M.U., 2019, January. IOT-based adaptive protection of microgrid. In *International Conference on Artificial Intelligence, Smart Grid and Smart City Applications* (pp. 123–130). Springer, Cham.

[37] Kumar, G.N. and Swathika, O.G., 2022. 19 AI Applications to. *Smart Buildings Digitalization: IoT and Energy Efficient Smart Buildings Architecture and Applications*, p. 283.

[38] Swathika, O.G., 2022. 5 IoT-Based Smart. *Smart Buildings Digitalization: IoT and Energy Efficient Smart Buildings Architecture and Applications*, p. 57.

[39] Lal, P., Ananthakrishnan, V., Swathika, O.G., Gutha, N.K. and Hency, V.B., 2022. 14 IoT-Based Smart Health. *Smart Buildings Digitalization: Case Studies on Data Centers and Automation*, p. 149.

[40] Chowdhury, S., Saha, K.D., Sarkar, C.M. and Swathika, O.G., 2022. IoT-based data collection platform for smart buildings. In *Smart Buildings Digitalization* (pp. 71–79). CRC Press.

[41] Abdulrahaman Okino Otuoze, M.W.M., Sofimieari, I.E., Dobi, A.M., Sule, A.H., Abioye, A.E. and Saeed, M.S., 2019. Electricity theft detection framework based on universal prediction algorithm. *Indonesian Journal of Electrical Engineering and Computer Science*, 15, pp. 758–768.

[42] Sharma, R., Kumawat, S.B. and Saini, M.K., 2016. IoT based theft detection and power optimization in electricity energy meter reading using PLC modem. *International Journal of Business & Engineering Research*, 10, pp. 1–7.

[43] Singh, S.K., Bose, R. and Joshi, A., 2019. Energy theft detection for AMI using principal component analysis based reconstructed data. *IET Cyber-Physical Systems: Theory & Applications*, 4(2), pp. 179–185.

[44] Kumar, S.N., Rao, S.K., Raju, M.S., Trimurthulu, S., Sivaji, K. and Reddy, T.R.M., 2017. IoT based control and monitoring of smart grid and power theft detection by locating area. *International Research Journal of Engineering and Technology*, 4(7), pp. 1923–1927.

[45] Shahid, M.B., Shahid, M.O., Tariq, H. and Saleem, S., 2019, July. Design and development of an efficient power theft detection and prevention system through consumer load profiling. In *2019 International Conference on Electrical, Communication, and Computer Engineering (ICECCE)* (pp. 1–6). IEEE.

[46] Gupta, A.K., Mukherjee, A., Routray, A. and Biswas, R., 2017. A novel power theft detection algorithm for low voltage distribution network. In *IECON 2017–43rd Annual Conference of the IEEE Industrial Electronics Society* (pp. 3603–3608). IEEE.

[47] Zheng, K., Chen, Q., Wang, Y., Kang, C. and Xia, Q., 2018. A novel combined data-driven approach for electricity theft detection. *IEEE Transactions on Industrial Informatics*, 15(3), pp. 1809–1819.

[48] Wei, L., Sundararajan, A., Sarwat, A.I., Biswas, S. and Ibrahim, E., 2017, September. A distributed intelligent framework for electricity theft detection using Benford's law and Stackelberg game. In *2017 Resilience Week (RWS)* (pp. 5–11). IEEE.

[49] Konstantinos, B. and Georgios, S., 2019. Efficient power theft detection for residential consumers using mean shift data mining knowledge discovery process. *International Journal of Artificial Intelligence and Applications (IJAIA)*, 10(1), pp. 69–85.

[50] Gupta, A.K., Routray, A. and Naikan, V.A., 2020. Detection of power theft in low voltage distribution systems: A review from the Indian perspective. *IETE Journal of Research*, 68(6), pp. 4180–4197.

# 19 Design and Implementation of Bluetooth-Enabled Home Automation System

*Nagavindhya Nagavindhya, Krithikka Jayamurthi,*
*V. Berlin Hency, and O.V. Gnana Swathika*
Vellore Institute of Technology

*Aayush Karthikeyan*
University of Calgary

*K.T.M.U. Hemapala*
University of Moratuwa

## CONTENTS

Abbreviations ........................................................................................................216
19.1   Introduction ............................................................................................216
19.2   Home Automation Using Different Modules.......................................217
19.3   Advanced Home Automation Systems ...............................................217
19.4   Home Automation Using Bluetooth Module HC-05 ........................220
19.5   Power Supply .........................................................................................221
19.6   Main Circuit............................................................................................221
19.7   Output .....................................................................................................222
19.8   Conclusion ..............................................................................................227
References..............................................................................................................228

### ABSTRACT

The main objective of the paper is home automation system using Bluetooth module and Arduino uno. In this study, we have made lights and fans be controlled by the app which is installed in our mobile phones. Single Pole Double Throw (SPDT) relays are used to control the amount of current flowing to the applications. The relays are connected to Arduino uno. We have done the coding part in Arduino IDE software which is installed in our device. Bluetooth module HC-05 is the connection between Arduino uno and the mobile phone. This module provides Universal Asynchronous Receive Transmit (UART) connection. LCD display is used as an indicator to the user about the working of the devices. This proposed system is useful for elders and physically handicapped persons to use it because it is tough for them to walk and use switches often. Since we have our phone at all time, we can switch ON/OFF the lights and fans remotely through our phone at all times.

### KEYWORDS

Home Automation System; Arduino Bluetooth App; Arduino uno; Bluetooth Module; AC Step-Down Transformer

DOI: 10.1201/9781003374121-19

## ABBREVIATIONS

| | |
|---|---|
| **BCI** | Brain–Computer Interface |
| **DTW** | Dynamic Time Warping |
| **EEPROM** | Erasable Programmable Read-Only Memory |
| **GSM** | Global System for Mobile Communication |
| **IoT** | Internet of Things |
| **LDR** | Light-Dependent Resistor |
| **Li-Fi** | Light Fidelity |
| **LoRa** | Long-Range Radio |
| **MQTT** | Message Queuing Telemetry Transport |
| **OLED** | Organic Light-Emitting Diode |
| **PLC** | Programmable Logic controller |
| **Wi-Fi** | Wireless Fidelity |

## 19.1   INTRODUCTION

Smart home automation system is the booming technology in terms of electronics. Physically handicapped and elder people who are weak and unable to use switches can use this new technology for their convenience. Lights and fans can be controlled by connecting light-dependent resistor (LDR) and passive infrared (PIR) sensor to the Arduino uno which is made of a microcontroller ATMEGA328P; based on the intensity of light sensed by LDR and the movement sensed by PIR sensor, lights and fans can be turned ON and OFF automatically [1]. Home automation system using Internet of Things (IoT) is very convenient and effective, and it can be used in many applications but it's high time to adopt them [2]. Web-based home automation system controls the appliances based on the web page. Raspberry Pi is used for controlling the appliances by receiving input from the web page and act accordingly [3]. A grid eye sensor is to see if people are present in house, and if present, then automatically all the appliances will work, and no need to turn ON/OFF the switch. And if want to turn ON/OFF a specific device, then Android app can be used for it. In case of emergency, the Global System for Mobile Communication (GSM) module sends message to the user and warns them. Also, if someone wants to come out of house, GSM module is used for this purpose [4]. The garden light, outside light, motor and garden motor are controlled by using programmable logic controller (PLC) module. It will store the instructions and perform the task accordingly. If watering the plants is scheduled at 7 AM and running the motor at 4 PM, then a microcontroller can be used to control these appliances; thereafter one can use the PLC module to register the time at which task is in the process. Then any task would be performed without manual intervention [5]. Graphical User Interface (GUI) controls the appliances via phone in remote areas. Microcontroller is used to get input from user. In case of power cut, erasable programmable read-only memory is used to store data. If a person wants to turn OFF the lights, which is turned on unnecessarily for a long period of time, then he/she can use the app where the lights in his home are connected to that app. One can use buttons in the app to turn OFF the lights. The input from that app is sent to microcontroller. And from the microcontroller, it can send to appliances [6]. Message Queuing Telemetry Transport (MQTT) protocol is used by IoT to control the devices. MQTT is used to transfer the information between two devices by Transmission Control Protocol/Internet Protocol (TCP/IP) protocol. When someone wants to turn OFF some appliances, then one can send message in our phone so that it can be transferred to main circuit by MQTT. With the help of MQTT, one can turn OFF the device easily [7]. This paper proposes a three-phase system which can classify, extract and feature the data. It is based on one-dimensional deep convolutional neural network, forecasting system for long short-term memory, a scheduling algorithm based on phase 2. These proposed schemes work efficiently which can control the devices and keep count of energy consumption [8]. Data transmission rate is an essential part in terms of efficiency and user's satisfaction. In order to fulfil this, Light Fidelity

(Li-Fi) technology is introduced to promote fast data transmission for instant function of the appliances. Comparing to Wireless Fidelity (Wi-Fi) speed, Li-Fi gives the 100 times faster internet connection as light travels at high speed to provide faster transmission of data [9]. High accuracy and less consumption are implemented in home automation system using dynamic time warping (DTW) where a guard sensor is selected which is common among all sensors. The function of the sensor is to reduce the energy consumption and extend the battery life to 137 days. There are many groups of neural networks which employ the recurrent neural network and DTW techniques [10].

## 19.2   HOME AUTOMATION USING DIFFERENT MODULES

Wireless home automation system using Bluetooth module is used to control the lights and prevent overflow of tank using photoresistor and transistor via Arduino uno [11]. Appliances can be turned ON and OFF by Google Assistant or text through phone, Wi-Fi module acts as an intermediate between the appliances and the phone; people present not only inside, but also outside the house can also use their phone to control the appliances in case if they forget to ON/OFF [12]. Safety and security play a critical role in this system, door can be locked and unlocked by typing the password in it and GSM module is used to send the status of the door to the user for safety purpose. During emergency situation, GSM module sends user about the condition and it also alarm the neighbour about the emergency condition if any abnormal sound is heard like screaming, noise, unusual movements, smoke and presence of fire in the home for immediate evacuation [13]. Wi-Fi communication acts as an interface between the phone and the appliances via voice control, Raspberry Pi controls the appliances based on the inputs given by the user via phone as it has in-built Bluetooth and Wi-Fi module [14,15]. Developing social media like Facebook Messenger for controlling the appliances like lights and fans even when we are not present in house is simple and effective as extending the social apps to an advanced level makes people easy to use it, as they are familiar in using those apps [16,17].

ARM Processor can be used to control the appliances by receiving instruction through phone via GSM module. Phones are used to send input to the user via Messager (SMS), and the ARM processor (LPC11U24) receives the input and controls the appliances [18]. Arduino mega 2560 is used for controlling the appliances by mobile appand the instructions are transferred by ESP 32 Wi-Fi module. More devices can be controlled using Arduino [19]. An economic wireless home automation system with more efficiency acts as a profit for users, as they spend less amount but get better results. Interfacing Zigbee and Raspberry Pi for low-power application yields better results [20,21]. Home automation system is performed by Xbee to provide wireless connectivity to device [22]. Cloud computing is the emerging technology in today's scenario. Command is sent through a website and the main circuit is controlled by Wi-Fi module [23]. Efficiency of smart home automation system can be increased by calculating the distance between the main device and the control; Long-Range Radio (LoRa) is used for calculating the distance, and based on the analysis, Bluetooth is used for short-range and server-based LoRa is used [24]. Protocols like RESTful and MQTT are used to integrate the small devices and sensors to provide modularity and flexibility, and a comparative analysis is made between developed smart home framework and smart things from Samsung in terms of user-friendliness, security and compactivity [25].

## 19.3   ADVANCED HOME AUTOMATION SYSTEMS

Using auditory steady-state response can control the home appliances for those who feel difficulty in speech and for motionless people. Connecting our brain and the external device can be done by Arduino uno and cortex headset. By receiving the impulse from brain, one can control the devices. If a person's arms and legs are motionless, then he can just remain himself to turn on the light, so that it's possible to analyse our brain signal by cortex headset, and it will analyse the signal type. Then the command is sent to Arduino uno, and Arduino uno will turn ON the fan accordingly [26]. By Arduino mega, we have connected many sensors across it which measures temperature, humidity, light and motion. The data collected by the sensor are sent for the pattern analysis and stored in it. By analysing the data, it will turn ON/OFF the appliances. If the surrounding light is less, then the LDR sensor senses the data to the pattern analyser, and according to the pattern, the lights are turned on [27]. Chat box API is used to control the appliances. If one can type anything in the chat box, then the message is sent to the controlling unit by Hypertext Transfer Protocol Secure transmission lines. The controlling unit analyzes the problem and sends an instruction to the main circuit unit through MQTT protocol to control the appliances. If he/she wants to turn off the fan when she is outside the house, then she can type the command in chat box API. Then it will analyse the command, and it is sent to the main circuit board through MQTT protocol and the fans are offed automatically [28]. Audio IoT device is used to record the sound of the home environment throughout the day; the data are sent to the machine learning where it is divided to small levels and classified into various levels. If some gunshots, screams and glass breaks happen, then the audio IoT senses it and turns on the emergency alarm so that the neighbour people may know about it. If there is a gunshot, then the sound is recorded and it is used to alarm the neighbour about the emergency so that the neighbours can be alerted [29]. Using a smart door locking system using radio-frequency identification card and password, the temperature and humidity of the home are displayed on screen to ensure that user can see it and turn ON/OFF the device by IoT; user also have gas and fire alarm to ensure safety. If someone wants to open the main door, then he/she should be aware of the password; without password, we can't open the door. If there is a fire in a house, then the alarm is turned to alert neighbour and the housemate [30].

Smart mirror is used to interact with users and displays information like weather, date, news, etc. It can also open a website to book cabs. We can also interact with smart mirrors by voice control or by hand gesture. Smart mirror is nothing but a big screen which is used to display the choices which we have to use by voice control, and by hand gesture we can choose our own convenience and do it accordingly [31]. Wireless home automation system is used to make the appliances to run effectively; user can use microcontroller to get input from the user and the mobile phone which is used control the appliances by the user. If someone wants to turn off the heater when user is out, then one can turn it off in mobile phone itself by sending the command via phone. The command sent by phone is received by a microcontroller. Then the microcontroller will turn off the heater [32]. Ontology-based system is used to represent the gesture and function of the appliances accordingly. If someone feels difficulty in speech, he/she can use gestures to turn ON/OFF the appliances. Then the main circuit understands the gesture and turn OFF the device accordingly. Gestures like using head and hands are allowed [33]. One can control the appliances by voice control, for example, the user can control the intensity or the speed of the appliances by voice. A wheel chair is controlled by voice so that it can be accessible anywhere in the house. An Android app is used to take the audio as input via Google Assistant. If the weather is moderate and medium, then we can adjust the fan speed to mode 3. Then we can adjust the fan speed to mode 3; if it is hot, then it will adjust to 5. If some handicapped people in home, then without anyone help they can use wheel chair by voice [34]. Home appliances and wheel chairs are controlled by air gestures. Like for turning light ON/OFF, one can use our hand gestures so that the algorithm can identify the type of gesture and do the operation accordingly. If someone wants to turn on the lights, they can use the gesture like nodding the head, so that the algorithm converts our action into command and send that to main circuit so that the lights will be on [35].

In this paper, with the help of smart cities, by combining the mobile application with the Arduino using Bluetooth or Wi-Fi, people can control their own home at any time with ease. It helps to level up the usability security and is helpful to achieve the main goal of the system. This paper also discusses about connecting intelligent systems to the new technology to make things faster and better. In this system, users have an option to choose either Bluetooth or Wi-Fi according to the range of distance and signal [36]. This paper proposes the design, deployment, development and a prototype for the secure, wireless home automation system which uses the latest version of OpenHAB for overall home security and to maximize productivity. In this system, Raspberry Pi Model B and Arduino mega 2560 interfaced with a 16-channel relay. Advanced Encryption Standard and JSON

Web Token interface procedures used for data encryption and authentication. This provides addition layer of security to the system. Both web and mobile applications were developed to view and control the status of home appliances [37]. In this paper, few examples of home automation and feasibility and effectiveness of IoT giving to the system are discussed. This paper discusses about how the household appliances are controlled using low-cost hardware board integrated with these software applications. This paper also proposes the internal mechanism of architecture of the system. Security, cost and challenges faced while implementation is also highlighted in the paper. Finally, they come up with solution of low-cost home automation [38]. A non-intrusive approach of collection of data from IoT devices is used for personalization. By using big data analysis and machine learning, one can integrate all the info and access it with ease. There are some open-source frameworks, for example Apache Spark, Azure, etc. This paper elaborated about how the big data analytics and machine learning (ML) are applied to predict behaviour of users [39]. Using the mobile application, IoT user can track the status of the device. This system provides response to physically disabled persons, and also caretakers are informed about their conditions simultaneously. Assistive demotics, a type of home automation system, has a wide range of features that can benefit people with specific accessibility issues in their houses. These technological systems and aiding equipment have become a feasible choice for elders who choose to remain in their homes, rather in an assisted living facility. Home appliances are controlled by hand gestures using glove-based home automation system. Simple movements used while wearing the glove can operate household equipment. A mobile app is created to keep track of the state and use of gadgets by members of house. As a result, the system delivers comfort to the differently abled while also keeping caregivers informed [40].

To connect IoT with household appliances, a voice-controlled artificial intelligence system is deployed. Putting in place a safe and intelligent voice-based system allows for seamless control of many home devices. This research focuses on three significant roadblocks: a lack of confidence due to security and privacy concerns, and consumers' inexperience with how to use machine intelligence to fully harness the smartness potential. Voice-based home automation systems are now implemented easily because of cloud-based systems that is present everywhere nowadays. However, there are demerits of using local speech channels and orders, as well as delays in fulfilling the needed reaction time for real-time services. To accomplish the intended results of simplicity, security and integration, adoption is also necessary. To address these limitations, we offer a paradigm for establishing safe home automation system using a voice control-based Artificial Intelligence (AI) system that incorporates IoT services and wireless technologies. This study addressed the embedding of IoT in a well-planned infrastructure paradigm applied over several platforms in faraway locations with AI voice control [41]. Home automation system using IoT interacts with different types of renewable energy resources and realizes about maximizing the security, convenience, cost, energy utilization and clean environment for the society. The main purpose of the study is to achieve the green and sustainable environment. A survey is conducted among the people to study the influence of home automation system and identify the problems. For the sustainable development, they took the results using IBM SPSS statistics version 23. It includes the present development of wireless system like GSM, Wi-Fi, Bluetooth, Voice Recognition, Zigbee which are widely used during practical implementation of the home automation approach [42]. This paper is about the study of advanced security solutions for IoT-based prototypes. It provides solutions for users and intercommunication with the technology and discusses about basic security requisites and secure intercommunication of IoT devices. According to surveys, it is important to secure storage systems as they are prone to hackers. Personal data are equally protected and managed by cloud service provider and data users. Cloud computing is linked to this home automation system, which takes data and detects the issues, and we further use this data to get solution for the hurdles [43]. Disabled person can be benefitted by steady-state visual-evoked potential (SSVEP) and eye blink signal-based home automation system by brain–computer interface (BCI) approach. It is a friendly channel configuration which has high accuracy, multiple commands and short response time, and might also offer a reference for the other BCI-controlled applications. They used CNN algorithms and Fourier transforms. Eye blink signals and SSVEP help the disabled. This experiment included two modules for analysing and processing electroencephalography (EEG) signals [44]. This paper discusses about comparison of 20 open-source home automation systems. It also provides ideas to change the existing system to make it a more efficient one. It would be helpful for developers who wish to improve and enhance the open-source home automation system.

This OpenHAB has a powerful result in all categories. They used two-phase study with extra special quality for home automation. In first phase study, it's about case-based analysis and the second phase is used to perform criteria-based analysis. This strengthens the open-source home automation [45]. Home automation system uses separate bridges for each network and OpenHAB, where each bridge links with a single master Wi-Fi gateway, providing a single window of control through an application or a web interface for an integrated Smart Home. In this paper, a cost-effective system is proposed; the one challenge faced by industrial phase is compatibility. So, they came up with two methods namely, threshold-based and Neural Network-based. The evaluation of this system is based on cost, life span of battery and recharge time [46]. The actuators and sensors were connected to the NodeMCU to continuously control and monitor it through the web interface. A versatile, low-cost and energy-efficient environmental monitoring tool is provided through an intuitive interface based on IoT. The suggested system may quickly and efficiently operate IoT-based devices for home automation and enhance home safety by operating autonomously, conserving energy and assuring occupants' desired comfort and safety. For safety and security reasons, an email is also sent to notify us if there is a hazard within the house. Devices are readily and effectively controlled via the web page, either by manual ON/OFF operation or voice control. Green energy is a fantastic choice for conserving energy. Rechargeable batteries, photovoltaic electricity, fuel cells, and wind turbines are just a few examples [47].

Speed of the fan and water level of the tank are adjusted automatically by temperature sensor and transistor, and PIC16F877A microcontroller is used to prevent fire accidents and protect the house owners and neighbours against fire accidents. Additional facilities are provided by notifying house owners, so they can investigate the problem [48]. Using the CNN model, we predict the gesture with the help of front- and back-end processing and assign the task to take place at the home interface. In this work, an Android application is used to create a household automation model that allows for easy management of home appliances. With this system, especially, the aged and physically handicapped people can carry out their daily duties with ease. Accelerometers have been used to track activity in the past, and while they are accurate, they are neither flexible nor portable. The suggested system recognizes the user's movements as input and controls the home appliance. The client interface is in charge of utilizing an Android application to capture the user's input gesture and transfer it to the Raspberry Pi server. It serves as a crucial pre-processor. Back-end processing includes picture pre-processing, CNN model training and image class category prediction for the input gesture image [49]. This paper presents an IoT-based hardware with reduced cost using voice command by the user using Google Assistant and the Blynk app. The ESP8266 NodeMCU microcontroller board serves as the system's key component. When utilized as smart home automation, the system is activated by a voice command delivered by the user via Google Assistant, in conjunction with a co-interfaced IFTTT (If-Then That) and the Blynk app platform, which enabled this capability. The central unit is connected to the multichannel relay module for loads. The data are processed by the central unit according to the user's input. Environmental factors such as soil moisture, air temperature and humidity (DHT11) have been assessed and shown on the organic light-emitting diode screen when utilized for agricultural monitoring. A water motor was used to manage the flow of water based on the soil moisture threshold value. The ESP8266 Wi-Fi-based NodeMCU uses the Hypertext Transfer Protocol [50].

## 19.4   HOME AUTOMATION USING BLUETOOTH MODULE HC-05

The main objective of the paper is home automation system using Bluetooth module HC-05. In this study, lights and fans can be controlled by the Android app which is installed in our mobile phone.

Single Pole Double Throw (SPDT) relays are used to control the amount of current flowing to the applications. The relays are connected to Arduino uno which controls the devices by the commands of user. Coding part is done using Arduino IDE software which is installed in our pc. Bluetooth module HC-05 is the connection between Arduino uno and the mobile phone.

It provides Universal Asynchronous Receive Transmit (UART) connection. LCD display is used as an indicator to the user about the working of the devices.

## 19.5 POWER SUPPLY

Figure 19.1 shows the block diagram of power supply. A step-down transformer is used to convert 230 V supply to 12 V to meet the requirement of the system. Diode-based bridge rectifier converts the supply to DC, as the appliances work in DC. Electrolytic capacitor is used to eliminate the repels in pulsating DC supply, and LM7805 regulator is used to maintain a standard voltage to the appliances as shown in Figure 19.2.

## 19.6 MAIN CIRCUIT

Figure 19.3 shows the hardware setup of proposed system. In this paper, by using our mobile phone, which is used to control the lights and fans through Bluetooth module as shown in Figure 19.4. The Bluetooth module is connected to the Arduino uno (as shown in Figure 19.5) which controls the function of the lights and fan. Figure 19.6 is the relay connection, which ensures a standard current flowing through it. LCD screen is used to indicate the status of the devices as shown in Figure 19.6.

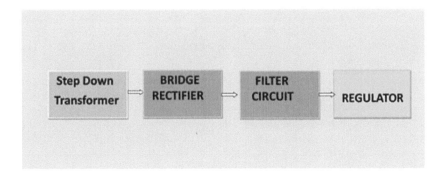

**FIGURE 19.1**   Block diagram of power supply.

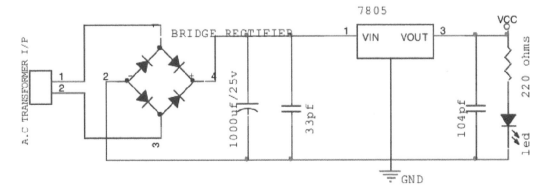

**FIGURE 19.2**   Circuit diagram of power supply.

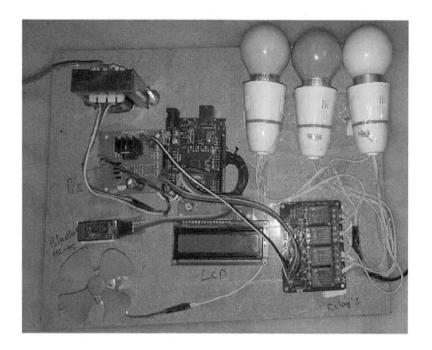

**FIGURE 19.3**   Hardware setup of the proposed system.

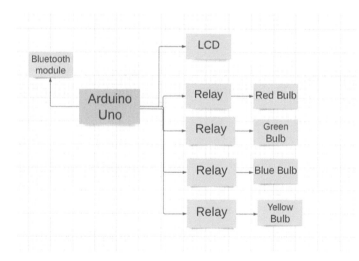

**FIGURE 19.4**   Block diagram of main circuit.

## 19.7   OUTPUT

Figure 19.7 indicates how the interface of the app occurs. In Figure 19.8, Bulb_1 is selected in the app. Figure 19.9 shows that bulb_1 is ON. Next, as shown in Figure 19.10, Bulb_1,3 and fan are selected in the app. Figure 19.11 shows that bulb_1,3 and fan are ON now. Figure 19.12 indicates that all the options are selected. Figure 19.13 indicates that Bulb_1,2,3 and fan are ON. Figure 19.14 shows that none of the options are selected. Further, Figure 19.15 shows that all the appliances are OFF and LCD is turned OFF.

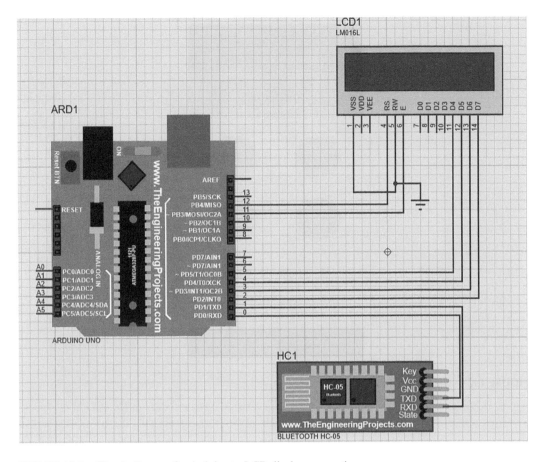

**FIGURE 19.5** Circuit diagram for Arduino to LCD display connection.

High Switching Current

SPDT Relay

normally closed relay

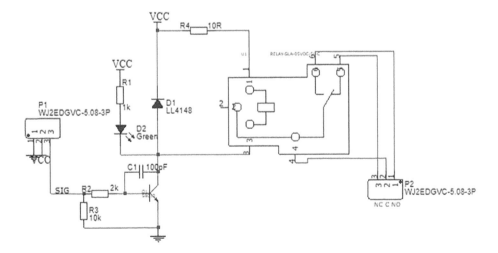

**FIGURE 19.6** Circuit diagram of relay connection.

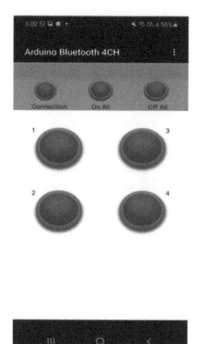

**FIGURE 19.7** Interface of the app.

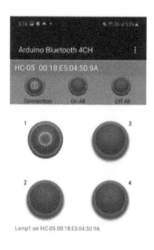

**FIGURE 19.8** Image showing Bulb_1 is selected in the app.

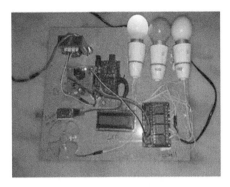

**FIGURE 19.9** Image showing bulb_1 is on.

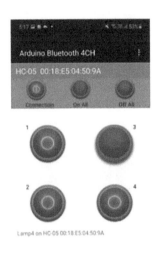

**FIGURE 19.10** Image showing Bulb_1,3 and fan are selected in the app.

**FIGURE 19.11** Image showing bulb_1,3 and fan are on.

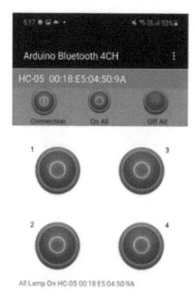

**FIGURE 19.12**   Image showing all the options are selected.

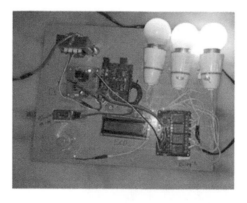

**FIGURE 19.13**   Image showing Bulb_1,2,3 and fan are on.

This paper explains how to use Arduino uno, IDE and Tinkercad software and also learn about the different types of sensors, Bluetooth module and relays and their specification. Initially Tinkercad software is used for simulation of the basic circuit, followed which the hardware model is realized which provided suitable results. And also able to note the energy consumption, this helps to save energy. Through this paper, one can learn how to do hardware and creative projects which helps to advance the present technology in an eco-friendly way.

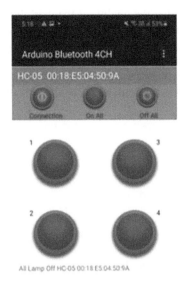

**FIGURE 19.14**    Image showing no option is selected.

**FIGURE 19.15**    Image showing all the appliances are OFF and LCD is turned off.

## 19.8   CONCLUSION

The overall concept of the ideas presented is to make the work efficient and useful for physically challenged and elder people in accessing the appliances. The proposed ideas use modules like Arduino uno, Arduino mega, Raspberry Pi, ARM architecture, ESP826 and Bluetooth module for controlling the appliances by receiving proper input from the user. Moreover, advanced research studies have been made in improving the quality of the work like pattern recognition and receiving input from the brain through impulse so that people with physical disabilities and cannot speak can use the home appliances without anyone help. LoRa technology is used to analyse the distance between the device and the user, which is used accordingly in other technologies like Li-Fi. Overall, this paper gives a good idea of different modules; technologies used in home automation system make it a more efficient, convenient, cost-efficient, and eco-friendly environment to humans.

## REFERENCES

[1] Naing, M. and Hlaing, N.N.S., 2019. Arduino based smart home automation system. *International Journal of Trend in Scientific Research and Development (IJTSRD)*, *3*(4), pp. 276–280.

[2] Abdulraheem, A.S., Salih, A.A., Abdulla, A.I., Sadeeq, M.A., Salim, N.O., Abdullah, H., Khalifa, F.M. and Saeed, R.A., 2020. Home automation system based on IoT. *Technology Reports of Kansai University*, *62*(5), pp. 2453–2464.

[3] Bepery, C., Baral, S., Khashkel, A. and Hossain, F., 2019. Advanced home automation system using Raspberry-Pi and Arduino. *International Journal of Computer Science and Engineering*, *8*(2), pp. 1–10.

[4] Hanif, M., Mohammad, N. and Harun, B., 2019, May. An effective combination of microcontroller and PLC for home automation system. In *2019 1st International Conference on Advances in Science, Engineering and Robotics Technology (ICASERT)* (pp. 1–6). IEEE.

[5] Khan, M.S., Ahmed, T., Aziz, I., Alam, F.B., Bhuiya, M.S.U., Alam, M.J., Chakma, R. and Mahtab, S.S., 2019, November. PLC based energy-efficient home automation system with smart task scheduling. In *2019 IEEE Sustainable Power and Energy Conference (iSPEC)* (pp. 35–38). IEEE.

[6] Haque, M.E., Islam, M.R., Rabbi, M.T.F. and Rafiq, J.I., 2019, December. IoT based home automation system with customizable GUI and low cost embedded system. In *2019 International Conference on Sustainable Technologies for Industry 4.0 (STI)* (pp. 1–5). IEEE.

[7] Cornel-Cristian, A., Gabriel, T., Arhip-Calin, M. and Zamfirescu, A., 2019, September. Smart home automation with MQTT. In *2019 54th International Universities Power Engineering Conference (UPEC)* (pp. 1–5). IEEE.

[8] Khan, M., Seo, J. and Kim, D., 2020. Towards energy efficient home automation: a deep learning approach. *Sensors*, *20*(24), p. 7187.

[9] Mansingh, P.M. and Yuvaraju, M., 2021. Improved data transmission using Li-Fi technology for home automation application. *Journal of Ambient Intelligence and Humanized Computing*, *12*(5), pp. 5581–5588.

[10] Khan, M., Seo, J. and Kim, D., 2021. Modeling of intelligent sensor duty cycling for smart home automation. *IEEE Transactions on Automation Science and Engineering*, *19*(3), pp. 2412–2421.

[11] Umer, M. and Khan, M.M., 2020. Smart home automation using ATMEGA328. *Advanced Journal of Science and Engineering*, *1*(3), pp. 86–90.

[12] Vishwakarma, S.K., Upadhyaya, P., Kumari, B. and Mishra, A.K., 2019, April. Smart energy efficient home automation system using IoT. In *2019 4th International Conference on Internet of Things: Smart Innovation and Usages (IoT-SIU)* (pp. 1–4). IEEE.

[13] Ajao, L.A., Kolo, J.G., Adedokun, E.A., Olaniyi, O.M., Inalegwu, O.C. and Abolade, S.K., 2018. A smart door security-based home automation system: an internet of things. *SciFed Journal of Telecommunication*, *2*(2), pp. 1–9.

[14] Zaidi SF, Shukla VK, Mishra VP, Singh B. Redefining home automation through voice recognition system. In *Emerging Technologies in Data Mining and Information Security 2021* (pp. 155–165). Springer, Singapore.

[15] Rajput, H., Sawant, K., Shetty, D., Shukla, P. and Chougule, A., 2018. Implementation of voice based home automation system using Raspberry Pi. *International Research Journal of Engineering and Technology*, *5*(5), pp. 2771–2776.

[16] Parthornratt, T., Kitsawat, D., Putthapipat, P. and Koronjaruwat, P., 2018, July. A smart home automation via Facebook Chatbot and Raspberry Pi. In *2018 2nd International Conference on Engineering Innovation (ICEI)* (pp. 52–56). IEEE.

[17] Ashraf, I., Umer, M., Majeed, R., Mehmood, A., Aslam, W., Yasir, M. N., & Choi, G. S. (2020). Home automation using general purpose household electric appliances with Raspberry Pi and commercial smartphone. *PLoS One*, *15*(9), p. e0238480.

[18] Manda, V.B., Kushal, V. and Ramasubramanian, N., 2018. An elegant home automation system using GSM and ARM-based architecture. *IEEE Potentials*, *37*(5), pp. 43–48.

[19] Susany, R. and Rotar, R., 2020. Remote control android-based applications for a home automation implemented with Arduino Mega 2560 and ESP 32. *Technium*, 2(2), pp. 1–8.

[20] Younis, S.A., Ijaz, U., Randhawa, I.A. and Ijaz, A., 2018. Speech recognition based home automation system using Raspberry Pi and Zigbee. *NFC IEFR Journal of Engineering and Scientific Research*, 5, pp. 40–45.

[21] Taiwo, O., Ezugwu, A.E., Rana, N. and Abdulhamid, S.I.M., 2020. Smart home automation system using Zigbee, Bluetooth and Arduino technologies. In *International Conference on Computational Science and Its Applications* (pp. 587–597). Springer, Cham.

[22] Sarah, A., Ghozali, T., Giano, G., Mulyadi, M., Octaviani, S. and Hikmaturokhman, A., 2020. Learning IoT: Basic experiments of home automation using ESP8266, Arduino and XBee. In *2020 IEEE International Conference on Smart Internet of Things (SmartIoT)* (pp. 290–294). IEEE.

[23] Raju, S.H., Rao, M.N., Sudheer, N. and Kavitharani, P., 2018. IOT based home automation system with cloud organizing. *International Journal of Engineering & Technology*, 7(2.32), pp. 412–415.

[24] Islam, R., Rahman, M.W., Rubaiat, R., Hasan, M.M., Reza, M.M. and Rahman, M.M., 2021. LoRa and server-based home automation using the internet of things (IoT). *Journal of King Saud University-Computer and Information Sciences*, 34(6), pp. 3703–3712.

[25] Parocha, R.C. and Macabebe, E.Q.B., 2019, November. Implementation of home automation system using OpenHAB framework for heterogeneous IoT devices. In *2019 IEEE International Conference on Internet of Things and Intelligence System (IoTaIS)* (pp. 67–73). IEEE.

[26] Shivappa, V.K.K., Luu, B., Solis, M. and George, K., 2018, May. Home automation system using brain computer interface paradigm based on auditory selection attention. In *2018 IEEE International Instrumentation and Measurement Technology Conference (I2MTC)* (pp. 1–6). IEEE.

[27] Iyer, R. and Sharma, A., 2019. IoT based home automation system with pattern recognition. *International Journal of Recent Technology and Engineering*, 8(2), pp. 3925–3929.

[28] Tseng, C.L., Cheng, C.S., Hsu, Y.H. and Yang, B.H., 2018, October. An IoT-based home automation system using Wi-Fi wireless sensor networks. In *2018 IEEE International Conference on Systems, Man, and Cybernetics (SMC)* (pp. 2430–2435). IEEE.

[29] Shah, S.K., Tariq, Z. and Lee, Y., 2018, December. Audio IoT analytics for home automation safety. In *2018 IEEE International Conference on Big Data (Big Data)* (pp. 5181–5186). IEEE.

[30] Nitu, A.M., Hasan, M.J. and Alom, M.S., 2019, May. Wireless home automation system using IoT and PaaS. In *2019 1st International Conference on Advances in Science, Engineering and Robotics Technology (ICASERT)* (pp. 1–6). IEEE.

[31] Kiran, S.R., Kakarla, N.B. and Naik, B.P., 2018. Implementation of Home automation system using Smart Mirror. *International Journal of Innovative Research in Computer and Communication Engineering*, 6(3), pp. 1863–1869

[32] Arora, Y. and Pant, H., 2019. Home automation system with the use of internet of things and artificial intelligence. In *2019 International Conference on Innovative Sustainable Computational Technologies (CISCT)* (pp. 1–4). IEEE.

[33] Maryasin, O., 2019, September. Home automation system ontology for digital building twin. In *2019 XXI International Conference Complex Systems: Control and Modeling Problems (CSCMP)* (pp. 70–74). IEEE.

[34] Rahman, M.I., Fahim, S.R., Avro, S.S., Sarker, Y., Sarker, S.K. and Tahsin, T., 2019, November. Voice-activated open-loop control of wireless home automation system for multi-functional devices. In *2019 IEEE International WIE Conference on Electrical and Computer Engineering (WIECON-ECE)* (pp. 1–4). IEEE.

[35] Rupanagudi, S.R., Bhat, V.G., Nehitha, R., Jeevitha, G.C., Kaushik, K., Pravallika Reddy, K.H., Priya, M.C., Raagashree, N.G., Harshitha, M., Sheelavant, S.S. and Darshan, S.S., 2018, December. A novel air gesture based wheelchair control and home automation system. In *International Conference on Intelligent Systems Design and Applications* (pp. 730–739). Springer, Cham.

[36] Alam, T., Salem, A.A., Alsharif, A.O. and Alhejaili, A.M., 2020. Smart home automation towards the development of smart cities. *APTIKOM Journal on Computer Science and Information Technologies*, 5(1), pp. 152–159.

[37] Sowah, R.A., Boahene, D.E., Owoh, D.C., Addo, R., Mills, G.A., Owusu-Banahene, W., Buah, G. and Sarkodie-Mensah, B., 2020. Design of a secure wireless home automation system with an open home automation bus (OpenHAB 2) framework. *Journal of Sensors*, 2020, p. 8868602.

[38] Shah, S.K.A. and Mahmood, W., 2020. Smart home automation using IOT and its low cost implementation. *International Journal of Engineering and Manufacturing (IJEM)*, 10(5), pp. 28–36.

[39] Asaithambi, S.P.R., Venkatraman, S. and Venkatraman, R., 2021. Big data and personalisation for non-intrusive smart home automation. *Big Data and Cognitive Computing*, 5(1), p. 6.

[40] Kshirsagar, S., Sachdev, S., Singh, N., Tiwari, A. and Sahu, S., 2020. IoT enabled gesture-controlled home automation for disabled and elderly. In *2020 Fourth International Conference on Computing Methodologies and Communication (ICCMC)* (pp. 821–826). IEEE.

[41] Venkatraman, S., Overmars, A. and Thong, M., 2021. Smart home automation—use cases of a secure and integrated voice-control system. *Systems*, 9(4), p. 77.

[42] Mahmood, Y., Kama, N., Azmi, A. and Ya'acob, S., 2020. An IoT based home automation integrated approach: impact on society in sustainable development perspective. *International Journal of Advanced Computer Science and Applications*, 11(1), pp. 240–250.

[43] Choudhury, T., Gupta, A., Pradhan, S., Kumar, P. and Rathore, Y.S., 2017, October. Privacy and security of cloud-based internet of things (IoT). In *2017 3rd International Conference on Computational Intelligence and Networks (CINE)* (pp. 40–45). IEEE.

[44] Yang, D., Nguyen, T.H. and Chung, W.Y., 2020. A bipolar-channel hybrid brain-computer interface system for home automation control utilizing steady-state visually evoked potential and eye-blink signals. *Sensors*, 20(19), p. 5474.

[45] Setz, B., Graef, S., Ivanova, D., Tiessen, A. and Aiello, M., 2021. A comparison of open-source home automation systems. *IEEE Access*, 9, pp. 167332–167352.

[46] Chaudhary, S.K., Yousuff, S., Meghana, N.P., Ashwin, T.S. and Guddeti, R.M.R., 2021. A multi-protocol home automation system using smart gateway. *Wireless Personal Communications*, 116(3), pp. 2367–2390.

[47] Ilyas, M., Ucan, O.N. and El Mohamad, Y., 2021. Smart home automation system design based on IoT device cloud. In *ICMI 2021* (pp. 116–123).

[48] Aboubakr, B., 2021. Study and implementation of a home automation and security system. *iKSP Journal of Computer Science and Engineering*, 1(1), pp. 17–22.

[49] Kheratkar, N., Bhavani, S., Jarali, A., Pathak, A. and Kumbhar, S., 2020, May. Gesture controlled home automation using CNN. In *2020 4th International Conference on Intelligent Computing and Control Systems (ICICCS)* (pp. 620–626). IEEE.

[50] Katangle, S., Kharade, M., Deosarkar, S. B., Kale, G. M., & Nalbalwar, S. L. (2020, February). Smart home automation-cum agriculture system. In *2020 International Conference on Industry 4.0 Technology (I4Tech)* (pp. 121–125). IEEE.

# 20 IoT-Based Smart Electricity Management

*R. Sricharan, E. Karthikeyan, K. Sethu Narayanan,
O.V. Gnana Swathika, and V. Berlin Hency*
Vellore Institute of Technology

## CONTENTS

20.1 Introduction ........................................................................................................... 231
20.2 Related Work ......................................................................................................... 232
20.3 Proposed System.................................................................................................... 233
    20.3.1 App Creation through MIT App Inventor Blocks ................................... 235
    20.3.2 App Development Screen ........................................................................ 235
    20.3.3 Firebase Realtime Database Code for Storing Username and Password
        for the Application ................................................................................... 236
20.4 Results.................................................................................................................... 236
    20.4.1 Arduino Output in Serial Monitor ......................................................... 236
    20.4.2 NodeMCU Output in Serial Monitor...................................................... 236
    20.4.3 ThingSpeak Output................................................................................. 238
    20.4.4 App Outputs............................................................................................ 238
    20.4.5 Firebase Realtime Database Storage for Username and Password ........ 240
20.5 Conclusion and Future Works................................................................................ 241
References........................................................................................................................ 241

**ABSTRACT**

In all our houses, we use the same standardized electricity meter, which is approved by the Electricity Board (EB) department. Currently, there are no live updates available for the amount of power we use daily. Previously, to solve the electricity crisis, the government has taken initiative on solving it by cutting the electricity down for an hour or two. With the help of the concept of Internet of Things (IoT), we can implement a smart energy meter, which can alert people on how much power they consume on a daily basis. The current and voltage values are collected with the help of sensors for a phase in the house. The collected inputs are then fed to the Arduino UNO microcontroller. The microcontroller then sends the data to NodeMCU ESP8266 Wi-Fi Module. The transmitted values then get updated in the ThingSpeak IoT website, thus helping the user for realtime analysis. This automated system is designed such that it can access the electricity meter of every consumer directly without any human intervention, and is also cheap, efficient and affordable.

## 20.1 INTRODUCTION

The usage of electricity at home and business owners has always been a concern. As resources to generate electricity are getting depleted day by day, the electricity costs are constantly rising. It is very much necessary that the general public need to be aware of their energy utilization and how efficiently it can be used. An easy way to give awareness is by deploying a management system where the user can understand when they use much of the energy and the efficient techniques that

DOI: 10.1201/9781003374121-20

can be implemented to conserve the electricity. Management can be done in a smart way by understanding how the systems can work most efficiently. The preliminary steps in energy administration are to diagnose manageable electricity losses and prevent troubles in typical residential, commercial and industrial energy systems. The basic idea is to implant a smart device on an appliance to read, analyse and improve the performance.

The objective of the paper is to create a cheap and efficient Internet of Things (IoT)-based smart energy management system with realtime monitoring using Web Application Peripheral Interface (API), and also, to make an app for smartphones where the user/consumer can see the power consumption with automated billing.

## 20.2   RELATED WORK

The discussion begins with a paper which focuses on Automated Systems for Energy Meter Reading and Billing. By implementing this system, the complexity in big apartments is reduced to a great extent. Also, this system can take a major share in the government's "DIGITAL INDIA" campaign [1].

Another paper focuses on batteries powered by solar modules that are used to sense the over current and to switch the medium loads of appliances like fans or lights. All data are monitored and stored in the server. The various parameters like real and reactive power, current and voltage can be depicted in graphical forms [2].

While discussing struggles of developing countries, one paper focuses on developing countries that face struggles in electricity meter reading. The aim of the paper is to study the measure of electricity consumption in household segment. It also aims to detect and control energy theft. Web API like ThingSpeak is used for realtime monitoring. If any theft is detected at the supplier's end, the user is alerted through a message and the energy supply is disconnected. It focuses on low-cost implementation [3].

With respect to fault indication, this paper focuses on developing an IoT-based smart energy meter reading system which also includes fault indications (happens every 15 days once), and also has the flexibility of automatic or manual options. Customers are alerted through SMS. In case of over power consumption in automatic mode, appliances are cut-off. Fault in meter is indicated to users through notifications [4].

Later, the discussion continues about the billing system, so this paper focuses on the automation of the ARM-based billing system. The proposed electronic meter will calculate the total energy consumed and automatically generates the bills and sends to the user's mobile as SMS or e-mail and thus helps in saving energy and cost [5].

IoT plays an important role in today's world, this paper focuses on different types of applications of IoT in everyday life. Through intelligent monitoring, we can achieve a smart system environment [6].

Another paper focuses on five parts: a voltage sensor, a current sensor, microcontroller, Bluetooth device and smartphone (Android app).The microcontroller reads the values of current and voltage from sensors and use the measured data to monitor the results of a three-phase electrical system using a new Android application in smartphone wirelessly through Bluetooth [7].

Shankara Narayanan and his colleagues give a proposal on how customers can monitor and analyse their usage using a user login process which can control the power supply during an emergency case [8].

Using Global System for Mobile communications (GSM) technology, Shaista Hassan Mir et al. gave a proposal which focuses on Arduino and GSM-based totally smart electricity meters for superior metering and billing structures that are capable to study and ship records by way of Wi-Fi protocol the use of GSM technological know-how via GSM modem, successful of the meters as properly as the line connections [9].

Using the same GSM technology in another proposal, a modification is made in the paper to make it smart. The paper uses a GSM module to send notifications as SMS to the user. Users can check the meters working from the web page as it shows the current reading with cost [10].

Another paper focuses on the design and deploying part of a single-phase energy meter which is used to measure the instantaneous and average realtime power consumed and remote monitoring capability using SMS with the help of Neoway M590 GSM module [11].

Similarly, another work also uses GSM. The work offers a single-section superior power meter based totally on a single-segment strength meter IC, a microcontroller and a GSM module to improve a computerized machine via which month-to-month energy consumption will be calculated precisely and at the identical time the ensuing unit will be dispatched to a faraway receiver for similar calculation and an up-to-date statistics additionally be acquired by means of the users about any information [12].

In another paper, small modification is made to the existing energy meters into a smart meter by using Arduino. In this paper, the aim is to make these meters smart with the use of Artificial Intelligence (AI) technology and IoT, so that the Energy Board can show the monthly consumption and also check for any energy theft detection [13].

An idea also focuses on monitoring the energy consumption of the user and generation of automatic billing systems, thus reducing the labour. The status of bill payment is updated to the user with due date, warning notifications. Thus, if the bill is not paid, the meter gets disconnected [14].

Smart energy meters need several parameters that need to be monitored. In a proposal, it allows the user to take a part in monitoring the energy management activities by monitoring various parameters like voltage, current, power factor, etc. The user also monitors the load in the house during the peak hours [15].

Although there are various ideas and proposals, this paper studies various proposed ideas and existing technology for smart billing systems and analyzes its use and drawbacks. Thus, in the proposed system in the paper, the electricity connection to each user may be given handiest to the registered person and the smart billing can be executed via IoT [16].

Energy shortages can occur during peak hours, so an idea which introduces smart electricity meters using wireless communication and LabVIEW is necessary. The smart meter uses something called Time Of Day tariff pricing, thus making the users to be an active part in it. This reduces the energy shortage during the peak hours [17].

IoT makes our lives easier. Smart electricity meters should not only benefit users but also meter readers who traditionally go to every house physically for meter analysing and recording the data. Thus, a proposal given in a paper discusses how the smart meter is preferred over the conventional meters, wherein the meter readers don't want to go to every user to record the power and to distribute the consignment slips [18].

Thus, discussion concludes here, as we have seen several proposals and work ideas on IoT-based smart electricity meters focusing on various scenarios and aspects.

## 20.3   PROPOSED SYSTEM

Parameters like current and voltage values are calculated using the current sensor (ACS712) and the voltage sensor (Voltage Detection sensor module). These values help us to calculate the power consumption in that phase. Power consumption from all the phases in the building is calculated and added for total energy consumption. The input parameters and calculated result are uploaded to Arduino UNO Microcontroller. NodeMCU ESP8266 module gives Arduino the access to Local Wi-Fi Network, to upload realtime data to ThingSpeak. Data on ThingSpeak are sent to MIT App Inventor through which the users can access the data real time through their phone devices. Consumers can also see their billing information in real time as well. The values are updated every 15 seconds; thus, the cost of power used in 15 seconds is calculated as follows:

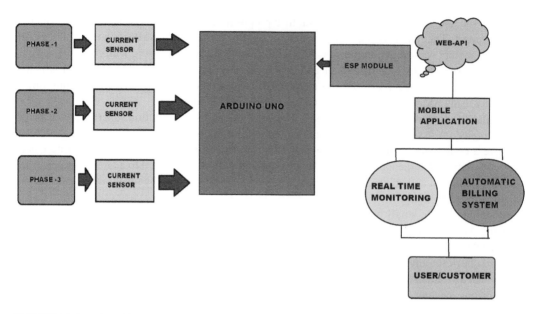

**FIGURE 20.1**  Block diagram consisting of important components.

$$₹/15 \ sec = (Power \times 10)/240 \ ₹.$$

Based on this formula, the consumers are billed in real time.

Real monitoring of power consumption and automated billing can help in reducing the energy wastage. In present time, saving energy is an important topic which helps in lesser usage of natural resources which in turn helps us to save the environment. Till now, monthly Electricity Board (EB) bills are computed by a person who visits the EB meters monthly and updates the power usage. Automated Billing can help in reducing the human errors and analysing the billing in real time.

Figure 20.1 explains the concept mentioned above. Most houses have three phases. Thus, separately obtaining the electricity consumed in each phase, these data are then sent to Arduino UNO microcontroller. Thus, by interfacing NodeMCU which contains ESP8266 Wi-Fi module, these data are sent to Web API (in this case, ThingSpeak). ThingSpeak allows us to enable realtime monitoring. Mobile application is created for users/customers for realtime monitoring and implementing automated billing processes in the IoT-Based Smart Electricity Management System.

Figure 20.2 shows the schematic diagram of the hardware components. The load is connected to the voltage sensor and ACS712 current sensor's +ve and −ve pins. Both sensors are powered by 5 V or 3.3 V pin and grounded by GND pin from the Arduino UNO. The signal pins of the voltage sensor and current sensors are connected to the analogue pins of Arduino UNO. Later, Arduino UNO is interfaced with NodeMCU through serial communication, i.e., two digital pins from Arduino UNO connected to the exact same two digital pins in NodeMCU. The GND pin of NodeMCU is connected to the GND pin of Arduino UNO. NodeMCU contains ESP8266 Wi-Fi module which helps us to connect with the internet. When the Arduino UNO is powered, all the sensors are also powered up.

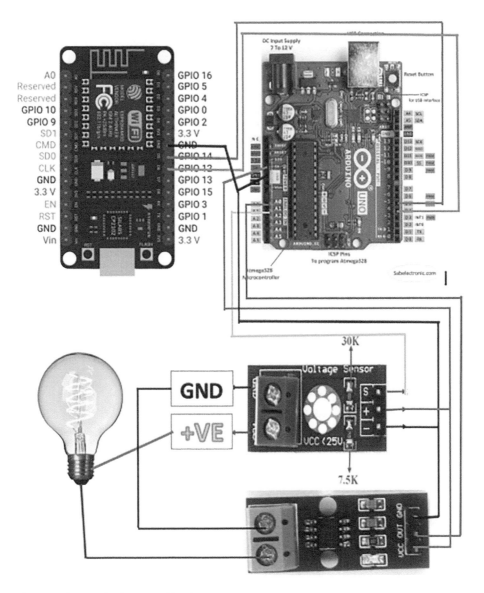

**FIGURE 20.2** Schematic diagram of hardware components.

### 20.3.1 App Creation through MIT App Inventor Blocks

Figure 20.3 shows the app being built using MIT App Inventor. The layout for both the login and main page is designed, and the block codes are assigned to each component such as buttons, text boxes and labels in their respective layout as shown in Figure 20.3.

### 20.3.2 App Development Screen

Figure 20.4 shows the design for the layout of the app is created using the MIT App Inventor. The complete user interface (UI) is created in such a way that the user does not find it challenging to operate as shown in Figure 20.4.

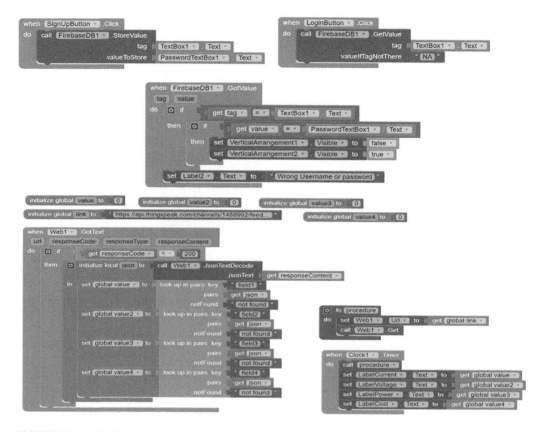

**FIGURE 20.3**    Block codes through MIT App Inventor.

### 20.3.3    Firebase Realtime Database Code for Storing Username and Password for the Application

The read and write rules are changed to true in the Realtime Database for storing and checking of the login credentials as shown in Figure 20.5.

## 20.4    RESULTS

### 20.4.1    Arduino Output in Serial Monitor

Figure 20.6 shows the output in the serial monitor configured for Arduino UNO. This displays the current and voltage values from a single phase.

### 20.4.2    NodeMCU Output in Serial Monitor

Figure 20.7 shows the output in the serial monitor configured for NodeMCU. This displays the status of internet connection and updates us about the data received from Arduino.

**FIGURE 20.4** App development screen in MIT App Inventor.

**FIGURE 20.5** Firebase database.

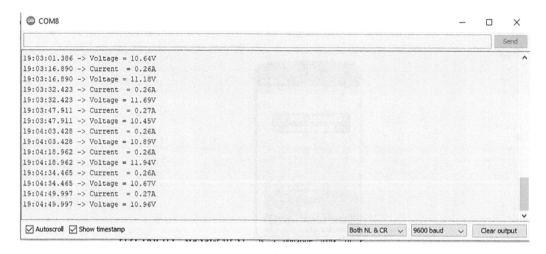

**FIGURE 20.6** Serial Monitor output 1.

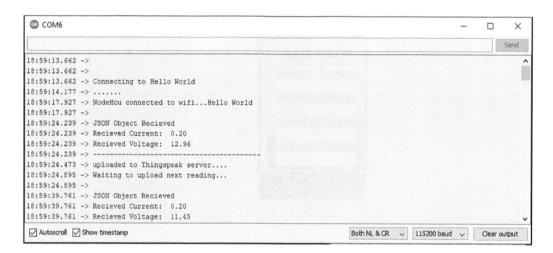

**FIGURE 20.7** Serial Monitor output 2.

### 20.4.3 ThingSpeak Output

Figure 20.8 shows the output in ThingSpeak. Here, each field shows us the value obtained from ESP8266 Wi-Fi module. Parameters like current, voltage, power and total cost are displayed both graphically and numerically.

### 20.4.4 App Outputs

Figure 20.9 represents the sample layout of the login and main page of the app. The login page is made user-friendly with a clear description of the text boxes to enter their username and password. The Login and Sign Up button is highlighted to give a clear view. The main page begins with a welcome message followed by four sections, current, voltage, power and cost.

Created: 3 months ago
Entries: 17

**FIGURE 20.8** ThingSpeak output.

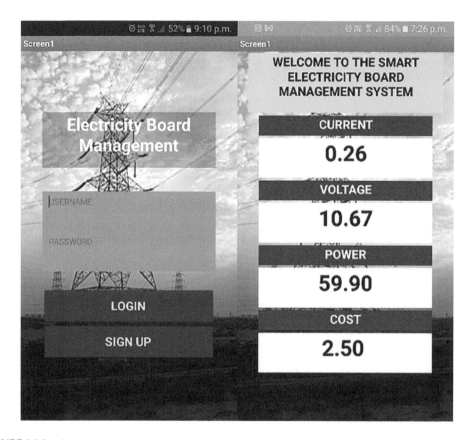

**FIGURE 20.9**   App outputs.

**FIGURE 20.10**   Firebase Realtime Database.

## 20.4.5   FIREBASE REALTIME DATABASE STORAGE FOR USERNAME AND PASSWORD

Firebase is used to do the task of storing usernames and passwords. On the sign up process, the username and password are sent to the Realtime Database in Firebase and stored in the format as shown in Figure 20.10.

## 20.5 CONCLUSION AND FUTURE WORKS

Our proposed smart meter achieves the required output needed and the current and voltage values are accurate. Power consumption values are calculated and displayed as required. The data are transmitted to the ThingSpeak IoT Web API with the help of NodeMCU ESP8266 module. Parameters can be viewed in real time in ThingSpeak IoT. Application for smartphone is created, which shows the parameters in real as well, thus ensuring remote access for consumer/user. Firebase Realtime Database is used for storing the username and passwords of the user/consumers. There is a scope for development in this project.

In future works, we are planning to find power usage for all phases in buildings and make their calculation and billing more accurate, by creating a user-friendly app, which is more interactive with the user/consumer, and to send notification to the user the information regarding the usage of power consumption and billings. Making realtime analysis and creating an invoice and table in the app help the user remotely access billing details.

## REFERENCES

[1] Pratik P. Bakshi, Akshay S. Patil, Akshay S. Gavali, Vibha U. Patel, "IoT Based Electric Energy Meter", *International Journal of Advanced Research in Computer and Communication Engineering (IJARCCE)*, Vol. 6, Issue 3, pp. 966–969 (March 2017).

[2] S.G. Priyadharshini, C. Subramani, J. Preetha Roselyn, "An IOT Based Smart Metering Development for Energy Management System", *International Journal of Electrical and Computer Engineering*, Vol. 9, Issue 4, pp. 3041–3050 (Aug 2019).

[3] M. M. Mohamed Mufassirin, A. L. Hanees, "Development of IOT Based Smart Energy Meter Reading and Monitoring System", in *Proceedings of 8th International Symposium-2018, SEUSL* (2018).

[4] Rishabh Jain, Sharvi Gupta, Chirag Mahajan, Ashish Chauhan, "Research Paper on IOT Based Smart Energy Meter Monitoring and Controlling System", *International Journal of Research in Electronics and Computer Engineering*, Vol. 7, pp. 1600–1604 (2019).

[5] P.C. Kishore Kumar, J. Antony Veera Puthira Raja, B. Ebenezer Abishek, S. Vishalakshi, G. Vidhya, "IOT Based Energy Meter Reading System with Automatic Billing", *International Journal of Engineering & Technology*, Vol. 7, pp. 431–434 (2018).

[6] N. Sathiyanathan, S. Selvakumar, P. Selvaprasanth, "A Brief Study on IoT Applications", *International Journal of Trend in Scientific Research and Development (IJTSRD)*, Vol. 4, Issue 2, pp. 23–27 (Feb 2020).

[7] Mohannad Jabbar Mnati, Alex Van den Bossche, Raad Farhood Chisab, "A Smart Voltage and Current Monitoring System for Three Phase Inverters Using an Android Smartphone Application", *Sensor*, Vol. 17, Issue 4, p. 872 (Apr 2017).

[8] R.N. Saravanan, U. Padmanaban, M. Santhosh Kumar, K. Shankara Narayanan, "IoT Based Smart Energy Meter", *International Journal of Advance Research, Ideas and Innovations in Technology*, Vol. 4, Issue 2 (2018), pp. 2071–2073.

[9] Shaista Hassan Mir, Sahreen Ashruf, Sameena, Yasmeen Bhat and Nadeem Beigh, "Review on Smart Electric Metering System Based on GSM/IOT", *Asian Journal of Electrical Sciences*, Vol. 8, Issue 1, pp. 1–6 (2019).

[10] Birendrakumar Sahani, Tejashree Ravi, Akibjaved Tamboli, Ranjeet Pisal, "IoT Based Smart Energy Meter" *International Research Journal of Engineering and Technology*, Vol. 4, Issue 4, p. 96 (Apr 2017).

[11] D.A. Shomuyiwa, J. O. Ilevbare, "Design and Implementation of Remotely-Monitored Single Phase Smart Energy Meter via Short Message Service (SMS)", *International Journal of Computer Applications*, Vol. 74, pp. 14–22 (July 2013).

[12] Afrin Hossain, Tajrin Jahan Rumky, Nursadul Mamun, "Implementation of Smart Energy Meter with Two Way Communication Using GSM Technology " *International Journal of Scientific & Engineering Research*, Vol. 4, Issue 7 (June 2013).

[13] Sagar Dadhe, Rohit Maske and Rohit Kalukhe, "IOT Based Smart Energy Meter", *International Journal of Innovations in Engineering Research and Technology*, pp. 1–5 (Mar 2021), 89–91.

[14] V. Arun, C.S. Aswathy, P. Asin Iqbal, S.R. Fathima Beevi, P. Sreejith, "Review of Smart Energy Meter", *International Journal of Advanced Research in Computer and Communication Engineering*. Vol. 9, Issue 1 (Jan 2020), 59–63.

[15] G. Vani, V. Usha Reddy "Application of Smart Energy Meter in Indian Energy Context" *IOSR Journal of Electrical and Electronics Engineering*, Vol. 10, Issue 3, pp. 7–13 (2015).

[16] A. M. Magar, Shradha Gaikwad, Nikita Katke, Yogeeta Karande and Ritu Raut, "A Survey on Various Smart Electricity Meter and Billing System", *International Journal of Advance Research and Innovative Ideas in Education*, Vol. 6, Issue 3, pp. 380–385 (2020).

[17] S. S. Al-Saheer, S. L. Shimi, S. Chatterji, "Scope and Challenges of Electrical Power Conservation in Smart Grids" *International Journal of Engineering Research & Technology*, Vol. 3, Issue 5, pp. 2212–2214 (May 2014).

[18] Zahid Iqbal Rana, M. Waseem, Tahir Mahmood "Automatic Energy Meter Reading using Smart Energy Meter", *International Conference on Engineering & Emerging Technologies (ICEET-2014)* (2014).

# 21 IoT-Based COVID-19 Patient Monitoring System

*S. Charan, K. Kaamesh, B. Aswin, O.V. Gnana Swathika, and V. Berlin Hency*
Vellore Institute of Technology

## CONTENTS

21.1 Introduction ....................................................................................................243
21.2 Literature Review ...........................................................................................244
21.3 Proposed Model..............................................................................................246
21.4 Block Diagram................................................................................................247
21.5 Circuit Diagram..............................................................................................249
21.6 Implementation ..............................................................................................250
21.7 Conclusion .....................................................................................................254
References....................................................................................................................254

### ABSTRACT

Coronavirus is a pandemic that changed the whole Earth upside down. Its outbreak has taken a toll on every aspect of life. Starting from the catastrophic death values to the huge drop in the stock market shares, some way or the other all humans in the planet Earth face an issue because of the ongoing pandemic. Though a solution to this pandemic is the need of the hour, we initiate to create an alternative platform that can play a vital role in treating the patients on time along with the required safety measures so as to save millions of lives and give hope for mankind to overcome this threat. The proposed work can be implemented in various health zones and/or public places which are in need: The proposed work is split into 3 A sections. The first one "Attach", which is developed to support hospital environment and maintain safety of doctors and support staff, is all about the patient's personal health monitoring with respect to various parameters like room temperature, humidity and pulse rate. The second is "Alert" system where we have designed and created an Automatic Sanitizer Dispenser along with a temperature checking system, which can be placed in entrances of every store or other public places. The sanitizer level is checked and alert is sent to the admin according to the remaining level via Bluetooth Communications. Finally, the "Administer" concept is indicated to show the availability of bed in each hospital/clinic in the local and public sector. This monitors the bed availability status in health centres and the system tracks the bed usage condition and transmits this data via Bluetooth to the users through an app developed.

## 21.1 INTRODUCTION

Centuries have witnessed major historical events that mark the construction or destruction of mankind and other lives on Earth. Coronavirus is one such pandemic that made the whole world topsy-turvy and human fear for many things till date, and the fear of uncertainty of life has become very common. Eventually every generation has survived with hopes and innovations of bringing up a better tomorrow, and we are now moving towards multiple safety measures to safeguard ourselves from this threat.

DOI: 10.1201/9781003374121-21

Medical and engineering fields have given tremendous inputs here to find a solution for a secured way of living and are still working on the betterment of the decision made [1]. It is very obvious that loss of lives has drastically come down, which given a ray of hope for us to be optimistic. There is always light at the end of the tunnel and with that note we have come to a stage to face this threat. Every department finds a mean to solve this problem and from the student's point of view, we have our own ideas to manage this situation. This may be of some help to meet out the needs as this is a trial-and-error scenario that we are forced in to. Vaccination procured in a short span of time proved helpful in many cases and people are trying to get new ideas every day to fight this hardship.

This paper aims to come up with an idea to add-on hope to the people based on the technology that is available as it can be used without physical contact and thereby helps in stopping virus mutation and multiplication of COVID cases. This paper is one such add-on to the innovations, and it has come out with solutions to address to three major concepts: Attach, Alert and Administer.

**Attach:** The first concept focuses on the primary issue of monitoring patients' personal health with respect to various parameters like room temperature, humidity and pulse rate [2]. This is advised through an instrument that constantly monitors, checks and uploads to the ThingSpeak Internet of Things (IoT) cloud platform. To have a better support to this system, we also equip it with an embedded ventilator sub-system which functions automatically based on the weather conditions. The Bluetooth data collected by the admin are also uploaded online in hospital websites that enables the patient's relatives to get the reports thus by avoiding physical contact. Doctors can also use this tool to monitor patient's condition, as it is so much more convenient, easier and safe.

**Alert:** Our next idea is mainly to attract the corporates and the retail store division, as these places constantly have a check on the temperature of the customers and also have a person to dispense sanitizer. To substitute this, we have designed an Automatic Sanitizer Dispenser along with a temperature checking system, which can be placed at entrances of every store or public places. The sanitizer level is also monitored from time to time and an alert is sent via Bluetooth Communications to the admin to refill the same. With this product, the long queues in front of a retail unit are thus avoided, and every customer is tested individually without any human intervention. This can reduce the transmission and mutation of the virus.

**Administer:** The third concept monitors the bed availability status in health centres. The system tracks the bed usage condition and transmits this data via Bluetooth to the users through an app. This is updated on a real-time basis on the APP making the public know about the current status of the hospital.

## 21.2  LITERATURE REVIEW

The discussion starts with an IoT node that monitors the fundamental readings which include a person's body temperature, heart rate, and blood oxygen saturation. It also checks the coughing pattern from time to time, and these parameters and individual risk factors are displayed on a smartphone app for further analysis. A mechanism via Bluetooth 4.0 technology is maintained for tracking the physical distance, and it does alert in case of any violation of safe social distancing and a fog server helps in collecting data from the IoT nodes. These users' information is sent using an application of machinelearning algorithm for automation [3]. Daniel Shu Wei Ting and Tien Y. Wong discuss all about IoT with respect to the telecommunication networks for the next generation. Examples of the next-generation networks like Artificial Intelligence, Big Data Analytics, Block Chain Technology and 5G. Hence, their establishments in hospitals and clinics with an interconnected digital ecosystem can enable real-time data collection at scale with a broader thought of involving AI and deep learning system in to it so as to understand the trends in healthcare, along with the model risk associated with them and predict outcomes [4]. In the case study stated by Cacovean Dan, Irina

Ioana, and Gabriela Nitulescu, there is a device that is worn by the person. This device reads the vitals of the respective patient, and the data are uploaded to the cloud. The physician can now access and analyse the data from the cloud and advise the patient on the necessary ailment. The wearable device is administered with a GSM module that alerts the patient in the event of any urgency and also consists of emergency contact numbers. In addition to that, these data are periodically uploaded to the regional health system, to monitor the current situation of the patient. By doing so, the COVID situation in the region can be analysed along with the symptoms and effects and that helps in getting an overall picture of the pandemic by collecting these data from various sources. The collected data are then used to figure out an idea for distribution of the effects and frequencies [5]. A wearable device designed to measure various vital signs related to COVID-19 is the new development every company is working on. Needless to mention, any violations of quarantine for potentially infected patients will be sent as an alert to the concerned medical authorities by monitoring their real-time GPS data. Primarily this is done by measuring the data from the IoT sensor layer which defines the health symptoms, and the next layer is used to store the information in the cloud database for pre- ventive measures, alerts and immediate actions. Notification and alerts for the potentially infected patient is done by the Android mobile application layer. Furthermore, the work disseminates digital remote [6]. Constructing an open-source mechanical ventilator at a very less price with a view to accommodate the need for the COVID-19 affected patients' region is the main objective taken by Vargas and Acho. It is designed to display numerical values on patient's pulmonary conditions and pressure measurements read from their inspiratory limb. By doing so, the doctors understand the real-time need for the patient and can evaluate if they are under a healthy or unhealthy situations [7].

Addressing all patients who live far away from the hospitals or in a remote zone where medi- cal treatment becomes tough for immediate needs is a major problem. Hence, the compact sensors with IoT aims to create a huge impact on every patient's life, as it monitors the patient's condition irrespective of the distance from their home to the hospital and hence reduces the fear in patients of any danger. This can be accessed from anywhere as it works with a sensor. This sensor helps the doctors to have the analytical values of the patients, and they can address to their needs from a far- off distance too. This will really be a great achievement in the medical field [8]. IoT makes a doctor's job easier as it opens up new healthcare opportunities and uplifts the information system to a higher level. The world-class results aim at the best treatment systems in the hospital, by equipping medical students to detect the disease better and train them in proper usage of IoT. This extends to help the medical practitioner resolve different medical challenges like speed, price and complexity. It can be customized to monitor caloric intake of the patient and treating COVID-19 patient for ailments like asthma, diabetes as well as arthritis [9]. The idea behind the IoT design for smart monitoring and keeping an emergency check on COVID-19 patients uses sensors like temperature sensor, sensor to detect blood oxygen level, heart rate sensor and Arduino Uno controller to monitor a patient in stage 1 of the disease. IoT server reads the patient's information instantly through these sensors. By doing so, the mortality rate is reduced considerably. With the alert received from IoT, hospital admissions could also be avoided and many emergency situations can be addressed well in advance. Further, the system supports in generating warning messages to the nearest hospital irrespective of the patient's health condition. Hence, this initiative of smart monitoring and emergency alert system will help in extending the monitoring support to COVID-19 patients [10]. An IoT framework is set to receive symptomatic data in the real time from users, and it serves in early identification of the suspected coronavirus cases. This also monitors the post-effect condition in virus-infected patients by col- lecting and analysing relevant data. It functions with the help of data from the following five com- ponents: Symptom Data Collection and Uploading (using wearable sensors), Quarantine/Isolation Centre, Data Analysis Centre (that uses machine learning algorithms), Health Physicians and Cloud Infrastructure [11]. Antonio Pietrosanto and his co-authors give a proposal of a device which could be worn by the patients for tracking their real-time body temperature. This functions with the IoT method. It also records the patient's indoor condition while in quarantine and indicates the data on emergency alert. This signal alert is given when the body temperature exceeds the allowed threshold temperature. In addition to that, a Repetition Spikes Counter based on the accelerometer algorithm is employed to monitor human activity. This checks if the quarantined person is performing his physical exercises regularly and maintains an auto-adjustment of threshold temperature. The real-time data are stored, and it helps the family members/doctors in regularly monitoring the updates on the quarantined people's body temperature behaviour in the tele-distance. The devices used for this include an M5stickC wearable device, a microelectromechanical system accelerometer, an infrared (IR) thermometer and a digital temperature sensor equipped with the user's wrist. This sensor keeps track with the room temperature and humidity and restricts the virus spread. The quarantined patient's room condition is also monitored, and all data are transferred to the cloud via Wi-Fi with the Message Queue Telemetry Transport broker. In the event of Wi-Fi failures or poor network connectivity, these data are sent through Bluetooth from the self-isolated person to the family member's electronic device [12].

## 21.3  PROPOSED MODEL

The proposal targets three different concepts to cover all possible industries. The "Attach" is a proposal to support hospital environment and maintain safety of doctors and support staff. "Alert" is a proposal to support the sectors in retail and public services where the product developed is useful for the public and the locality members. Finally, "Administer" is a proposal to indicate the availability of bed in each hospital/clinic in the local and public sector.

A. **"Attach":** Proposal to support hospital environment and maintain safety of doctors and support staff.

In view of the hospital environment and safety of the patients, this proposal stands to check on the health condition of the patients affected with COVID-19 who are isolated in different wards in a hospital. The rooms are regularly administered with temperature and humidity sensors to monitor the day-to-day room temperature. In addition to this, there is also an automated exhaust fan installed in these wards which automatically turns on whenever the temperature or humidity falls below the threshold value.

As per the instructions from the World Health Organization (WHO), hospitals have to maintain an optimum room temperature of 20°C–24°C in cold regions and 24°C–35°C in hot regions to reduce the transmission of the novel coronavirus. Hence, the relative humidity should be maintained between 40% and 60% [13]. In adherence to these said values, the system is configured with a temperature threshold of 40°C and with a relative humidity level of 65%.

Oximeter heart rate sensor values are as well monitored and the $SPO_2$ average value in humans should lie between 90% and 95%. In case of person with COVID, studies show a decrease in $SPO_2$ measurement. To have a regular check on this value and to analyse the patients better, the system updates these values to the cloud. This can be viewed in the dashboard of ThingSpeak Environment for the doctors, and it enhances their treatment even in patient's conditions remotely.

The sensors used for implementing the above-stated statement: DHT-11 Temperature and Humidity Sensor Module and MAX30100 Pulse Oximeter Heart Rate Sensor Module.

B. **"Alert":** Proposal to support the sectors in retail and public services where the product developed is useful for the public and the locality members.

On the outbreak of COVID in severe condition globally, big countries started to impose lockdowns to prevent the movement of the public in social gathering to stop the spread of the virus. This affected many industries globally and many were forced to even shut down their units. With the decline of the first wave, lockdown was taken back, and people returned to normalcy with a compulsion to maintain social distance and were asked to follow sanitization procedures. But monitoring this has become a strenuous process, and the risk factors became more to insist people on sanitization and temperature check on daily basis. At last, this became time-consuming and also had a threat of spreading.

To avoid this, "Alert" proposal was introduced with an automatic non-contact temperature check system with automatic provision of sanitizer. Every retail store or a common place disposes an access for these, which comprises a LCD Module, Sanitizer System and a non-contact Temperature Module. As the client or the public of interest approaches near the system with an interval of 10 cm between the unit and the person [14], the module reads the temperature and also dispenses the Sanitizer with automatic door open mechanisms. These values are regularly updated in an APP provided to the owner or the admin. Depending on the feedback provided by the APP, the admins can analyse the situation and act accordingly.

The sensors used for implementing the above-stated statement: MLX90614 Non-Contact IR Temperature Module, IR Sensors and Ultrasonic Sensor Module.

C. **"Administer"**: Proposal to indicate the availability of bed in each hospital/clinic in the local and public sector.

Sudden surge in COVID has limited the resources available around the globe. Full strength in hospitals has pushed the management to a state where they couldn't admit the affected patients into wards. This situation is reflected in a confused public behaviour, and lots of deaths were recorded just because the patients were not admitted into hospitals at the right time.

To provide transparency in the bed status available at every hospital in the locality, "Administer" aims to showcase the status of the bed occupancy in an APP environment. Force sensors are placed under the hospital beds and these read the force exerted. If a bed is occupied, the reading of the force value is high, and if it's vacant, there is no force exerted. To minimize the wrong readings, an interval of 15 minutes is provided between the free and occupied status update. This information displayed on the APP can be of great use to the public and depending on the availability in a particular hospital, patients can approach that hospital for treatment.

Force sensor with a specification of 39.1 mm sensing boundary is used to implement the above case.

## 21.4 BLOCK DIAGRAM

Figure 21.1 explains the Attach concept as mentioned above. The MAX30100 Pulse Oximeter is used to measure the heart beat and $SPO_2$ values from the patients, and these values are sent to the NODEMCU for processing and updating. Similarly, humidity sensors are placed in hospital wards and depending on the feedback analysed in NODEMCU, the exhaust fan is controlled remotely with commands from the NODEMCU. The values are also stored in the ThingSpeak cloud platform from NODEMCU for future analysis and predictions.

Figure 21.2 explains the Alert concept as mentioned above. Once the presence of a visitor is detected by the ultrasonic sensor, depending on the feedback from the sensor the Arduino starts the system. Also, there has to be sufficient sanitizer available in the dispenser for the system to start. Hence, an IR sensor measures this, and depending on the feedback the system is started. This feedback from the sensor is displayed in the LCD and also indicated with LEDs placed. If ultrasonic sensor detects a person in the range limit, the MLX90614 sensor is triggered to measure the temperature of the person. This temperature value is compared with the standardized value in Arduino and servo motor is triggered to dispense the sanitizer from the dispenser.

Figure 21.3 explains the Administer concept as mentioned above. Force sensors are placed under the beds in hospital and Arduino is used to check for the minimum threshold. Depending on this feedback, the status is updated on an APP platform designed which connects with the Arduino using Bluetooth Technology [15].

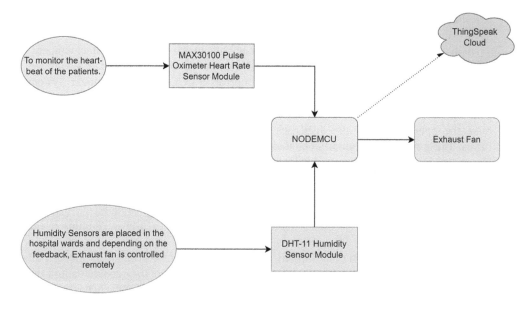

**FIGURE 21.1** Attach block diagram.

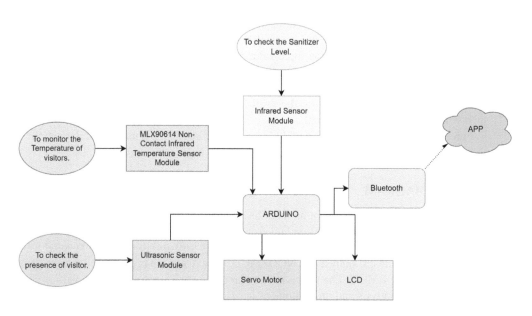

**FIGURE 21.2** Alert block diagram.

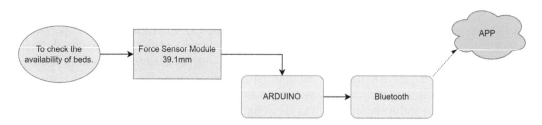

**FIGURE 21.3** Administer block diagram.

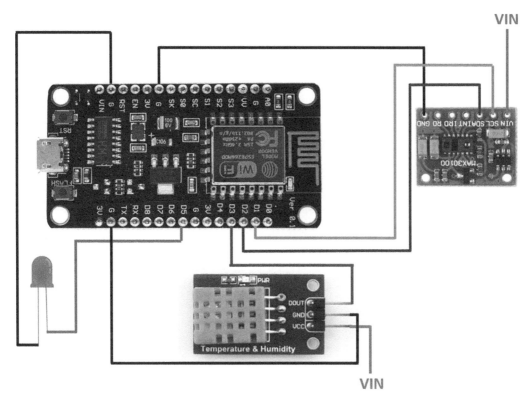

**FIGURE 21.4** Attach circuit diagram.

## 21.5 CIRCUIT DIAGRAM

The MAX30100 Pulse Oximeter is powered on by connecting the VIN pin to VIN pin of NODEMCU and GND pin to GND pin of NODEMCU. The SCL pin is connected to D1 and SDA is connected to D2. The DHT-11 sensor is powered by connecting the VCC pin to VIN and GND pin to GND. The DOUT pin is connected to D3. In this case, an LED is replaced in the place of an exhaust fan, and the positive of LED is connected to D5 and the negative is connected to GND.

Figure 21.5 discusses the circuit diagram of Alert concept, the MLX90614 sensor, IR sensor, ultrasonic sensor and the servo motor powered on by connecting the VIN/VCC pin to 5 V pin of Arduino and GND pin to GND pin of Arduino. The SCL pin and SDA pin of MLX90614 are connected to the SCL and SDA pins in the Arduino, respectively. The Echo pin of ultrasonic is connected to D10 and Trigger Pin is connected to D9 [14]. The servo motor output is connected to D6 and the IR sensor output is connected to A0 pin. The Bluetooth Module is connected by connecting its TX pin with the RX pin in Arduino and by connecting its RX pin with the TX pin in Arduino. The VIN is connected to 3.3v pin and GND is connected to GND. The LCD is conventionally connected by following the standard connections [15]. DB4 to DB7 are connected to D5 to D2 pins in Arduino. The LED+ is connected to 5 V and LED− is connected to GND. The Enable is connected to D11 and RS is connected to D12. The RW pin is grounded and VCC is connected to 5v and GND is connected to GND pin. The contrast is connected to a potentiometer with its one terminal connected to 5 V and the other to GND.

In Figure 21.6, the Bluetooth Module is connected by connecting its TX pin with the RX pin in Arduino and by connecting its RX pin with the TX pin in Arduino. The VIN is connected to 3.3v pin and GND is connected to GND. The force sensor is connected to the A0 pin and the same

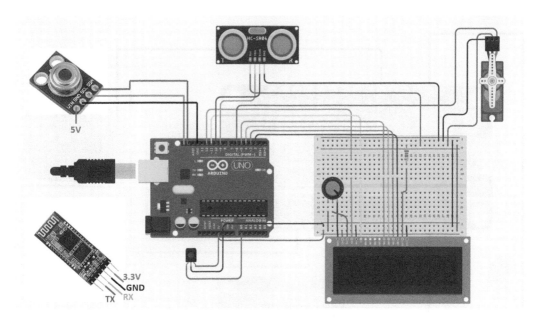

**FIGURE 21.5**  Alert circuit diagram.

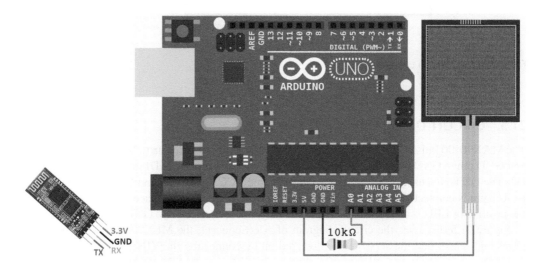

**FIGURE 21.6**  Administer circuit diagram.

terminal is connected to GND through a resistor with 10 kilo-ohm resistance. Another terminal of sensor is connected to 5 V Pin.

## 21.6  IMPLEMENTATION

A. **"Attach":** Proposal to support hospital environment and maintain safety of doctors and support staff.

Figure 21.7 represents the Attach circuit implementation for collecting the responses from user; these responses are processed and depicted as data points in the cloud window. Figures 21.8–21.10 represent the cloud window implementation.

**FIGURE 21.7** Attach circuit implementation.

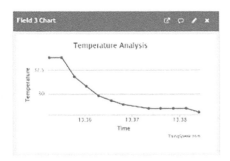

**FIGURE 21.8** Attach cloud implementation A.

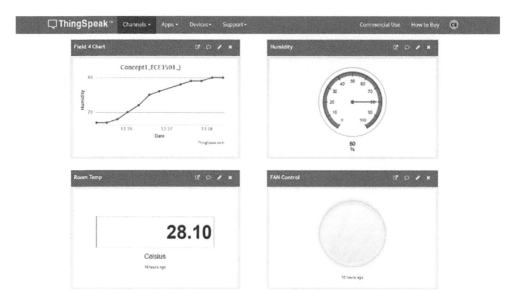

**FIGURE 21.9**    Attach cloud implementation B.

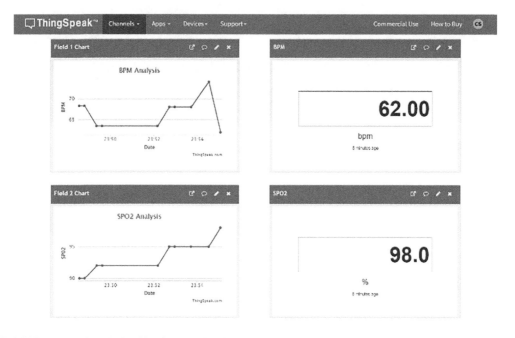

**FIGURE 21.10**    Attach cloud implementation C.

B. **"Alert":** Proposal to support the sectors in retail and public services where the product developed is useful for the public and the locality members.

    Figure 21.11 gives an insight into the Alert concept circuit implementation, and Figure 21.12 displays the APP implementation screenshot.

C. **"Administer":** Proposal to indicate the availability of bed in each hospital/clinic in the local and public sector.

    Figure 21.13 depicts the Administer APP implementation screenshot.

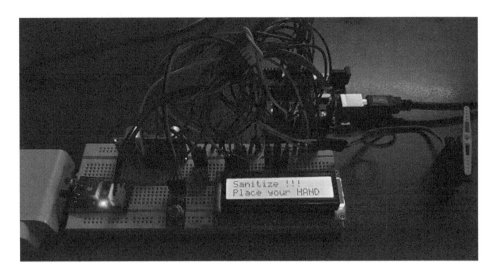

**FIGURE 21.11**   Alert circuit implementation.

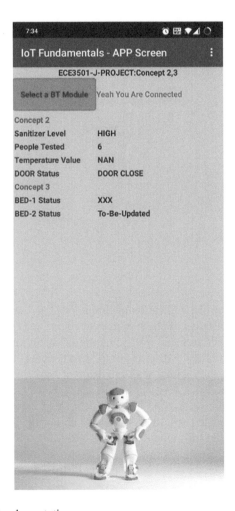

**FIGURE 21.12**   Alert APP implementation.

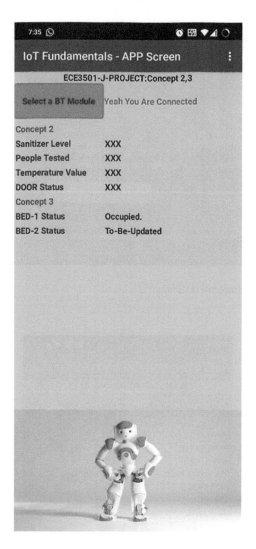

**FIGURE 21.13** Administer APP implementation.

## 21.7 CONCLUSION

To conclude, we ensure to list the application of three concepts that were developed to set up a COVID-19 Patient Monitor System with the resources available and listed above. By doing so, now one can collect the room temperature and relative humidity values and display in a cloud platform with an automatic exhaust fan linked with it as per the first concept. To satisfy the second concept, an automatic temperature checker system with provision of sanitizer is set up and these data are sent to an APP platform for analysis. Last but not the least, the bed availability status in hospitals is collected real-time and updated in an APP platform that was developed. Hence, the aim of the paper is achieved, and we hope to serve the patients and ensure the helpers' security alongside by our innovative mechanism using IoT module.

## REFERENCES

[1] Nandy, Sudarshan, and Mainak Adhikari. "Intelligent Health Monitoring System for Detection of Symptomatic/Asymptomatic COVID-19 Patient." *IEEE Sensors Journal* 21, no. 18 (2021): 20504–20511.

[2] Wurzer, David, Paul Spielhagen, Adonia Siegmann, Ayca Gercekcioglu, Judith Gorgass, Simone Henze, Yuron Kolar et al. "Remote Monitoring of COVID-19 Positive High-Risk Patients in Domestic Isolation: A Feasibility Study." *PLoS One* 16, no. 9 (2021): e0257095.

[3] Vedaei, Seyed Shahim, Amir Fotovvat, Mohammad Reza Mohebbian, Gazi ME Rahman, Khan A. Wahid, Paul Babyn, Hamid Reza Marateb, Marjan Mansourian, and Ramin Sami. "COVID-SAFE: An IoT-Based System for Automated Health Monitoring and Surveillance in Post-Pandemic Life." *IEEE Access* 8 (2020): 188538–188551.

[4] Ting, Daniel Shu Wei, Lawrence Carin, Victor Dzau, and Tien Y. Wong. "Digital Technology and COVID-19." *Nature Medicine* 26, no. 4 (2020): 459–461.

[5] Cacovean, Dan, Irina Ioana, and Gabriela Nitulescu. "IoT System in Diagnosis of Covid-19 Patients." *Informatica Economica* 24, no. 2 (2020): 75–89.

[6] Al Bassam, Nizar, Shaik Asif Hussain, Ammar Al Qaraghuli, Jibreal Khan, E. P. Sumesh, and Vidhya Lavanya. "IoT Based Wearable Device to Monitor the Signs of Quarantined Remote Patients of COVID-19." *Informatics in Medicine Unlocked* 24 (2021): 100588.

[7] Acho, Leonardo, Alessandro N. Vargas, and Gisela Pujol-Vázquez. "Low-Cost, Open-Source Mechanical Ventilator with Pulmonary Monitoring for COVID-19 Patients." In *Actuators*, vol. 9, no. 3, p. 84. MDPI, 2020.

[8] Sathya, M., S. Madhan, and K. Jayanthi. "Internet of Things (IoT) Based Health Monitoring System and Challenges." *International Journal of Engineering & Technology* 7, no. 1.7 (2018): 175–178.

[9] Javaid, Mohd, and Ibrahim Haleem Khan. "Internet of Things (IoT) Enabled Healthcare Helps to Take the Challenges of COVID-19 Pandemic." *Journal of Oral Biology and Craniofacial Research* 11, no. 2 (2021): 209–214.

[10] Sabukunze, Igor Didier, Djoko Budiyanto Setyohadi, and Margaretha Sulistyoningsih. "Designing an IoT Based Smart Monitoring and Emergency Alert System for COVID19 Patients." In *2021 6th International Conference for Convergence in Technology (I2CT)*, pp. 1–5. IEEE, 2021.

[11] Otoom, Mwaffaq, Nesreen Otoum, Mohammad A. Alzubaidi, Yousef Etoom, and Rudaina Banihani. "An IoT-Based Framework for Early Identification and Monitoring of COVID-19 Cases." *Biomedical Signal Processing and Control* 62 (2020): 102149.

[12] Hoang, Minh Long, Marco Carratù, Vincenzo Paciello, and Antonio Pietrosanto. "Body Temperature— Indoor Condition Monitor and Activity Recognition by MEMS Accelerometer Based on IoT-Alert System for People in Quarantine Due to COVID-19." *Sensors* 21, no. 7 (2021): 2313.

[13] Hoang, Minh Long, Marco Carratù, Vincenzo Paciello, and Antonio Pietrosanto. "Body Temperature— Indoor Condition Monitor and Activity Recognition by MEMS Accelerometer Based on IoT-Alert System for People in Quarantine Due to COVID-19." *Sensors* 21, no. 7 (2021): 2313.

[14] Robu.in. "Ultrasonic Sensor" [Online]. Available: https://robu.in.

[15] eTechnophiles. "HC-05 Bluetooth Module" [Online]. Available: https://www.etechnophiles.com/hc-05-pinout-specifications-datasheet/.

# 22 Interleaved Cubic Boost Converter

C. Sankar Ram, Aditya Basawaraj Shiggavi, A. Adhvaidh
Maharaajan, R. Atul Thiyagarajan, and M. Prabhakar
Vellore Institute of Technology

## CONTENTS

22.1   Introduction .................................................................................................... 257
22.2   Circuit Description ......................................................................................... 258
22.3   Modes of Operation ....................................................................................... 258
    22.3.1   Mode 1 (S1-ON; S2-ON) ................................................................ 258
    22.3.2   Mode 2 (S1-ON; S2-OFF) ............................................................... 259
    22.3.3   Mode 3 (S1-ON; S2-ON) ................................................................ 259
    22.3.4   Mode 4 (S1-OFF; S2-ON) ............................................................... 259
22.4   Voltage Gain and Design Details ................................................................... 260
22.5   Switch Ratings ............................................................................................... 261
22.6   Diode Ratings ................................................................................................ 263
22.7   Design Expressions ........................................................................................ 264
22.8   Simulation and Inference ............................................................................... 264
22.9   Conclusion ..................................................................................................... 270
References ................................................................................................................ 270

### ABSTRACT

This paper delineates a novel high-gain DC-DC converter designed for photovoltaic (PV) applications. The proposed converter yields an overall gain of 22.22. The converter is synthesized by interleaving two similar cubic cell structures using a voltage lift capacitor. The two switches in each cubic cell operate at a 180-degree phase shift, thus canceling out the ripple currents. The switches in the proposed converter are subjected to 50% of Vo. The prototype is simulated using PSIM software to verify the proposed voltage gain concept. The converter specifications are as follows: 18 V input, 400 V output, 50 kHz switching frequency, and 200 W power rating. The efficiency of the prototype is 93.64%. Further, under closed-loop conditions, a regulated output voltage of 400 V is obtained.

## 22.1   INTRODUCTION

The modern world is moving toward sustainable development. As an alternative to conventional sources, renewable energy sources are being increasingly used to meet the ever-increasing electrical energy demand. A solar panel is one such source that supplies power to DC homes using a power electronic converter like boost and boost-derived converters [1–3]. Since the output from the solar panels is very low, high-gain boost converters are preferred. For applications demanding a voltage gain >10, boost converters suffer from diode reverse recovery issues, incremental power loss, and high voltage stress on the semiconductor devices when operated at extreme duty ratios. The idea of developing hybrid boost-derived converters stems from the issues faced by the classical boost-derived converters for high-voltage-gain applications [4–9].

DOI: 10.1201/9781003374121-22

Interleaved boost converters are mainly used in photovoltaic (PV) applications to obtain a ripple-free input current [10–13]. In PV applications, input current ripples should be very minimal as it affects the overall life span of the panels [14,15]. In Ref. [16], an interleaved quadratic converter is presented. A cubic cell is a single-switch multi-stage high-gain converter. The proposed converter has a single structure in which two cubic cells are interleaved using a lift capacitor which increases the overall gain. As coupled inductors (Cis) occupy lesser volume when compared to discrete inductors and result in a compact converter when implemented in hardware, coupled inductors are used. The converter is designed to yield a voltage gain of 22.22.

## 22.2   CIRCUIT DESCRIPTION

Figure 22.1 portrays the circuit diagram of the presented interleaved cubic boost converter (ICBC). The circuit comprises two similar cubical cell structures (voltage gain is cube times the gain of classical boost converter) interleaved using a $C_{Lift}$ capacitor to increase the output voltage. The two cubic cell structures have two switches operating at the same frequency with a 180-degree phase shift. Interleaving the two structures results in high efficiency, reduced ripple currents, and improved reliability. Additionally, the two switches being situated near the input side, experience low voltage stress only. The resultant output voltage gain is twice the gain of an individual cubic cell. The structure employs two coupled inductors and four discrete inductors. The inductors L1 and L2 are coupled in one cubic cell, and L4 and L5 are coupled in another cubic cell.

## 22.3   MODES OF OPERATION

### 22.3.1   MODE 1 (S1-ON; S2-ON)

In this mode, both the switches are turned ON. Inductors L1 and L4 start storing energy and linearly charge up to $V_{in}$ and forward-bias diodes D1 and D5, respectively. Diodes D2 and D6 will be in reverse blocking state. Capacitor C1 and C3 will transfer their stored energy to L2 and L5, respectively, which forward-bias the diodes D3 and D7, and diodes D4 and D8 will be reverse-biased.

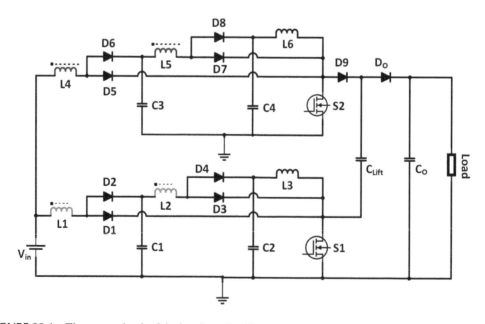

**FIGURE 22.1**   The power circuit of the interleaved cubic converter.

Similarly, C2 and C4 will be transferring the stored energy to L3 and L6, respectively, and $C_O$ expends its stored energy to the load. Mode 1 ends when inductors L4, L5, and L6 reach $I_{max}$.

$$i_{L1}(t) = \frac{V_{L1}(t)}{L1} t \quad i_{L4}(t) = \frac{V_{L4}(t)}{L4} t \tag{22.1}$$

$$i_{L2}(t) = \frac{V_{C2}(t)}{L2} t \quad i_{L5}(t) = \frac{V_{C3}(t)}{L5} t \tag{22.2}$$

$$i_{L3}(t) = \frac{V_{C2}(t)}{L3} t \quad i_{L6}(t) = \frac{V_{C4}(t)}{L6} t \tag{22.3}$$

### 22.3.2 Mode 2 (S1-ON; S2-OFF)

In Mode 2, switch S1 will be in ON state and S2 will be in OFF state. Inductors L1, L2, and L3 continue to charge. Inductors L4, L5, and L6 forward-bias diodes D6, D8, and D9 to transfer their stored energy to the capacitors C3, C4, and $C_{Lift}$, and diodes D6, D8, and $D_O$ will be reverse-biased. The capacitor $C_O$ transfers its stored energy to the load. Mode 2 ends when inductors L4, L5, and L6 reach $I_{min}$.

$$i_{L1}(t) = \frac{V_{L1}(t)}{L1} t \quad i_{L4}(t) = \frac{V_{L4}(t)}{L4} t \tag{22.4}$$

$$i_{L2}(t) = \frac{V_{C2}(t)}{L2} t \quad i_{L5}(t) = \frac{V_{C4}(t) - V_{C3}(t)}{L5} t \tag{22.5}$$

$$i_{L3}(t) = \frac{V_{C2}(t)}{L3} t \quad i_{L6}(t) = \frac{V_{CLift}(t) - V_{C4}(t)}{L6} t \tag{22.6}$$

### 22.3.3 Mode 3 (S1-ON; S2-ON)

In Mode 3, both the switches are turned ON. Hence, the circuit operation is similar to Mode 1.

### 22.3.4 Mode 4 (S1-OFF; S2-ON)

In Mode 4, switch S1 will be in OFF state and S2 will be in ON state. Inductors L4, L5, and L6 continue to charge. Inductors L1, L2, and L3 discharge their energy stored to the capacitors C1, C2, and $C_O$ through the diodes D2, D4, and $D_O$, respectively. Diodes D1, D3, and D9 will be reverse-biased. The capacitor $C_O$ transfers its stored energy to the load. Mode 4 concludes when inductors L1, L2, and L3 reaches $I_{max}$.

$$i_{L1}(t) = \frac{V_{L1}(t)}{L_1} t \quad i_{L4}(t) = \frac{V_{L4}(t)}{L4} t \tag{22.7}$$

$$i_{L2}(t) = \frac{V_{C2}(t) - V_{C1}(t)}{L2} t \quad i_{L5}(t) = \frac{V_{C3}(t)}{L5} t \tag{22.8}$$

$$i_{L3}(t) = \frac{V_{Clift}(t) - V_{C2}(t)}{L3} t \quad i_{L6}(t) = \frac{V_{C4}(t)}{L6} t \tag{22.9}$$

The equivalent circuit of each operating mode is depicted in Figures 22.2–22.4, while the characteristic waveforms are presented in Figure 22.5.

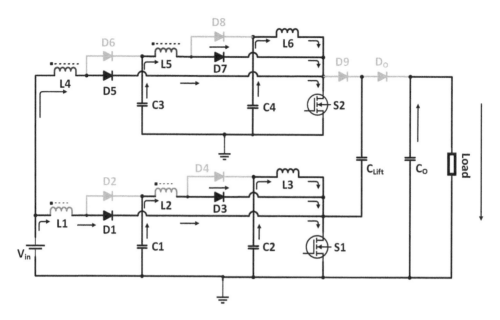

**FIGURE 22.2**  Equivalent circuit of the interleaved cubic converter during Mode 1 and Mode 3.

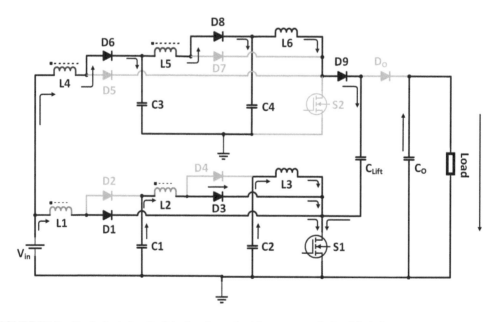

**FIGURE 22.3**  Equivalent circuit of the interleaved cubic converter during Mode 2.

## 22.4  VOLTAGE GAIN AND DESIGN DETAILS

The proposed circuit is designed by interleaving two identical cubic cell structures using a voltage lift capacitor. Using the volt–second balance in the circuit, the overall voltage gain expression is obtained; the voltage gain expression of a classical boost converter is given by

$$M_{IBC} = \frac{1}{(1-D)} V_{in}. \tag{22.10}$$

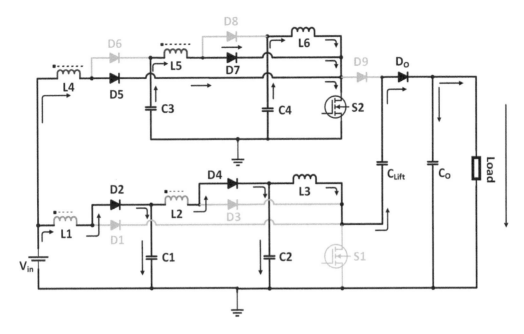

**FIGURE 22.4**  Equivalent circuit of the interleaved cubic converter during Mode 4.

Voltage gain expression of an interleaved boost converter with lift capacitor is given by

$$M_{IBC} = \frac{2}{(1-D)} V_{in}.$$ (22.11)

As both cubic cell structures are identical in nature, the voltage gain across cubic cells remains the same which is given by

$$M_{Cubic} = \frac{1}{(1-D)^3} V_{in}.$$ (22.12)

Now the overall voltage gain expression of the circuit calculated using volt–second balance is given by

$$\frac{V_o}{V_{in}} = \frac{2}{(1-D)^3},$$ (22.13)

## 22.5  SWITCH RATINGS

As the name suggests, voltage stress across a switch is the potential difference across the anode and cathode terminals of the switch when it is in the OFF state. The switches S1 and S2 function in 180° phase shift at 0.52 duty ratio. Due to the identical cubic cell structure, the voltage stress across the switches S1 and S2 remains the same. The voltage stress expression is given by

$$V_{S1} = V_{S2} = V_{C2} = \frac{1}{(1-D)^3} \times V_{in}.$$ (22.14)

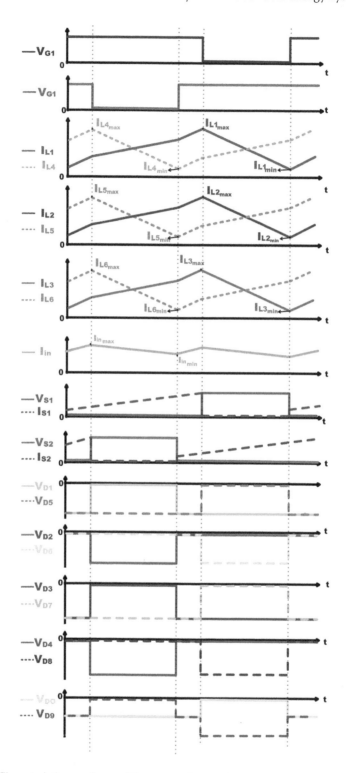

**FIGURE 22.5** Characteristic waveforms of the proposed converter.

## 22.6  DIODE RATINGS

Similarly, in diodes, the voltage stress is the potential difference across the anode and cathode terminals of the diode in reverse-biased condition.

During Mode 1 and Mode 3, diodes D2, D6, D4, and D8 are reverse-biased. The anode terminal of the diode is grounded via switch S1, and the cathode is connected to C1. Hence, the voltage across D2 is the same as the potential at the upper plate of C1. Similarly, the voltage across D4, D6, and D8 is the voltage across capacitors C3, C2, and C4, respectively. The voltage gain expressions are given by

$$V_{D2} = V_{D6} = V_{C1} = \frac{1}{(1-D)} \times V_{in} \qquad V_{D4} = V_{D8} = V_{C2} = \frac{1}{(1-D)^2} \times V_{in}. \qquad (22.15)$$

In Mode 4, the diodes D1 and D3 are reverse-biased. The cathode of D1 is connected to $C_{Lift}$ and the anode of D1 is clamped by the potential developed at the top plate of C1. Hence, the voltage across D1 is the potential difference between C1 and the lower plate of $C_{Lift}$. This is because

$$V_{D1} = V_{C_{Lift}} - V_{C1} = \frac{2D-D^2}{(1-D)^3} \times V_{in}. \qquad (22.16)$$

Similarly, the voltage across D3 is the potential difference between top plate of C2 and the lower plate of $C_{Lift}$.

$$V_{D3} = V_{C_{Lift}} - V_{C2} = \frac{D}{(1-D)^3} \times V_{in} \qquad (22.17)$$

In Mode 2, the diodes D5 and D7 are reverse-biased. The cathode of D5 is connected to $C_{Lift}$ and the anode is connected to C3. Hence, the voltage across D5 is the potential difference between C3 and the upper plate of $C_{Lift}$.

$$V_{D5} = V_{C_{Lift}} - V_{C3} = \frac{2D-D^2}{(1-D)^3} \times V_{in} \qquad (22.18)$$

Similarly, the voltage across D7 is the potential difference between C4 and the upper plate of $C_{Lift}$.

$$V_{D7} = V_{C_{Lift}} - V_{C4} = \frac{D}{(1-D)^3} \times V_{in} \qquad (22.19)$$

The diode D9 operates in reverse bias condition in Modes 1, 3, and 4. In Mode 1 and Mode 2, as the cathode of D9 is connected to the upper plate of $C_{Lift}$, whereas the anode is grounded via S2, the voltage across D9 is the potential difference across $C_{Lift}$; whereas in Mode 4, the cathode of D9 is connected to the upper plate of $C_O$ and the anode is grounded via S2. Hence, the voltage across D9 is the potential difference across $C_O$.

- In Mode 1 and Mode 3,

$$V_{D9} = V_{C_{Lift}} = \frac{1}{(1-D)^3} V_{in}. \qquad (22.20)$$

- In Mode 4,

$$V_{D9} = V_{C_O} = \frac{2}{(1-D)^3} V_{in}. \qquad (22.21)$$

Voltage across $D_O$ is the potential difference between $C_O$ and $C_{Lift}$. The cathode is connected to the upper plate of $C_O$ and the anode is connected to the upper plate of $C_{Lift}$.

$$V_{Do} = V_{CO} - V_{CLift} = \frac{1}{(1-D)^3} V_{in}$$

(22.22)

## 22.7  DESIGN EXPRESSIONS

There are two cubic structures interleaved together using a voltage lift capacitor. The respective inductors and capacitors in both structures have the same values. Inductors L1, L2, and L3 are connected to obtain one cubic cell, whereas inductors L4, L5, and L6 make the other cubic cell. Both cubic cells are operated by switches S1 and S2, respectively. They have a duty ratio of D and frequency of f.

The voltage across L1 and L4 will be the input voltage, and the inductance values are the same since the cubic structures are identical. The input current divides equally into the two branches and hence current through L1 and L4 is the same.

Inductors L2 and L5 are both energy-storing elements and their respective voltages are the same due to identical cubical structures. Similar to L1 and L4, inductors L2 and L5 are connected to switch S1 and S2.

The current through L2 and L5 is the same as that through capacitors C1 and C3, as C1 and C3 discharge through L2 and L5, respectively. Similarly, the voltage across inductors is equal to the voltage across their respective capacitors.

Similarly, the inductors L3 and L6 are directly connected to the switches. The current through L3 and L6 and C2 and C4 are same, when C2 and C4 discharge through L3 and L6, respectively. Similarly, the voltage across inductors will be equal to the voltage across their respective capacitors.

$$L_1 = L_4 = \frac{V_{in} \times D}{f \times \Delta i_{L_x}} \qquad L_3 = L_6 = \frac{V_{C2} \times D}{f \times \Delta i_{L_x}}$$

where $x = 1, 4$        where $x = 3, 6$

(22.23)

$$C_1 = C_3 = \frac{I_{C1} \times D}{f \times \Delta V_{C_x}} \qquad C_2 = C_4 = \frac{I_{C2} \times D}{f \times \Delta V_{C_x}}$$

where $x = 1, 3$        where $x = 2, 4$

(22.24)

$$C_{lift} = \frac{I_{Clift} \times D}{f \times \Delta V_{CLift}}$$

(22.25)

Capacitors C1, C2, C3, and C4 are part of cubic cell structure. C1 and C2 are connected to switch S1 which is part of one cubic cell, while C3 and C4 are connected to S2 which is part of the other cubic cell. $C_O$ is the capacitor across load. $C_{Lift}$ capacitor is used for cascading both cubic cell structures in parallel and lifts the output voltage from the cubic cell.

## 22.8  SIMULATION AND INFERENCE

The proposed converter is simulated in PSIM software using values specified in Table 22.1. The overall efficiency of the power circuit at a rated load of 200 W can be determined using the waveform in Figure 22.6, which turns out to be 93.64%. For the input voltage specified in Table 22.1, the desired output of 400 V is obtained. Hence, a voltage gain of 22.22 is obtained and verified by plugging in the values in the formulated voltage gain.

**TABLE 22.1**
**Parameters Used in the Proposed Converter**

| Parameter | Values |
|---|---|
| Input voltage ($V_{in}$) | 18 V |
| Output voltage ($V_0$) | 400 V |
| Output power | 200 W |
| Switching frequencies | 50 kHz |
| Inductance L1=L4, L2=L5, L6=L3 | 178.9 µH, 444.98 µH, 2.215 mH |
| Mutual inductance L1 and L2=L4 and L5 | 253.93 µH |
| Duty ratio | 0.521 |
| Capacitor C1=C2, C3=C4 | 136.8 µF, 27.49 µF |
| Capacitor $C_{Lift}$, $C_O$ | 5.54 µF, 1.38 µF |

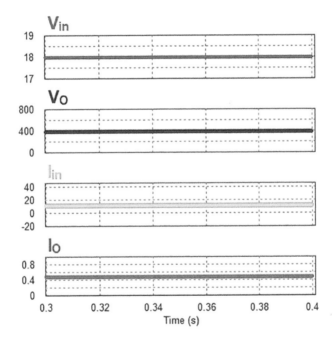

**FIGURE 22.6**  Input and output voltage, and the current waveform of the proposed converter.

The voltage stress across S1 and S2 are projected in Figure 22.7. The complementary operation of S1 and S2 are clearly observed. Further, their voltage stresses are equal and same as the voltage developed in a cubic boost converter. In percentage terms, the voltage stress of each switch is only 50% of $V_O$.

In Figure 22.8, the voltage stress on D1, D2, D3, and D4 is projected. The complementary operation of D1 and D3, and D2 and D4 is observed. The distortions in the waveforms of D3 and D4 are due to the leakage inductance.

Similarly, in Figure 22.9, the stress on D5, D6, D7, and D8 is projected. The complementary operation of D5 and D7, and D6 and D8 is observed. The distortions in the waveforms of D7 and D8 are due to the leakage inductance.

In Figure 22.10, voltage waveforms of gate pulses of S1 and S2, D9, and $D_O$ are depicted. When both the switches are in ON state, the voltage across D9 is 200 V which is half of $V_O$, while in

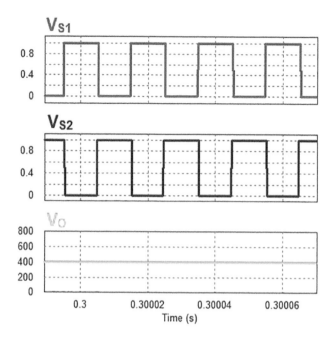

**FIGURE 22.7**   Voltage across switches of the proposed converter.

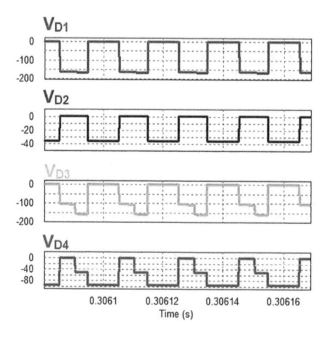

**FIGURE 22.8**   Voltage across diodes of the first cubic cell of the proposed converter.

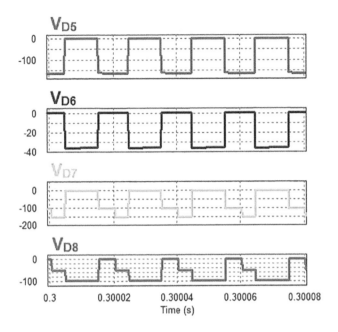

**FIGURE 22.9** Voltage across diodes of the second cubic cell of the proposed converter.

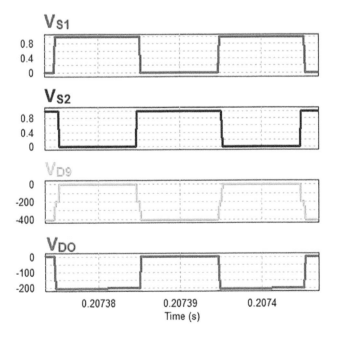

**FIGURE 22.10** Voltage across switches and diode D9 and $D_O$ of the proposed converter.

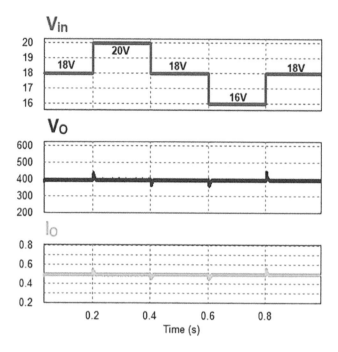

**FIGURE 22.11**   Voltage regulation due to variation of input voltage of the proposed converter.

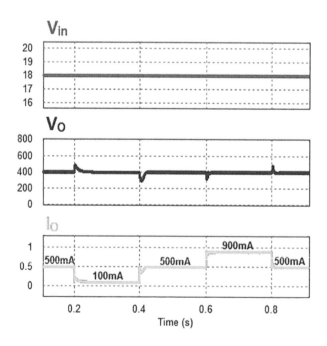

**FIGURE 22.12**   Voltage regulation due to variation of input current of the proposed converter.

Mode 4, maximum voltage stress of 400 V across D9 is observed. The diode $D_O$ is positioned between $C_O$ and $C_{Lift}$ and its voltage stress is obtained from the voltage difference between the top plates of $C_{Lift}$ and $C_O$. The complementary operation of $D_O$ and S1 is also observed.

In Figures 22.11 and 22.12, a regulated output voltage can be observed. As closed-loop control is employed, even when the input voltage is varied, a nearly constant output voltage of 400 V is obtained. In Figure 22.12, the load is indirectly varied by varying the output current using step current sources in PSIM. Even when the load is varied, a regulated output voltage is obtained due to the closed-loop control mechanism. Further, the initial overshoots during the transition period are also within acceptable limits.

In Figure 22.13, current waveforms at the input port and through L1 and L4 are visualized. Since the switches operate at duty ratio values different from 0.5, the input current ripples do exist but are minimal.

Figure 22.14 depicts the efficiency curve of the proposed ICBC. The maximum efficiency value of 93.6% is obtained at 200 W power level.

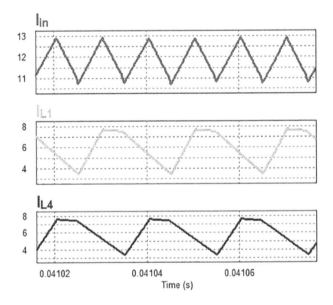

**FIGURE 22.13**   Input current and current across inductors in cubic cell of the proposed converter.

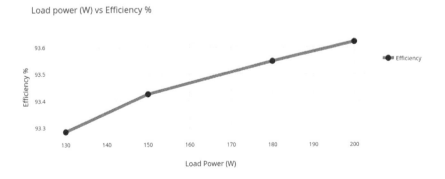

**FIGURE 22.14**   Efficiency of the proposed converter at various load power values.

## 22.9   CONCLUSION

In this chapter, an interleaved cubic boost converter was developed by interleaving two cubic cells with the voltage lift technique. The developed converter significantly enhances the voltage conversion ratio with low voltage stress across the switches. A cubic cell is a single-switch multi-level high-gain boost-derived converter. In the proposed converter, two cubic cells are interleaved to obtain an almost ripple-free input current waveform. By the combined actions of the lift capacitor and the cubic cells, a high gain of 22.22 is obtained. The switches of the proposed ICBC experienced minimal voltage stress of only 50% of output voltage. Simulation results obtained from an 18 V/400 V, 200 W converter confirm the voltage gain extension principle. Further, the converter operated under a full load efficiency of 93.64%, with a regulated output voltage of 400 V.

## REFERENCES

[1] Tofoli, Fernando Lessa, Denis de Castro Pereira, Wesley Josias de Paula, and Demercil de Sousa Oliveira Junior (2015). Survey on non-isolated high-voltage step-up dc-dc topologies based on the boost converter. *IEEE Transactions on Power Electronics*, 8(10), pp. 2044–2057.

[2] Forouzesh, Mojtaba, Yam P. Siwakoti, Saman A. Gorji, Frede Blaabjerg, and Brad Lehman (2017). Step-up DC–DC converters: A comprehensive review of voltage-boosting techniques, topologies, and applications. *IEEE Transactions on Power Electronics*, 32(12), pp. 9143–9178.

[3] Schmitz, Lenon, Denizar C. Martins, and Roberto F. Coelho. (2017). Generalized high step-up DC-DC boost-based converter with gain cell. *IEEE Transactions on Circuits and Systems I: Regular Papers*, 64(2), pp. 480–493.

[4] Samuel, Vijay Joseph, Gna Keerthi, and Prabhakar Mahalingam. (2019). Coupled inductor based DC-DC converter with high voltage conversion ratio and smooth input current. *IET Power Electronics*, 13. doi: 10.1049/iet-pel.2019.0933

[5] Samuel, Vijay Joseph, Gna Keerthi, and M. Prabhakar. (2021). High gain converters based on coupled inductors and gain extension cells. In *Recent Trends in Renewable Energy Sources and Power Conversion: Select Proceedings of ICRES 2020* (pp. 13–24). Springer, Singapore. doi: 10.1007/978-981-16-0669-4_2

[6] Samuel, Vijay Joseph, Gna Keerthi, and M. Prabhakar. (2019). High gain interleaved quadratic boost DCDC converter. In 2019 2nd International Conference on Power and Embedded Drive Control (ICPEDC) (pp. 390–395). doi: 10.1109/ICPEDC47771.2019.9036565

[7] P. Poovarasan, M. Saraswathi and R. Nandhini. (2015). Analysis of high voltage gain DC-DC boost converter for renewable energy applications. In *2015 International Conference on Computation of Power, Energy, Information and Communication (ICCPEIC)* (pp. 0320–0324). doi: 10.1109/ICCPEIC.2015.7259484

[8] Alzahrani, Ahmad Saeed Y. (2018). Advanced topologies of high-voltage-gain DC-DC boost converters for renewable energy applications. Missouri University of Science and Technology.

[9] Chakraborty, Uddesh, Pratinav Kashyap, Harmeet Kapoor, and Prabhakar Mahalingam. (2020). Interleaved high step-up DC-DC converter for photovoltaic applications. In *2020 IEEE First International Conference on Smart Technologies for Power, Energy and Control (STPEC)* (pp. 1–6). IEEE. doi:10.1109/STPEC49749.2020.9297728

[10] Kashyap, Pratinav, and Prabhakar Mahalingam. (2021). Non-isolated high step-up DC-DC converter with low input current ripple. In *2021 IEEE International Power and Renewable Energy Conference (IPRECON)* (pp. 1–6). doi: 10.1109/IPRECON52453.2021.9640764

[11] Lee, Sin-Woo, and Hyun-Lark Do. (2019). Quadratic boost DC–DC converter with high voltage gain and reduced voltage stresses. *IEEE Transactions on Power Electronics*, 34(3), pp. 2397–2404. doi: 10.1109/TPEL.2018.2842051

[12] Muhammad, Musbahu, Matthew Armstrong, and Mohammed A. Elgendy. (2015). A non-isolated interleaved boost converter for high-voltage gain applications. *IEEE Journal of Emerging and Selected Topics in Power Electronics*, 4, pp. 352–362. doi: 10.1109/JESTPE.2015.2488839.

[13] Balapattabi, Sri Revathi, and Mahalingam Prabhakar. (2016). Non-isolated high gain DC-DC converter topologies for PV applications - a comprehensive review. *Renewable and Sustainable Energy Reviews*, 66, pp. 920–933. doi: 10.1016/j.rser.2016.08.057.

[14] Elkhateb, Ahmad, Nasrudin Abd Rahim, Jeyraj Selvaraj, and Barry W. Williams. (2013). The effect of input current ripple on the photovoltaic panel efficiency. In *CEAT 2013-2013 IEEE Conference on Clean Energy and Technology*. doi: 10.1109/CEAT.2013.6775680.

[15] Alzahrani, Ahmad, Pourya Shamsi, and Mehdi Ferdowsi. (2020) Interleaved multistage step-up topologies with voltage multiplier cells. *Energies*, 13, p. 5990. doi: 10.3390/en13225990

[16] Zhu, Binxin, Lulu Ren, and Xi Wu. (2016). A kind of high step-up dc/dc converter using a novel voltage multiplier cell. *IET Power Electronics*, 10. doi: 10.1049/iet-pel.2016.0354.

# 23 Emerging Role of AI, ML and IoT in Modern Sustainable Energy Management

*Arpan Tewary*
Central University of Jharkhand

*Chandan Upadhyay*
Dr. R. M. L. Avadh University

*A.K. Singh*
R. P. P. G. College Sultanpur

## CONTENTS

23.1 Introduction ........................................................................................................ 274
23.2 Role of Artificial Intelligence (AI) in the Renewable Energy Sector ................ 275
    23.2.1 Merits ...................................................................................................... 276
    23.2.2 Demerits .................................................................................................. 276
    23.2.3 Applications of Artificial Intelligence .................................................... 276
        23.2.3.1 Improved Integration of Microgrids ...................................... 276
        23.2.3.2 Intelligent Centralization of Control Centres ........................ 277
        23.2.3.3 Smart Grid with Intelligent Energy Storage (IES) ................ 277
        23.2.3.4 Supply Chain Management in Biogas Plants .......................... 277
        23.2.3.5 Improved Safety and Reliability ............................................ 277
        23.2.3.6 Market Expansion .................................................................. 277
23.3 Role of Machine Learning (ML) in the Renewable Energy Sector .................... 277
    23.3.1 Merits ...................................................................................................... 278
        23.3.1.1 Solar Energy ........................................................................... 278
        23.3.1.2 Wind Energy ........................................................................... 278
    23.3.2 Demerits .................................................................................................. 278
    23.3.3 Applications of Machine Learning .......................................................... 279
        23.3.3.1 Accurately Predict Energy Demand ....................................... 279
        23.3.3.2 Optimize Energy Consumption .............................................. 279
        23.3.3.3 Prediction of Accurate Prices of Energy ............................... 280
        23.3.3.4 Predict Merit Order of Energy Prices .................................... 280
        23.3.3.5 Price Optimization through Better Trading ............................ 280
        23.3.3.6 Modelling of Bioenergy Plants .............................................. 280
        23.3.3.7 Predicting Malfunction of Wind Turbines .............................. 281
        23.3.3.8 Increase Power Plant Profitability with Optimized Scheduling
                and Pricing ............................................................................. 281
        23.3.3.9 Reduce the Potential for Human Error ................................... 281
        23.3.3.10 Predict Customer Lifetime Value (CLV) ............................... 281
        23.3.3.11 Predict the Probability of Winning a Customer ..................... 281

DOI: 10.1201/9781003374121-23

          23.3.3.12  Make Incredibly Specific Offers to Customers ................................ 282
          23.3.3.13  Reduction in Consumer Turndown ...................................................... 282
23.4    Role of the Internet of Things in the Renewable
      Energy Sector ......................................................................................................... 282
      23.4.1  Types of Internet of Things ......................................................................... 282
          23.4.1.1   Actuators .................................................................................. 283
          23.4.1.2   Sensors ...................................................................................... 283
          23.4.1.3   Communication Technologies................................................. 283
      23.4.2  Merits.............................................................................................................. 283
          23.4.2.1   Intelligent Grid........................................................................ 283
          23.4.2.2   Sustainability ........................................................................... 283
          23.4.2.3   Cost-Cutting and Data Management ...................................... 284
          23.4.2.4   New Business Prospects........................................................... 284
          23.4.2.5   Advanced Analytics................................................................. 284
      23.4.3  Demerits.......................................................................................................... 284
      23.4.4  Applications of Internet of Things.............................................................. 285
          23.4.4.1   Smart Grid Management ......................................................... 285
          23.4.4.2   Efficient Smart Meters ............................................................ 285
          23.4.4.3   Balanced Distribution Systems............................................... 285
          23.4.4.4   Higher Cost-Efficiency .......................................................... 285
          23.4.4.5   Better Transparency in the Way People Use Electricity.................. 286
          23.4.4.6   Supreme Residential Solutions................................................ 286
          23.4.4.7   Monitoring Biogas Production................................................ 286
          23.4.4.8   Better Control, Automation and Futuristic Possibilities................. 286
23.5    Future Scopes in AI, ML and IoT in the Renewable Energy Sector ................................. 287
23.6    Conclusion ............................................................................................................... 287
Bibliography ...................................................................................................................... 287

**ABSTRACT**

The relevance of modern computing tools and techniques is becoming quite significant nowadays. With time, such techniques will play a massive role in almost all spheres of human life to make it more comfortable. The renewable energy sector is also not an exception in this case. With the growing demand for clean energy and sustainable living, the need to integrate modern computing tools and techniques like Artificial Intelligence, Machine Learning and Internet of Things into the renewable energy sector is becoming more substantial. Such integration leads to meeting the demand satisfaction, improved output performance and reduced losses, which finally results in simultaneous as well as inclusive management of multiple energy generation, distribution and storage systems far and near. In this chapter, we put forward a wide-ranging discussion of these computing tools and how their integration can desirably upgrade our renewable energy systems as well as provide us with a robust platform for sustainable energy management. We have also included the pros and cons of these modern techniques and a few applications for each one of them in the renewable energy sector. We have wrapped up our discussion with an appurtenant conclusion and future scope that lies ahead.

## 23.1 INTRODUCTION

Presently, the global energy scenario is transforming very rapidly. With the rapid increase in human population worldwide and the huge exploitation of natural resources concurrently to satisfy the ever-increasing energy need, unprecedented pressure on the ecosystem is being strongly felt. Due to the overuse of conventional energy resources like coal, oil and natural gas to fulfil the necessary energy needs, a huge amount of pollution has already engulfed the global ecosystem. This has already started to show its consequences like climate change which is raising the average global temperature resulting in global warming. The burning of fossil fuels causes a lot of air pollution and also leaves endangered particles that affect the health of the people. Looking at the present state we see no signs of abatement in pollution over the next few years but rather more possibility of further deterioration in the current environment. Given this alarming situation, it is quite wiser to quickly shift towards a clean and green economy as well as adopt a sustainable lifestyle for ourselves and our planet.

The energy sector is facing significant problems and requires immediate attention along with strong revival actions. Policy pledges like the Paris agreement for a net-zero future, need a quick changeover towards a carbon-free economy. While renewable energy is expected to thrive in this environment, its sporadic nature will make it essential to find methods to keep systems steady. To integrate renewable energy, the manufacturing sector is also shifting from a market based on product pricing to a market based on technical resolutions. Reasonable predictions of power production and net load are becoming crucially influential, as the energy industry uses more variable generating options, ensuring system dependability, reducing carbon emissions and making the most of nonconventional energy resources.

Renewable energy techniques are one of the best options to combat the critical environmental challenges that we are facing today. They are non-polluting sources of energy generation units which are almost ecofriendly and easily available around us in one way or the other. However, their services are limited by constraints like less efficiency, performance losses, the smaller magnitude of production and most important intermittency in production. Therefore, these constraints outplay the renewable energy systems when compared with the conventional energy generation systems causing them to be less preferred among the masses. However, with the advent of technology, the integration of modern techniques like AI, ML and IoT is the need of the hour. The demand for renewable energy has soared rapidly in the last few years and the need for smart energy solutions is highly essential with this unparalleled soaring rate. Such integration will lead to inclusive management and control over all the assets simultaneously wherever located. This will help to prevent undesirable losses, hence better output. This will also make renewable energy cheaper and more reliable. For example, automatic or programmable solar photovoltaic/thermal systems can change their position as per the availability of sunlight during the day leading to optimized performance. Hence, such systems do not require human assistance for alignment or management as per need during operation, thus making them more efficient. Similarly, the health of any renewable energy system can also be thoroughly monitored round the clock even remotely to keep a close vigil on its performance and position. This will help to provide better control over the operation and maintenance of the system. Fault detection and any other malfunctioning of the systems and their components can also be identified and timely remedies can be provided. Such things can be done easily using robust and computerized systems which are run by modern computing tools. Such features have strongly supported the cause of installing renewable energy systems worldwide and now it can be said that these systems can challenge the conventional energy generation systems within the next few years, when completely installed and integrated with modern techniques in multiple places suitably with more output than the traditional systems. Grid integration and development of microgrids for well-organized and effective power distribution can also be facilitated using these tools. This work is a comprehensive discussion of the emerging computerized energy management tools and techniques. It also describes how these tools are potentially strong enough to robustly manage modern sustainable energy systems. Each one of these major tools is individually discussed below along with realistic examples in detail.

## 23.2   ROLE OF ARTIFICIAL INTELLIGENCE (AI) IN THE RENEWABLE ENERGY SECTOR

AI is a very futuristic tool and its integration with renewable energy systems is highly ultramodern. As we are determined to move towards more industrialized yet smarter societies, the need for sustainable power cannot be ignored. Power producers, distributors and consumers are looking forward to a system which can easily fulfil the huge energy demand with less pollution while operating from a common platform for operation and concern redressal. AI can be the suitable answer to all our modern requirements. AI's feature of predicting things much before their occurrence can help us to provide improved forecasting on-demand and failures, cost and asset management, etc. is a huge advantage.

AI when supported with other techniques such as ML, IoT and big data analytics makes the system more robust with better output. Such things help us to achieve automation capabilities, operational excellence, competitive results, etc. Therefore, AI can help us to unlock the huge potential stored in renewables. AI is much superior to humans' fault detection and addressing the complaint can be done within a shorter period. It can easily predict and accordingly suggest suitable solutions to a situation. AI algorithms are used for the said purpose. This increases the speed of operation with effective results. Seamless integration of complex electricity grids and real-time data collection helps in better innovation and development of more advanced techniques of system management in the future. This will make the grid and so the supply stable as well as reliable, therefore reducing carbon footprints indirectly through reduced losses.

### 23.2.1 MERITS

AI is said to be the most powerful tool in the modern energy sector. It can completely change the energy sector like it was never before, thereby shifting it into a completely digital platform. AI can be used for predictive analysis, and help to improve demand and supply forecasts in electricity trading. It can be used for policy framework, grid integration, supply security, etc. Nowadays, the concept of setting up a virtual power plant and regulation of energy storage can also be realized using this technology. This has reduced human dependence, which is a great achievement in itself.

### 23.2.2 DEMERITS

The potential for AI to transform the nonconventional energy sector is apparent, but that does not mean that expanding its use across the sector will be without obstacles. Poor data, user distrust and regulatory obstacles might all be hurdles for the technology. Questions have been voiced about relying too heavily upon AI could facilitate energy networks more prone to cyber-attacks. Operational technology structures must be segregated from IT systems and have no network connections between them, leading them much harder to enter. In terms of technology, reliance on cellular technologies would limit AI's potential in many emerging nations, particularly low-income ones. A lack of reliable connectivity is a significant barrier in areas where cellular network coverage is scant or limited. Insufficient theoretical background, practical expertise and lack of finance hurt the implementation of AI on a large scale in the renewable energy sector.

Users are likely to be dubious at first, as with any other new technology. Owners and occupants of buildings are likely to be concerned about the technology's ability to cut energy use and expenses while preserving energy services and comfort. To persuade customers to trust the technology, extensive awareness and marketing efforts will be required.

### 23.2.3 APPLICATIONS OF ARTIFICIAL INTELLIGENCE

### 23.2.3.1 Improved Integration of Microgrids

Integration of microgrids and distributed energy management can both benefit from AI. When community-scale renewable energy-producing units are added to the major grid, balancing the energy flow within the grid becomes difficult. The AI-powered control system has the potential to solve both quality and congestion problems.

### 23.2.3.2   Intelligent Centralization of Control Centres

To capture a significant amount of data, the electricity grid can be networked with devices and sensors. When combined with AI, this data can provide grid controllers with fresh insights for improved control operations. It gives energy companies the ability to intelligently adapt supply to demand. Advanced load control systems can be integrated with equipment such as industrial furnaces or huge air conditioners, allowing them to automatically shut down when the power supply is low. The flow of supply can also be modified using intelligent storage devices. Smart machines and modern sensors can also predict load and weather, which can help to upsurge integration and efficiency.

### 23.2.3.3   Smart Grid with Intelligent Energy Storage (IES)

AI combined with IES can offer long-term dependable results to the renewable energy market. This smart grid will be capable of analysing a large amount of data received from several instruments and making favourable energy allocation choices. This will also assist microgrids in managing local energy needs while maintaining power exchange with the main grid.

### 23.2.3.4   Supply Chain Management in Biogas Plants

With the use of AI, threats or disorderly happenings that might occur across the supply chain can be quickly assessed and more effectively minimized. Resource evaluation, logistics planning and power plant design are the three main factors that influence the bioenergy-producing plant's entire supply chain management. By keeping their multi-scaling feedstock database, which includes feedstock qualities, availability, demand, logistic data, etc., AI offers various other significant applications in supply chain management.

### 23.2.3.5   Improved Safety and Reliability

While managing intermittency is the primary purpose of AI in green energy systems, it can also increase reliability, safety and efficiency. It can assist you in determining energy usage patterns, as well as identifying energy wastage and device well-being. Data from wind turbine sensors, for example, can be collected using AI-powered advanced analytics to track depreciation. The system will keep track of the machine's general health and alert the operator when maintenance is required.

### 23.2.3.6   Market Expansion

By providing new service models and encouraging greater participation, AI integration can help green energy suppliers to further develop and enlarge their business. Such systems will also assist to study the energy collection data as well as deliver valuable perceptions into energy consumption. Such information would aid vendors in optimizing current services and launching new ones. It can also assist retailers in reaching out to new consumer markets.

## 23.3   ROLE OF MACHINE LEARNING (ML) IN THE RENEWABLE ENERGY SECTOR

Renewable energy is becoming more popular around the world. Predicting future situations becomes increasingly difficult as renewables become a larger part of the world's energy generation. ML applications, thankfully, can improve renewable energy forecasts in various ways. ML is basically a subtype of AI in which algorithms learn to find patterns from data with little or no human intervention. Many businesses utilize it to find ways to improve their operations or anticipate forthcoming changes. Renewable energy can also benefit from pattern-based prediction. Because renewables are dependent on nature, their efficiency and output can vary greatly. Improved prediction technologies can assist energy providers and users in getting the most out of these installations. ML is a type of technology that can filter through algorithms and data and learn and improve its ways as it gains more experience. It's a subset of AI, which is referred to as a network capable of thinking like a person. AI will play a critical role in the future of energy, according to analysts and experts throughout the industry. This is especially true in the subject of sustainable energy, where, while there are many uses of ML in this field, its potential to effect changes and innovation has yet to be completely realized. Some of the challenges that ML is beginning to tackle for energy businesses are an elegant fit. This is especially true for solar and wind generation, which have historically been hampered by difficult-to-predict weather patterns with numerous variables to consider.

### 23.3.1 MERITS

ML has a distinctive role to play and has a lot of advantages in the renewable energy sector. The use of ML in the renewable energy sector is quite new, but within this short span of time it has already started to show outstanding and promising results. Its efficient role in the field of two major renewable energy generation sources is described below:

#### 23.3.1.1 Solar Energy

AI has already aided solar generation in various areas, most notably in weather prediction. Grid supply may be forecast more accurately with more precise weather predictions. Grid operators place great importance on this. In the United States, the Department of Energy teamed with IBM to develop Watt-Sun, an ML technology that sorts through data acquired from a large database of weather reports. The goal was to reduce the variability of solar energy's variable output, with the project's expenditure justified by the ability to drastically reduce expenses associated with excess energy storage and deficit energy production. Forecasting accuracy was increased by up to 30% because of this technology. The unpredictability of sunshine is unavoidable, but it may be managed. Another fascinating area in which ML is being used is AI integration with microgrids. ML is increasingly being used to address the difficulties of intermittent electrical supplies and varying loads. Operators can strive to alleviate grid bottlenecks by assessing current flows and deciding whether energy should be stored in a battery, distributed or sold in an electrical exchange at a specific location and time, resulting in more efficient energy use. The battles over how to best distribute electricity that has surrounded the microgrid since its conception now appear to be relying on advances in AI and ML to be once again practicable. Such initiatives offer enticing benefits that aren't confined to solar energy.

#### 23.3.1.2 Wind Energy

Wind energy suffers many of the same issues as solar energy when it comes to predicting weather conditions. In 2014, Google along with the AI expertise of the company DeepMind was able to anticipate 36 hours of production ahead of time using ML techniques and generally available weather reports. This helped them reach their goal of raising the value of their energy by up to 20%.

Another application of ML that is shown to be capable of dramatically lowering costs for energy producers is condition monitoring, particularly when applied to wind turbines. Recent models have been able to monitor blade defects and generator temperature, enabling quick repairs and maintenance utilizing ML and data. Because energy demand and production fluctuate, operations at offshore wind farms must work smoothly. A fundamental component of this is having dependable and precise monitoring and regulation. The effect of smoothing out the bumps in the energy supply chain and boosting efficiency cannot be overstated; nonetheless, this alone is insufficient to accurately anticipate the power output of a particular turbine or even farm.

### 23.3.2 DEMERITS

For correctly integrating ML approaches into energy system workflows, several methodological orientations for ML are necessary. These include efforts to address the physical requirements of energy systems through hybrid physical modelling and reliable ML, to better integrate ML into deployment workflows by using interpretable and uncertainty-aware methods, and to address problems like dispersion shift and irregular data accessibility through transfer learning and domain adaptation. To support the adoption of ML models, the energy industry will need to solve many blockages, including data accessibility and the digital split, as well as offer testbeds or collaborative stages to fill the opening between investigative studies and implementation. Since ML is inherently a system amplifier within which it operates, strong regulatory steps will be needed to guarantee that energy system inducements are adequately harmonized with sustainable goals.

### 23.3.3 Applications of Machine Learning

The growth of renewable energy, shifting towards a smarter grid, and aggressive marketing are all transforming the energy distribution industry and reducing the utilities' margin of profit. The necessity to make wiser decisions is greater on a scale. Fast decisions must be taken to compete. The most important instrument for choosing better prices and creating more effective consumer interactions is quickly evolving to be dependent on ML. The following are some potential areas where ML can be used.

1. Estimate prices and demand.
2. Optimize wholesale and retail prices.
3. ML may help us to streamline retail costs, design the best attractions and draw in more consumers as well as lower the consumer turndown.
4. Estimate the lifespan of the customers.
5. Ultimately help both service provider and customer in reducing undue interruptions through smooth and adequate services.

#### 23.3.3.1 Accurately Predict Energy Demand

Any utility firm that serves its customers should be capable of anticipating their energy requirements with accuracy. Presently, due to the unavailability of any viable means for large-scale energy storage, energy must be supplied as well as immediately used as soon as it is produced. To increase these projections' accuracy, ML is now being applied. The amount of energy utilized is influenced by previous energy consumption data, prediction of weather and the kinds of consumers operating on a given day. Office buildings require more energy when the air conditioning systems are running at full capacity, as on a hot summer day during the week, for instance. To prevent rolling blackouts brought on by air conditioners in the summer, forecasting weather and previous data can help spot these patterns early on. For example, the day of the week, time, anticipated wind and solar radiation, historical demand, air temperature, mean demand, moisture and pressure, along with wind direction are some of the influencing factors that ML can uncover complex patterns. This information can be used to explain demand fluctuations. Due to its ability to identify more complex patterns, ML can predict outcomes better than humans. As a result, it is possible to purchase energy more effectively and save money without making expensive adjustments.

#### 23.3.3.2 Optimize Energy Consumption

Long-time users of energy are aware of how much they consume both at home and work. We are unaware of which devices/appliances consume the highest energy because we are unable to gather an overall picture of energy use without performing numerous human calculations. The expansion of IoT and smart meters are to blame for the changes in all of that. Disaggregation, also known as non-intrusive appliance load monitoring, uses ML to analyse energy use at the device level. Which appliances cost the most to run? That is something which can be helped with. Both commercial and residential clients will benefit from this in terms of fine-tuning their consumption patterns to reduce their costs and energy use. The choice is theirs to use expensive gadgets infrequently or swap them out for energy-efficient alternatives.

### 23.3.3.3 Prediction of Accurate Prices of Energy

Predictably domestic renewable (wind and solar) energy production will become more convenient and affordable, as consumers and companies are now insisting more on producing their required energy. Personal power generation enables individuals to produce, use and store energy. Depending on the place they reside, they could even be able to sell extra electricity to the energy production or distribution company. When it comes to choosing the most favourable moment to generate, store or sell energy, ML can help. Energy should be consumed or stored while costs are low, and supplied back into the system when costs are high. When we employ ML models to assess previous data, consumption patterns, and forecast the weather, we can produce hourly projections that are considerably more accurate. This supports people and companies who have power generation systems in their strategic resource allocation decisions. An example of this is the employment of the Adaptive Neural Fuzzy Inference System in predicting short-term wind patterns for electricity production. Such things enable businesses to enhance energy output while also reselling it at peak prices to the grid.

### 23.3.3.4 Predict Merit Order of Energy Prices

Energy supplies available to utilities include nuclear electricity, fossil fuels and renewables like wind and solar. At the time of selling power, all these diverse energy means are arranged in a proper order depending on price. The order in which the different power sources are sold is affected by this. Considering that we have data available from numerous means, we may use ML to analyse both current and past data. By considering all of the various factors that affect pricing, including weather, demand, the amount of energy that is available from different sources, historical usage and so forth, ML algorithms are also good at anticipating optimum merit. You are then able to decide on your power source with greater knowledge. Since it can be challenging to predict when electricity will be available from renewable sources, like wind, this is particularly helpful in those markets. If you oversee wind turbines or another kind of power plant, you are well aware of the following: Every year, millions of dollars may be lost due to maintenance, human error, downtime and ineffective planning. ML is proving to be a potent tool for the following areas in which conventional statistics have stalled: Predict malfunctions more precisely and sooner, identify human errors before they become a major issue, and increase your profitability by optimizing power plant scheduling.

### 23.3.3.5 Price Optimization through Better Trading

Offering the most competitive costs for electricity in a market that is open to competition is essential to surviving. Price comparison is inevitable when consumers have the opportunity to choose who supplies their electricity. To remain competitive, utility companies use ML to identify the proper times to purchase electricity depending on the lowermost prices. In terms of commodity tracing, a plethora of factors, like the time of day and the weather, affect energy pricing. Utility firms will be able to make better decisions about when to acquire and sell energy thanks to the application of ML to evaluate minor changes in these parameters. It can also help in better predicting the day-ahead prices for the consumers, combining gradient boosting, general additive and deep learning models. ML can also help energy distributors in choosing trading strategies to lower the cost of future purchases.

### 23.3.3.6 Modelling of Bioenergy Plants

Building models for accurate predictions based on previous experience or theory is challenging due to the intricate nature of bioenergy systems and the constraints of human understanding. New opportunities may now be available as a result of recent advancements in data science and ML. ML can assist in bioenergy technology, including the use of lignocellulosic biomass for energy, the growth of microalgae and the conversion and use of biofuels. A new generation of bioenergy and biofuel conversion technologies may advance due to the capabilities and possibilities of sophisticated ML approaches while dealing with a wide range of activities in the future.

### 23.3.3.7    Predicting Malfunction of Wind Turbines

Despite being an excellent source of renewable energy, maintaining wind turbines is infamously costly. It may make up as much as 25% of the overall cost per kWh. Problems that are already present may be much more expensive to fix. By identifying problems before the turbine breaks down, ML can help in keeping ahead of the issue and cut down on maintenance expenses. This is crucial since repair costs are significantly higher when wind farms are constructed in remote areas, like the middle of the ocean. Supervisory Control and Data Acquisition software which is used in gathering real-time data could assist to identify probable issues with the system early enough to prevent disaster. To train ML models to anticipate failure precursors like low lubricant levels, for instance, we can use data from turbine sensors, including oil, grease and vibration sensors. This method allows failures to be predicted a considerable number of days in advance.

### 23.3.3.8    Increase Power Plant Profitability with Optimized Scheduling and Pricing

Because energy costs are as volatile, even something as simple as the time of day can affect the profitability of a power plant. However, due to the fast-paced nature of the utility industry, manually tracking all of the data needed to make these judgements can be difficult. ML in this case can be of great assistance. You can estimate the optimal times to run your plant and generate more money by feeding previous prices and consumption data into a ML algorithm. ML can identify periods when energy consumption is high but raw material prices are low. These incredibly precise estimates result in a more profitable generation schedule.

### 23.3.3.9    Reduce the Potential for Human Error

Human mistake is responsible for up to 25% of power plant failures each year. This results in service interruptions for customers, as well as the loss of up to millions of megawatt-hours of annual energy generation. It also means that the costs of correcting the problem and bringing the system back online are unnecessary. To overcome this, we can employ ML to assist control room workers in making judgements. ML allows for continuous system monitoring, which aids in the detection of anomalies. We also advise a course of action to keep the situation from getting worse. It can even solve an issue without the need for human intervention. This decreases the danger of human error owing to distraction, lack of understanding, or reaction time—control room operators can't always react quickly enough to stop a disaster.

### 23.3.3.10    Predict Customer Lifetime Value (CLV)

Indicators like CLV must receive more attention from utility owners and providers in an open utility market. They can determine how much each customer will spend throughout their contract thanks to this ML technology. You can achieve more with ML than just increased CLV prediction accuracy. By incorporating information about the client, their consumption patterns, location, past purchases and payment behaviour, we may be able to anticipate the entire value of a certain client using ML models like deep neural networks. With ML, we are even able to recommend strategies to increase consumer value. This can entail making extremely specific offers to clients who behave similarly or enhancing customer service through the use of natural language processing.

### 23.3.3.11    Predict the Probability of Winning a Customer

To stay ahead of the competition, energy providers operating in open markets must have a complete picture of their prospective clients. But ML is capable of giving you more than just this overall picture. Additionally, it could provide you with the information needed to make data-driven marketing decisions. Such things imply that you'll be able to identify right away whether a visitor to your website will convert to a buyer. With the use of ML, we can accurately portray the individual as a customer based on the information they provide, including their location, the type of computer they are using, their surfing and search histories and the number of times they have visited your website.

After that, ML can determine whether or not that person is likely to become a customer using a process commonly known as scoring as well as the best way to convert them.

### 23.3.3.12   Make Incredibly Specific Offers to Customers

Consumers have a choice of utility suppliers in open energy markets. Particularly, in times when the loyalty of the brand isn't as strong as it was once, personalized offers are essential for bringing in new customers and keeping hold of existing ones. With the aid of ML, we can assist you in gaining the knowledge required to create enticing offers that specifically address the demands of a given customer. By analysing purchasing patterns and client information, ML can help you decide what kind of offer is most appropriate to make to a certain consumer at any particular time. You may offer to waive the connection fee at a customer's new address if the data indicate that they are about to move. With a personalized offer like this, you can outperform the competition and reduces customer dissatisfaction.

### 23.3.3.13   Reduction in Consumer Turndown

In open energy markets, wherein consumers have a selection of utility suppliers, it is imperative to comprehend which consumers will leave the market. The proportion of customers who stop using a service in a given year can reach 25% in the energy industry. You need to be able to predict consumer turndown accurately and stop it from happening if you want to survive. Utility providers can forecast when a consumer is going to leave with the use of ML. Utility companies can use techniques like Cross-Industry Standard Process for Data Mining, AdaBoost and Support Vector Machines, as well as previous consumption data, to identify crucial indicators that determine whether a customer would turn down or not. Examples of these include client happiness, employment status, energy use, home ownership and rental status. Any of these symptoms may change if a client is getting ready to discontinue service. If we recognize these symptoms early enough, we can stop consumer turndown by assisting clients in solving any issues they may be experiencing.

## 23.4   ROLE OF THE INTERNET OF THINGS IN THE RENEWABLE ENERGY SECTOR

The IoT has successfully altered the current world landscape. It's critical to develop future-proof energy systems, and the IoT can help you get there quickly. IoT applications that are gaining traction in the renewable energy industry, include grid management, distributed systems, smart meters and residential solutions, as well as the benefits that IoT provides, such as better control, cost efficiency and transparency. IoT devices, in their simplest form, assist in the remote management and monitoring of equipment in real time. They can also automate specific operations in a system and give workers more control, resulting in lower operational costs and less reliance on fossil fuels in the energy industry. Finally, enhanced load control and process efficiency result in greater cost-effectiveness. As a result, it is critical to implement smart energy production systems as soon as possible. The current market dominance of IoT is inconceivable. IoT in the green energy industry alone can probably play a crucial role in raising corporate productivity, driving operational efficiency and improving employee safety, which is the need of the hour. As you can see, the tremendous shift to renewable energies necessitates optimization and digitization. This is especially useful when turbines are installed in inaccessible locations.

### 23.4.1   Types of Internet of Things

In general, IoT is categorized into five different types:

1. Consumer IoT,
2. Commercial IoT,

3. Industrial IoT,
4. Infrastructure IoT and
5. Internet of Medical Things (IoMT).

Commercial IoT applications are essentially a large-scale version of Consumer IoT mostly utilized by supermarkets and commercial buildings. Consumer IoT includes typical IoT devices and applications used by consumers. In the meantime, Industrial IoT applications are employed in industrial settings such as factories, manufacturing units and so forth. Infrastructure IoT, on the other hand, refers to the integration of IoT technologies with intelligent infrastructures.

IoMT ecosystem allows for the gathering, analysis and transmission of health data to increase efficiency, enhance outcomes and reduce healthcare costs. However, more people are now highly interested in using IoT in the renewable energy sector.

### 23.4.1.1 Actuators

Actuators are machines that turn one type of green energy into motion. Oscillatory, linear and rotational motions are among the movement patterns they produce. Actuators in the energy sector communicate with other nearby devices to provide useful services.

### 23.4.1.2 Sensors

These units are used to collect and transmit real-time data. The major goal is to produce, transfer and distribute green energy in real time, allowing for energy optimization and load control.

### 23.4.1.3 Communication Technologies

Sensors are linked to IoT gateways via wireless systems, allowing for end-to-end data exchange in the environment. Wind and solar energy plants, as previously indicated, are frequently located in remote locations that are difficult to access. One can ensure real-time data flow while conserving energy by deploying IoT devices.

## 23.4.2 MERITS

Optimization of resources and effectively monitoring the processes in a power plant, using sensor devices automates processes and provides superior, usually error-free services. IoT technology is a clever notion that protects against the overuse of resources while also assisting with consistency. IoT enables smart process monitoring, which provides data on every aspect of the plant's operations.

### 23.4.2.1 Intelligent Grid

The IoT enables a smart grid system to gain control over power flow or significantly reduce energy use. It reduces the energy load, even more, to match generating that is real-time or nearly real-time. IoT is a computerized concept that provides a cost-effective method of interconnecting users for efficient power consumption. This also contributes to the economy by conserving energy as a resource and decreasing energy waste. To combat the environmental effects of increased vehicle traffic, plug-in hybrid cars and autos are currently being offered. These vehicles also necessitate a smart grid for cost-effective charging, which allows customers to save money. As a result, adopting an intelligent grid enables managers to get real-time monitoring healthier energy management and speedier data restoration.

### 23.4.2.2 Sustainability

All assets have been designed to communicate with one another via IoT. The energy sector is a primary driver of responsibility in terms of finding innovative solutions to address environmental challenges. IoT also allows for automatic maintenance and reporting, smart grid optimization, renewable energy generation and real-time carbon consumption measurement. Through its sophisticated ways, technology is enabling sustainability throughout the industrial world, allowing managers to make educated decisions for improved corporate success.

### 23.4.2.3   Cost-Cutting and Data Management

In the energy sector, the IoT is a sophisticated process that combines consumption pattern planning and energy management across different domains. It gives managers complete control over energy data and allows them to drastically improve the process. Sensor-based technologies are used to establish the automated functioning of the energy sector when using an IoT-powered solution. With real-time data processing and analytical decision-making, the industry has reaped the greatest benefits. It effectively manages data and makes every piece of information matter by safely keeping it on the cloud platform. As a result, IoT has a tremendous impact on the energy sector, adding intelligence to every critical feature.

### 23.4.2.4   New Business Prospects

The IoT introduces new business opportunities as well as newer and more advanced concepts. Sensor devices, gateway connectivity and communication protocols are all combined to form an IoT architecture for a variety of enterprises. IoT technology can be used to help businesses and provide smart strategies for increased production and growth.

### 23.4.2.5   Advanced Analytics

The power industry's sensor-based operation is ushering in a revolution. It makes use of innovative approaches to meet corporate needs and produce high-quality output. Industrialists are maximizing the value of advanced analytics in their operations. It extracts data from assets using sensor-enabled data and makes better decisions than before. The power sector can benefit from data analytics to optimize generation and planning.

### 23.4.3   DEMERITS

Even though IoT is becoming more prevalent among consumers and businesses, the absence of security is a major concern. Just a few years back a very well-known European country faced a grid cyberattack causing complete disruption due to a malware attack in its power distribution system leading to power cuts for longer hours. This kind of attack on the system stopped the normal functioning of the substations and affected millions of people, because IoT devices exchange data over the internet. As a result, the compromised data have a significant impact on consumer and industry privacy. IoT devices must build top-notch encryption for sharing information over the internet to overcome the difficulties. Apart from that, technology causes unemployment in poor countries by substituting thing-to-thing communication for human interaction. People also want to complete their jobs faster and with less effort. While the IoT makes their jobs easier and more efficient, it also makes them lazy and reliant on technology. The main constraints at the moment are financial costs and some countries' incapacity to keep up with modern technology. However, with a collective worldwide effort, IoT may be implemented and a sustainable future can be achieved. It is unsurprising that in modern times, we consume enormous amounts of natural, nonrenewable resources, particularly energy. Energy production from nonrenewable resources is costly and has significant environmental consequences. Coal consumption, for example, results in the emission of additional greenhouse gases, particularly carbon dioxide, into the atmosphere ($CO_2$). Many people are unaware, however, that 1.2% of the Sahara desert may be used to meet world energy demands by installing solar panels. According to research, the cost of installing solar panels in 2016 was around $5 trillion. Aside from the expense, the power loss owing to the remote position can be as high as 10%. As a result, it is becoming increasingly difficult to put global solutions into practice. Furthermore, most countries, particularly developed and developing countries, lack the necessary infrastructure to undertake large-scale investments in renewable energy resources. In many circumstances, they lack the financial resources to do so. As a result, they are unable to realize their ambition of constructing a smart city. Smart meters provide organizations with the difficult issue of integrating them with their existing infrastructure. They must also make it a valuable solution package so that their firm may reap major benefits. Most businesses employ antiquated infrastructure that cannot be coupled with contemporary, digital technology. As a result, they will need to completely redesign their operations to meet their objectives. Unfortunately, the majority of them lack the necessary funds.

### 23.4.4 Applications of Internet of Things

#### 23.4.4.1 Smart Grid Management

Human labour makes continuous real-time data collection nearly difficult. IoT, on the other hand, may make it so simple. Companies will be able to collect real-time consumer usage data by integrating IoT technologies and sensors into smart grid distribution lines and power substations. It enables businesses to make smarter, more efficient decisions about voltage regulation, network configuration and load switching between distribution lines. They can also detect problematic lines quickly thanks to real-time data collection. As a result, the risk of wildfires, electrocution and other hazards is reduced. Furthermore, because IoT offers an automated procedure, allowing sensors to take decisive steps instantly if the need arises, overall grid management becomes easier. In the event of a large-scale outage, smart switches, for example, can effectively shut down troublesome areas. The distribution lines are, therefore, protected. IoT can effectively control and regulate power excess without disrupting overall operations.

#### 23.4.4.2 Efficient Smart Meters

Many negative consequences can be avoided by using IoT smart meters. If your meters are connected to the internet via IoT, for example, you can keep a close eye on your energy consumption. You may also track your intake in great detail. There are various advantages to using smart meters. They are affordable and have the potential to eliminate power disruptions. They may be able to diagnose and correct the problem remotely in some circumstances. Power is soon restored as a result of this. Smart meters can also help people conserve water. They may track their water consumption in real time and adjust their water allocation accordingly. The most important advantage of smart meters may be precise invoicing. Nobody will be overcharged. There will be no more ludicrous behaviour. There will be no more astronomical invoices or unknown hidden costs. In an industrial setting, these meters provide the same benefit.

#### 23.4.4.3 Balanced Distribution Systems

Smart energy grids have seen remarkable growth in recent years due to an increase in residential and business demand. As a result, controlling the main generators and the smaller units dispersed throughout the smart grid is getting increasingly challenging. It is, however, very simple to control and monitor these vast smart grids using IoT apps. Between the manufacturing and transmission lines, strategically placed sensors gather real-time data. The smart grid can be handled more efficiently the more control it has. In distributed power networks, this results in lower costs and improved balance.

#### 23.4.4.4 Higher Cost-Efficiency

IoT technologies are the key to better power usage control and monitoring. Electricity producers, providers and energy firms can collect critical data via IoT technologies. They will be able to monitor energy consumption, peak demand times and so on. As a result, people will have more information to make better selections. For example, they can balance power demand and supply over a set period to reduce waste. Furthermore, IoT is more dependable and cost-effective than having humans perform the activity. They can thereby save money in both circumstances.

### 23.4.4.5 Better Transparency in the Way People Use Electricity

One can acquire insights into their power usage habits and make adjustments using IoT-driven data analytics and visualization solutions. Raw usage data can be transformed into illustrative graphs and charts, which can then be presented in an easy-to-read dashboard. As a result, both families and utility providers may benefit from the data and take action that matters.

### 23.4.4.6 Supreme Residential Solutions

The advantages of renewable energy sources are becoming more widely recognized. Although their initial setup costs are substantial, their ongoing operating and maintenance costs are modest when compared to what they pay utility companies. According to several reports, the average power wasted per person per month is nearly 250 kW or more in a few developed countries. Installing solar panels means you won't have to pay for such massive waste, lowering your costs. As a result, renewable energy implementation is not limited to public or private sector utility firms. Small-scale solar panels can be placed in people's backyards to generate electricity. Frequently, the product meets the energy needs of the houses. Sensors can be added to residential systems, and IoT technology will make management easier. They can simply operate their smart roofs, windows, solar panels and other home electronics via a desktop or mobile application, allowing them to control the electronics in their home remotely. Thermostats are an excellent example of one of the most widely utilized IoT devices in home settings.

### 23.4.4.7 Monitoring Biogas Production

IoT can be used for monitoring biogas generation. Both the quantity and quality of biogas produced in the biogas plant can be monitored. To monitor, forecast and improve biogas output and thus get beyond the current difficulties and limitations of biogas production, IoT-managed systems paired with several other types of controllers and software are utilized. The initial work is to detect and then forecasting is done followed by resolving the issue. All parameters like temperature, pressure, pH, the quantity of raw material loaded and the amount of gas produced are all measured. Based on the antecedents, further forecasting is done with the proper outcomes. The anticipated results are further visualized layer to identify the variables influencing gas production and the appropriate actions to be done to increase gas production.

### 23.4.4.8 Better Control, Automation and Futuristic Possibilities

Wind and solar energy are the two most well-known renewable energy sources in the world today. We are all aware that large-scale electricity can be generated using many wind turbines and solar panels when connected together. Such systems can be built for domestic uses as well as for large-scale commercial and industrial uses. Though small systems to satisfy domestic needs can be set up in our homes or locality, these large-scale systems require good amount of space, and hence they are installed in dedicated areas called power plants or renewable energy parks. Nowadays, the concept of generating electricity through hybridization of two or more renewable energy technologies is also gaining momentum. Most of the developing countries are not far behind, with a high percentage of commercial, industrial and residential solar power and wind power in use. The cost of constructing such power plants, as well as the cost of producing electricity, has reduced dramatically over time. As a result, technologically advanced nations are also now adopting this strategy. A huge amount of renewable energy in developed countries is produced from solar and wind. The entire operation will be more dependable if both production methods are combined with IoT sensor applications. For example, an IoT system can manage the dual-axis trackers on solar panels remotely. Renewable energy workers may easily employ IoT sensors to vary their angle according to the sun's beams to enhance radiation intake. In addition, IoT devices can be used in windmills to closely monitor a variety of important characteristics for generating electricity. As a result, total manufacturing efficiency is improved. This can also further lead to possible technological bloom in this sector in the future.

## 23.5   FUTURE SCOPES IN AI, ML AND IoT IN THE RENEWABLE ENERGY SECTOR

Given the above study, we can say that these tools are highly futuristic and hence hold a lot of promise. We see that these tools can address things much better than they could be done manually in a time-bound manner. AI, ML and IoT will make things more user-friendly and customer-centric. In the energy sector, AI has a lot of promise to improve power generation, transmission, distribution and consumption. ML can act concurrently with AI and obtain results for policy framing. IoT can help in the efficient operation and maintenance of power-generating systems. Hence, both the government and non-governmental agencies can invest more in developing and integrating such systems to make this sector much more advanced, profitable, yet affordable and immensely people-centric.

## 23.6   CONCLUSION

From the above discussion, it can be easily inferred that the integration of modern computing tools into renewable energy systems is highly futuristic. Such facilities help us to realize the vast potential present in these renewable energy technologies which can capably satisfy all our energy needs. The role of these computing techniques is indeed prolific towards achieving sustainability, net zero and providing a respite from the evils of global catastrophes like pollution, unhealthiness and climate change. There are several other future scopes and promises as employment opportunities for engineers, and technicians, reduced pressure on the global ecosystem and many more that do exist in the implementation of these tools which can certainly make human life more comfortable in the future.

## BIBLIOGRAPHY

1. Srivastava, S. K., Application of artificial intelligence in renewable energy. In: *2020 International Conference on Computational Performance Evaluation (ComPE)*, pp. 327–331, 2020.
2. Jha, S. K., Bilalovic, J., Jha, A., Patel, N., Zhang, H. Renewable energy: Present research and future scope of artificial intelligence. *Renew. Sust. Energ. Rev.*, 77, 297–317, 2017.
3. Hannan, M. A., Al-Shetwi, A. Q., Ker, P. J., Begum, R. A., Mansor, M., Rahman, S. A., Dong, Z. Y., Tiong, S. K., Mahlia, T. M. I., Muttaqi, K. M. Impact of renewable energy utilization and artificial intelligence in achieving sustainable development goals. *Energy Rep.*, 7, 5359–5373, 2021.
4. Dubey, A., Narang, S., Srivastav, A., Kumar A., García-Díaz, V. *Artificial Intelligence for Renewable Energy Systems*, 1st Edition. View series: Woodhead Publishing Series in Energy, 2022.
5. Vyas, A. K. Balamurugan, S., Dhiman, H. S., Hiran, K. K. *Artificial Intelligence for Renewable Energy Systems*. Scrivener Publishing, Wiley (2022).
6. Tripathi, S.L., Dubey, M.K., Rishiwal, V., Padmanaban, S. *Introduction to AI Techniques for Renewable Energy Systems*, 1st Edition. CRC Press. 2021.
7. Karad, S., Thakur, R. Efficient monitoring and control of wind energy conversion systems using Internet of things (IoT): A comprehensive review. *Environ. Dev. Sustain.*, 23, 14197–14214, 2021.
8. Ponnalagarsamy, S., Geetha, V., Pushpavalli, M., Abirami, P. Impact of IoT on renewable energy. In: Singh, I., Gao, Z., Massarelli, C. (eds), *IoT Applications Computing*. IntechOpen. 2021.
9. Birleanu, F.G., Bizon, N. Control and protection of the smart microgrids using internet of things: Technologies, architecture and applications. In: Mahdavi Tabatabaei, N., Kabalci, E., Bizon, N. (eds) *Microgrid Architectures, Control and Protection Methods. Power Systems*. Springer, Cham. 2020.
10. Shaw, R. N., Mendis, N., Mekhilef, S., Ghosh, A. *AI and IOT in Renewable Energy*. Springer, Singapore, 2021.
11. Batcha, R. R., Geetha, M. K. A survey on IOT based on renewable energy for efficient energy conservation using machine learning approaches. In: *2020 3rd International Conference on Emerging Technologies in Computer Engineering: Machine Learning and Internet of Things (ICETCE)*, pp. 123–128, 2020.
12. Rangel-Martinez, D., Nigam, K. D. P., Ricardez-Sandoval, L. A. Machine learning on sustainable energy: A review and outlook on renewable energy systems, catalysis, smart grid and energy storage. *Chem. Eng. Res. Des.*, 174, 414–441, 2021.

13. Daniel, C., Shukla, A. K., Sharma, M. Applications of machine learning in harnessing of renewable energy. In: Baredar, P.V., Tangellapalli, S., Solanki, C.S. (eds) *Advances in Clean Energy Technologies. Springer Proceedings in Energy.* Springer, Singapore, pp. 177–187, 2021.

14. Alvarez, L. F. J., González, S. R., López, A. D., Delgado, D. A. H., Espinosa, R., Gutiérrez, S. Renewable energy prediction through machine learning algorithms. In *2020 IEEE ANDESCON*, pp. 1–6, 2020.

15. Gu, G. H., J., Noh, Kim, I., Jung, Y. Machine learning for renewable energy materials. *J. Mater. Chem. A,* 7, 17096–17117 2019.

16. Kumar, K., Rao, R. S., Kaiwartya, O., Kaiser, S., Padmanaban, S. K. *Sustainable Developments by Artificial Intelligence and Machine Learning for Renewable Energies.* Academic Press, 2022.

17. Khedkar, M.K., Ramesh, B. AI and ML for the smart grid. In: *Intelligent Renewable Energy Systems,* pp. 287–306, 2022.

18. Bose, B. K., Artificial intelligence applications in renewable energy systems and smart grid – some novel applications. In: *Power Electronics in Renewable Energy Systems and Smart Grid: Technology and Applications,* pp. 625–675, 2019.

19. Dutta, S., Sadhu, P.K., Cherikuri, M., Mohanta, D.K. Application of artificial intelligence and machine learning techniques in island detection in a smart grid. In: *Intelligent Renewable Energy Systems,* pp. 79–109, 2022.

20. Xiaoyi, Z., Dongling, W., Yuming, Z., Manokaran, K. B., Antony, A. B., IoT driven framework based efficient green energy management in smart cities using multi-objective distributed dispatching algorithm. *Environ. Impact Assess. Rev.,* 88, 106567, 2021.

21. Meena, M., Shubham, S., Paritosh, K., Pareek, N., Vivekanand, V. Production of biofuels from biomass: Predicting the energy employing artificial intelligence modelling. *Bioresour. Technol.,* 340, 125642, 2021.

22. Wang, Z., Peng, X., Xia, A., Shah, A. A., Huang, Y., Zhu, X., Zhu, X., Liao, Q. The role of machine learning to boost the bioenergy and biofuels conversion. *Bioresour. Technol.,* 343, 126099, 2022.

23. Chamundeswari, V., Niraimathi, R., Shanthi, M., Mahaboob Subahani, A. Renewable energy technologies. In: *Integration of Renewable Energy Sources with Smart Grid.* 2021.

24. Pandian, I., Begum, S., S. Kumaravel, S.P. An integrated IoT and fuzzy logic controller system for biogas digester to predict methane. *Environ. Dev. Sustain.,* 2021, https://doi.org/10.1007/s10668-021-01943-7.

# 24 Automated Water Dispenser – A Hygiene Solution for Pandemic

*G.G. Raja Sekhar and D. Kalyan*
Koneru Lakshmaiah Education Foundation

*R. Ramkumar and M. Lakshmi*
Dhanalakshmi Srinivasan University

## CONTENTS

24.1 Introduction .............................................................................................................289
24.2 Methodology.............................................................................................................290
24.3 Simulation Model: Automated Filling of Water in Tanks .......................................291
24.4 Circuit Modelling in TinkerCAD ............................................................................293
24.5 Experimental Setup ..................................................................................................295
24.6 Results and Discussion ............................................................................................295
24.7 Conclusion and Future Scope ..................................................................................295
References........................................................................................................................296

### ABSTRACT

World Health Organization suggested the primary steps to eradicate illness episodes including COVID infection 2019 (COVID-19, coronavirus), such as using disinfectants, safe drinking water, waste treatment, cleanliness conditions, and most importantly ensuring human well-being. Ensuring proper waste management and washing practices in homes, workplaces, medical services, and commercial gatherings will help reduce contamination and human-to-human transmission of microorganisms including SARS-CoV. This paper demonstrates a hygiene solution for the current pandemic due to COVID-19. An automated water dispenser dispenses water automatically and is controllable based on its position and level and the quantity required. It has programmed, embedded controller using Arduino integrated with ultrasound sensor and infrared sensors, and the actuators used are solenoid valve and single channel relay. The achieved volume of 350 mL per serve is calibrated by the proposed model. The operation of working model is simulated using TinkerCAD, a virtual automation tool.

## 24.1 INTRODUCTION

Water shortage is one of the serious issues which the significant urban areas face across the world. We as a whole realize that approximately 70% of the Earth's surface is covered with water, but only a minor percent of it is utilized for day-to-day activities. Researchers have been looking for some strategies to make significant well-springs of water, for example, ocean and sea water, to be utilized for human use. Apart from making dams and generating power, availability of water to all is a must. There are numerous diseases, which have come into prominence, mostly resulting from the pollution of water. As per reports, approx. 4,000 children across the world are biting the dust every year, because of diarrheal sicknesses due to perilous water which they use in everyday life. 70% of the overall freshwater is designated for agriculture, and a large portion of these cultivating water system frameworks works at just 40% effectiveness. According to our predecessors, the following universal conflict if happens will occur due to water issues, as there have been numerous contentions for waterways, dams, and other water bodies [1–3].

The programmed water distributor has a number of functions, including managing the water level, displaying Total Dissolved Solids (TDS) estimations, displaying temperature estimations, and releasing water according to a schedule. In today's world, there should be a few components that are regulated, and the Programmed Water Allocator provides individuals with water that is of acceptable quality. Insightful frameworks are used in a wide range of situations in everyday life, and they are included into the plan. In order for some real components to carry out their typical tasks, they must be managed in daily life. According to a study, a programmed water distributor is a system that has the capacity to maintain every water boundary, such as water level, temperature estimate, and TDS assessment [4,5]. As a result, a control framework may be defined as a device or a group of devices that supervises, orders, coordinates, or directs the behaviour of several devices. This paper is engaged in introducing how a Programmed Water Gadget can be installed. The thing by which we get purred is the wastage of water and the contamination of water. Hence realize that it will help the climate and water cycle by which we can save water for our future [6,7].

## 24.2 METHODOLOGY

The functioning of ultrasonic sensors is appropriate for the situation of water-level monitoring, since it is a module that uses sound waves to measure the distance of an obstruction. It works by sending a sound wave at a specific frequency and listening for it to ricochet back. It is possible to calculate the distance between the sensor and the object by recording the lapse in time between the sound wave emitted and the sound wave bounced back. In this work, the distance of the item is estimated through an ultrasonic distance sensor and the sensor yield is associated with signal processing unit and after that, it is passed through an Arduino microcontroller. The distance measured is calibrated into a level of water of the glass [8,9]. Figure 24.1 shows the ultrasonic sensor with sonic burst and echo generation and IR module features. The infrared (IR) transmitter emits incessant infrared beams that are collected by an infrared receiver module. The acceptance of IR beams determines how an IR yield terminal of the beneficiary differs. Because this variety can't be broken down this way, this yield may be handled by a comparator circuit, which uses an LM339 operation speaker (activity enhancer) as a comparator. When the IR receiver does not get a signal, the risk of receiving distressing information is greater than the comparator IC LM339's non-altering contribution [10,11]. Figure 24.2 shows the block diagram of automated water dispenser.

As a result, the comparator's yield falls dramatically. The drive, on the other hand, lacks radiance. When the IR beneficiary module receives signals from the modifying input, the potential decreases. The yield of the comparator LM339 rises as a result of this, and the drive shines brightly. For initiating the flow of water depending on glass detection and volume requirements, solenoid and relay actuators were employed [12,13].

The flowchart from Figure 24.3 describes the logical steps behind the automation process by which water dispenses into the glass as per the user requirements. The ultrasonic sensor and IR sensor's integrative work for solenoid triggers through the relay. When glass is detected with IR1 sensor and the level is below set value 15 cm using ultrasound and IR2 sensor did not detect the obstruction, then Arduino programmer generates a low pulse to trigger a relay to turn ON the solenoid valve for water filling [14,15]. When the ultrasound reaches above 15 cm or IR2 sensor detects the obstruction, Arduino programmer generates a HIGH pulse to trip the solenoid valve which stops the flow of water.

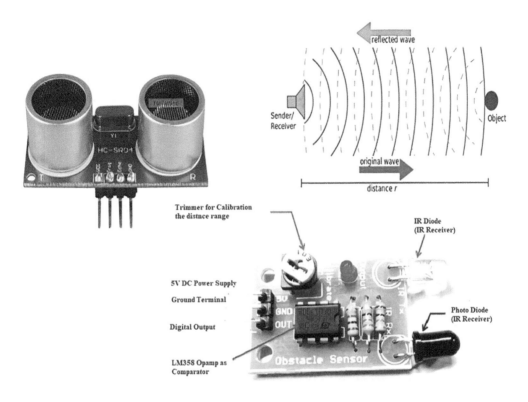

**FIGURE 24.1**  Ultrasonic sensor with sonic burst and echo generation and IR module features.

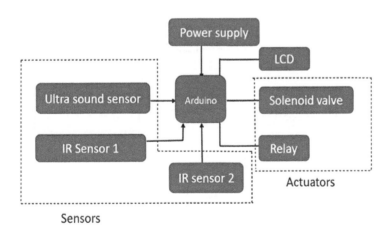

**FIGURE 24.2**  Block diagram of automated water dispenser.

## 24.3  SIMULATION MODEL: AUTOMATED FILLING OF WATER IN TANKS

More often than not, we will turn ON the syphon when the tank is empty and neglects to remember the same. This causes large amount of wastage of water as well as burn-through loads of electric force. For a shopper, to foresee the level of the water and control the valve ultimately is absurd constantly. A programmed syphon and a water-level regulator are considered as solutions to these

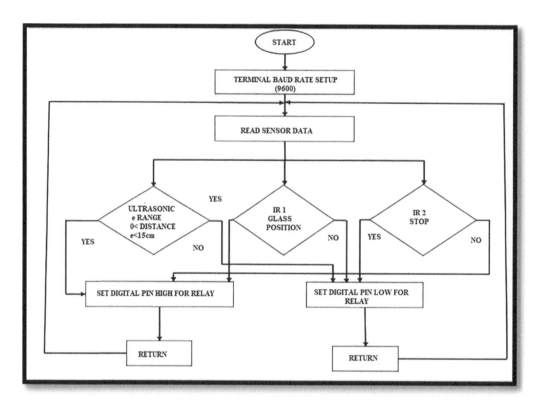

**FIGURE 24.3** Flowchart for automated water dispenser system.

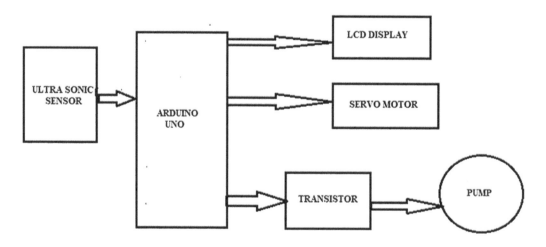

**FIGURE 24.4** Block diagram for TinkerCAD level control.

problems. The basic principle behind this concept is that when the tank is empty, the syphon should activate, and when the tank is full, the syphon should automatically shut off. In this paper, we demonstrate the use of a programmed syphon and water-level regulator with an ultrasonic sensor and a few actuators utilizing Arduino [16,17]. Figure 24.4 shows the block diagram for TinkerCAD level control.

**TABLE 24.1**

**Components for Automatic Tank-Filling System**

| Name | Quantity | Component |
|------|----------|-----------|
| M1 | 1 | DC Motor |
| U2 | 1 | Arduino Uno R3 |
| U3 | 1 | LCD 16 × 2 |
| R1 | 1 | 220 Ohm Resistor |
| R2, R3, R4 | 3 | 1 k Ohm Resistor |
| SLIDE | 1 | Slide Switch |
| SERVO1 | 1 | Micro Servo |
| TRANS1 | 1 | Transistor |

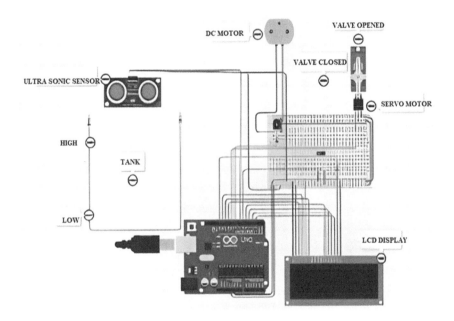

**FIGURE 24.5**   Automatic tank-filling system.

The components of the proposed system are listed in Table 24.1, as well as the quantity of each component is mentioned.

## 24.4   CIRCUIT MODELLING IN TINKERCAD

The first process is to start the tank-filling cycle by looking at the sensor value, which determines the amount of water in the tank. The interaction will begin when the tank falls below sensor value, in which case the engine will turn on and water will be poured into the tank. After some time, the water will reach that level of the sensor [18,19]. Naturally, the motor will be turned off, and when the tank is empty, an indicator such as a bell sound signals that the tank is empty. The interaction will then resume, with a valve installed to drain the water. Observations are made for the following two cases 1 and 2 [20]. Figure 24.5 shows the Arduino connection diagram of automatic tank-filling system.

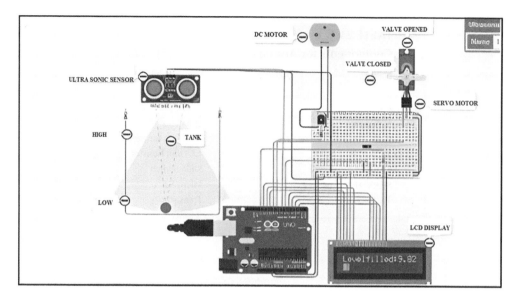

**FIGURE 24.6**   Output when tank is empty.

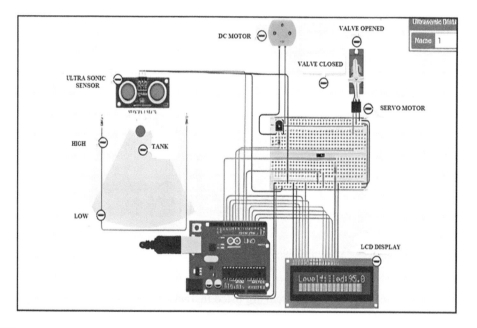

**FIGURE 24.7**   Output when tank is filled.

**Case 1**: Tank level at minimum

Figure 24.6 shows the Arduino connection diagram of tank. The figure shows the output for an empty tank. When the tank's level is low, it means that the tank is not filled. The valve is closed and the engine turns on. The tank is now at 9.82% capacity. A programmed syphon and a water-level regulator are considered as solutions to these problems.

**Case 2**: Tank level at maximum

When the tank is empty, the syphon should turn on, and when the tank is full, the syphon should be turned off. Figure 24.7 shows the output for tank-filled condition. When

**FIGURE 24.8**  Proposed model of automated water dispenser.

the water level in the tank reaches a certain level, it means that the tank is full. Water is no longer syphoned, since the motor/solenoid valve is stopped ("0" rpm) [21].

## 24.5  EXPERIMENTAL SETUP

The output voltage waveforms for five-level and seven-level inverters for different loads are obtained. Figure 24.8 shows the hardware setup of automated water dispenser. Worst-case harmonics for both the inverters were determined and dominant lower order frequencies were obtained to be of order seven and nine, respectively. An LCL filter is designed for resonance to dominant lower order harmonics, and its performance is evaluated in terms of complying IEEE 519 standards for harmonics in voltage and current [22]. A reduction in overall system cost owing to filter size reduction for seven-level inverter compared to five-level inverter- based system is achieved.

## 24.6  RESULTS AND DISCUSSION

The working model of an automated water dispenser performed as per the expectation. Figure 24.7 shows the RED LED indication for glass detection. The level set through ultrasound sensor is achieved for 15 cm, which provides 350 mL of water in the glass. IR2 interrupts the flow even when the user would like to stop the flow of water into the glass, i.e., when the user places his hand in front of the IR2 sensor. The proximities of ultrasound and IR sensors can be tuned using the trim potentiometers available on the IR modules [23].

## 24.7  CONCLUSION AND FUTURE SCOPE

The automated water dispenser is designed and simulated and demonstrated its operation in this paper. The operation of sensors is reliable due to noncontact-type transducers, and grid power driving made the module flexible and continuously operable [24]. The prototype is cost-effective and durable, and can also be operated with solar-driven inverter setup [25].

## REFERENCES

[1] Dusarlapudi, K., Kiran, K.U., Narayana, M.V. and Suman, M.V., 2018. Flow based valve position control using NI my DAQ. *International Journal of Engineering and Technology (UAE)*, 7, pp. 955–957.

[2] Bhargavi, R., Ganesh, P., Raja Sekhar, G.G., Prakash, R.B.R. and Muni, T.V., 2020. Design and implementation of novel multilevel inverter. *Journal of Advanced Research in Dynamical and Control Systems*, 12(2), pp. 1322–1328.

[3] Narayana, M.V., Dusarlapudi, K., Kiran, K.U. and Kumar, B.S., 2017. IoT based real time neonate monitoring system using Arduino. *Journal of Advanced Research in Dynamical and Control Systems*, 9(14), pp. 1764–1772.

[4] Prasad, J.S., Obulesh, Y.P. and Babu, S., 2017. FPGA controlled five-level soft switching full bridge DC-DC converter topology. *Journal of Electrical Systems*, 13(2), pp. 266–284.

[5] RajaSekhar, G.G. and Basavaraja, B., 2019. Solar PV fed non-isolated DC-DC converter for BLDC motor drive with speed control. *Indonesian Journal of Electrical Engineering and Computer Science*, 13(1), pp. 313–323.

[6] Sudharshan Reddy, K., Sai Priyanka, A., Dusarlapudi, K. and Vijay Muni, T., 2019. Fuzzy logic based iUPQC for grid voltage regulation at critical load bus. *International Journal of Innovative Technology and Exploring Engineering*, 8(5), pp. 721–725.

[7] Pandian, A., Rishika, M., Sudakshina, M. and Karthik, M.S.V.R., 2020. Control technique for power electronic converter in hybrid electric vehicle. *Journal of Advanced Research in Dynamical and Control Systems*, 12(2), pp. 35–42.

[8] Ravi Kumar, K., Pandian, A. and Sastry, V., 2020. Modeling simulation and performance comparison of bidirectional dc dc converters. *Journal of Advanced Research in Dynamical and Control Systems*, 12(2), pp. 1292–1298.

[9] Dusarlapudi, K. and Narasimha Raju, K., 2020. Embedded prototype of 3 DOF parallel manipulator for endoscope application using 3 axis MEMS accelerometer. *Journal of Advanced Research in Dynamical and Control Systems*, 12(2), pp. 1225–1235.

[10] Bade, R., Ramana, B.V. and Dusarlapudi, K., 2019. A six servo six DOF position controlled platform for sorting application. *International Journal of Engineering and Advanced Technology*, 8(4), pp. 1215–1219.

[11] Dukkipati, S., Marthanda, A.V.G.A., Rajasekhar, G.G. and Kumar, M.K., 2019. Design analysis and implementation of a three phase seven level PV & wind based microgrid. *International Journal of Innovative Technology and Exploring Engineering*, 8(10), pp. 4495–4498.

[12] Baig, K., Raj, K.P., Sekhar, G.R., Muni, T.V. and Kumar, M.K., 2020. Power quality enchancement with active power control. *Journal of Critical Reviews*, 7(9), pp. 739–741.

[13] Muppavarapu, M., Rajasekhar, G.G., Muni, T.V. and Prakash, R.B.R., 2019. Enhancement of power quality in a grid connected UDE based PV inverter. *Journal of Critical Reviews*, 7(2), p. 2020.

[14] Harshith, I., Raj, B.P., Sekhar, G.R. and Muni, T.V., 2019. A novel methodology for single phase transformerless inverter with leakage current elimination for PV systems application. *International Journal of Innovative Technology and Exploring Engineering*, 8(6), pp. 1017–1021.

[15] RajaSekhar, G.G. and Basavaraja, B., 2018. Brushless DC motor drive with single current sensor fedfrom PV with non-isolated interleaved converter. *Journal of Advanced Research in Dynamical and Control Systems*, 10(9).

[16] Sai, P.S., Rajasekhar, G.G., Muni, T.V. and Chand, M.S., 2018. Power quality and custom power improvement using UPQC. *International Journal of Engineering & Technology*, 7(2.20), pp. 41–43.

[17] Praveenkumar, B., Srikanth, K.S., Kiran Kumar, M. and Raja Sekhar, G.G., 2018. ANN control based variable speed PMSG-based wind energy conversion system. *International Journal of Engineering & Technology*, 7(2.7), pp. 526–531.

[18] Sekhar, G.R. and Banakara, B., 2018. Performance of brushless DC drive with single current sensor fed from PV with high voltage-gain DC-DC converter. *International Journal of Power Electronics and Drive System (IJPEDS)*, 9(1), pp. 33–45.

[19] Sekhar, G.R. and Banakara, B., 2018. An internal current controlled BLDC motor drive supplied with PV Fed high voltage gain DC-DC converter. *International Journal of Electrical and Computer Engineering*, 8(2), p. 1262.

[20] RajaSekhar, G.G. and Banakara, B., 2006. Photo-voltaic system fed high voltage gain DC-DC converter feeding BLDC drive with simplified speed control. *ARPN Journal of Engineering and Applied Sciences*, 12(24), pp. 7088–7095.

[21] Zedak, C., Belfqih, A., Boukherouaa, J., Lekbich, A. and Elmariami, F., 2022. Energy management system for distribution networks integrating photovoltaic and storage units. *International Journal of Electrical & Computer Engineering*, *12*(4), pp. 3352–3364.

[22] Al-Rawashdeh, A.Y., Dalabeeh, A., Samarah, A., Alzyoud, A.E. and Alzyoud, K.Y., 2022. Modeling and analysis of energy losses under transient conditions in induction motors. *International Journal of Electrical and Computer Engineering*, *12*(4), p. 3388.

[23] Sharma, P., Salkuti, S.R. and Seong-Cheol, K., 2022. Advancements in energy storage technologies for smart grid development. *International Journal of Electrical and Computer Engineering*, *12*(4), p. 3421.

[24] Satta, S., Bayadi, A. and Boudissa, R., 2022. Alternating current flashover voltage of a uniform polluted glass flat insulator under parallel electric discharges effect. *International Journal of Electrical and Computer Engineering*, *12*(4), p. 3454.

[25] Akshath, N.S.S., Naresh, A., Barman, M., Nandan, D. and Abhilash, T., 2022. Analysis and simulation of even-level quasi-Z-source inverter. *International Journal of Electrical and Computer Engineering*, *12*(4), pp. 3477–3484.

# 25 Review of IoT-Based Smart Waste Management Systems

*T. Sakthi Ram, S. Vetriashwath, V. Mruthunjay, Rahul Srikanth, Yuvan Shankar, L. Yogesh, and O.V. Gnana Swathika*
Vellore Institute of Technology

*Aayush Karthikeyan*
University of Calgary

## CONTENTS

25.1 Introduction ...................................................................................................299
25.2 Smart Segregation of Waste Using Embedded System Solutions .....................300
25.3 Cloud Integration and IoT-Based Solutions ...................................................303
25.4 Conclusion .....................................................................................................305
References.................................................................................................................305

### ABSTRACT

This paper aims to study in detail and review the existing and proposed methods for effective waste management coupled with Internet of Things (IoT) and cloud-based management systems. This paper also proposes a smart waste segregation system for smart cities improvised with IoT and cloud integration. The waste is segregated into different categories like dry, wet, biodegradable, non-biodegradable, metal, and non-metal. The waste level in each bin can be monitored using ultrasonic distance sensors, and this data can then be sent to the cloud. This allows various optimisations like finding the shortest route, filling only the necessary bins, etc. This paper studies proposed methods for such optimisations and smart waste management systems in detail.

## 25.1 INTRODUCTION

Segregation and smart management of waste can vastly improve utilisation of waste resources. It can also reduce the footprint of the waste materials on the environment. We have studied the existing methods for both smart segregation of waste using embedded systems and sensors and also the cloud integration in waste collections. Internet of Things (IoT)-based waste management is employed to a rate of 85% accuracy. Here different sensors are used to identify and segregate different objects such as metals, plastic and paper containers but fail to identify glass objects. The sensors used are proximity sensors to locate any things without making direct contact. Inductive sensors are used to differentiate metallic and non-metallic objects and capacitive sensors are used to differentiate solid and liquid wastes. These sensors are integrated to programmable logic controllers and an Arduino is used to operate robotic arm for systematic operations. Global System for Mobile communication (GSM)/General Packet Radio Services (GPRS) modules are used to communicate the expert regarding filled bin and status. A Liquid Crystal Display (LCD) is used in the framework to see the progress and status of the framework. The proposed system also has an image processing algorithm using a camera attached to differentiate biodegradable and non-biodegradable waste. The

DOI: 10.1201/9781003374121-25

IoT mechanism is dealt with by a fixed sensor. Then the cloud is updated with the trash level and status-related information [1].

This paper proposes an eco-friendly IoT-based waste segregation, separating waste into metallic, plastic, and biodegradable categories utilising sensors for moisture, induction, capacitance, gas, level, and bacteria. The main heart of the proposed system is the STM32 microcontroller which controls most of the operations by getting the input data from different sensors. The bin has three segments for three different categories. To determine the level of segregated garbage, a level sensor is utilised; if it hits a threshold level, a message is sent to the user. A bacteria sensor is used to detect the microbial activities and odour sensor to stop the unpleasant odour by spraying chemicals [2].

Classification of biodegradable and non-biodegradable waste cannot be easily done using classical methods but there are machine learning (ML)-based approaches to classify the waste based on image classifications. This paper adds on to prove that Support Vector Machines (SVMs) can be modelled in a way to classify images where the basic structure can be featured into dimensional histograms. This mechanism can be used as a viable alternative to Radial Basis Function (RBF) kernels. A simple recurrence of the supplied input vector can be observed in the process as well. The superior performance of the SVM can be narrowed down to the high-calibre generalisation in high-dimensional regions. The steps involved are separating the hyperplanes and generalising the learning technique to classify said images. Selection of kernels involves using the Gaussian-plane separation formulae and experiments to sort out the histograms obtained. It is recommended to use nonlinear transformation of the input vectors rather than introducing kernels to maximise efficiency [3].

The cloud integration with waste management unlocks lots of new potential for effective segregation and reuse of the waste and can also reduce the resources required in the process of waste management such as selective emptying of bins based on waste load patterns and also finding the shortest distance for the waste collectors. The paper proposes cloudSWAM, a cloud-connected waste management for smart cities. Pre-separated wastes are stored in different bins, with each of them containing sensors to monitor the level of waste filled in the bins with the categories of organic/plastic/metal wastes. The waste collectors are also notified of the levels of bins through the cloud. The real-time monitoring of the data allows various possibilities like route optimisation for waste collectors or predicting the amount of potential recyclable material using cloud computing. This system introduces several possibilities for management of the collected waste through connecting the live status of bins and all the stakeholders [4].

This paper explores the possibilities to improve waste management through IoT and deep learning. The primary novelty is a custom design for smart bin which has separate compartments for different wastes including one for general wastes which don't fit in any of the categories. The waste placed in the bin can be sent to the appropriate bin using retractable platform and lids at different levels. It also uses Raspberry Pi and Arduino Uno-based system for controlling of servo motors in the retractable platform and lids. There is also a Radio Frequency Identification (RFID)-based authorisation which only opens the top-most platform only if a valid RFID card is used, enhancing security. They have also used LoRa for sending data to the server. LoRa is suitable for applications which require low power and less bandwidth requirement. Tensor flow-based deep learning algorithm is used in the classification of wastes, which work on the basis of the images captured by the sensor and sent to the cloud via Raspberry Pi [5].

## 25.2   SMART SEGREGATION OF WASTE USING EMBEDDED SYSTEM SOLUTIONS

In this proposed system, dry, wet, and metallic wastes are separated. In this model, servo motors operate a metal plate that drops trash onto a conveyor belt. The electromagnet affixed to the servo-driven arm separates metallic garbage from other rubbish, as it moves over the conveyor belt. A

DC air blower is used to remove dry waste, which is then directed into a different container. The remainder of the wet waste simply empties into a container, as it drops off the conveyor belt's edge. The heart of this framework is ATmega328P microcontroller used to control servo motors, motor driver, and transceiver. The device is driven by DC helical geared motors, and servo motors are used for arm movements. Variable pulse width modulation is used to regulate servo motors. A proximity sensor is used to differentiate the waste so that the whole setup gets activated. A strong electromagnet is used to identify metallic waste. The air blower module is a squirrel-cage air blower. To guard against controller damage, it is connected to a microcontroller using a transistor switch and diode circuit. In this system, there are three drivers: two (2) for movement and one (1) for conveyor belt movement. Xbee series 2 radio modules are used for the segregator's and the user's wireless interface. It wirelessly transmits and receives the data [6].

The proposed framework is divided into three layers, the top to drop the waste, the middle to segregate, and the bottom to store the wastes separately. Power supply for the framework is provided by solar panels which generate 12 V DC supply or AC supply is converted to 12 V DC supply using a full-wave rectifier. Voltage regulator IC is used to convert 12 V–9 V. The 9 V current will be converted into 5 V DC current using the microcontroller power input system, and this 5 V is used for GSM, a servomotor, an ultrasonic sensor, a motion sensor, and a rain sensor. A motion sensor is employed to find if waste is dropped; a rain sensor (moisture sensor) is employed to find the wetness: if it is more than 30%, then classified as wet waste, and if not, then classified as dry waste. Using a servo motor, the dry and wet waste are tilted to their respective bins. Two ultrasonic sensors are used in separate bins for indication of level to the microcontroller and to the LCD to display the level. Once 80% of level reaches, the microcontroller commands the GSM module to send notification to the respective agent [7].

In this proposed system, automatic waste segregation bins namely SmartBin are employed. Smart waste management is an essential step towards smart cities. Arduino Mega 2560 acts as the brain of this framework that controls all workflow from segregation to sending messages via GSM module. A tray is fixed to keep the trash and attached to servo motors. This tray consists of a moisture sensor to identify wet and dry waste and an infrared (IR) sensor which acts as a proximity sensor to detect objects so energy can be saved. The Arduino issues orders based on the signal from the moisture sensor to the servo motor for tilting the waste into wet and dry waste bins. Each sub-bin is attached with an IR sensor at 80% level so that when trash reaches 80% level, it activates a GSM module to alert the nearby waste collection centre [8].

In this paper, electronically assisted automatic waste segregation is achieved using robotic arm and five layered convolution layer. Here, an Arduino Uno microcontroller is used to control the whole workflow and the robotic arm. The robotic arm is fixed and has a 360-degree range of motion which gives flexible workspace for the arm to move and pick the waste. The robot doesn't have any intelligence and is controlled by an Arduino which also has several DC motors that use a driver circuit for the position control. Webcams are used to record the photos of the trash when it is deposited in a segregator which is fed to five layered convolution layers to identify biodegradable and non-biodegradable wastes [9].

In this proposed system, three types of wastes are segregated: wet, dry, and metal. Here, a conveyor belt is used to collect all the wastes, and this belt is driven by a microcontroller. The dry wastes are removed by a powerful blower, so because of the blower dry waste gets blown away and gets segregated in a separate bin, whereas the wet stays on the conveyor belt. The metal waste is removed by the help of electromagnet, so due to the magnetic field created metallic waste gets collected separately [10].

In this paper, the waste segregation mechanism is basically a conveyor belt sensor detection model, that is, the waste is placed on a conveyor belt, initially IR sensor detects whether the object is metal or non-metal depending on the material, the conveyor belt is operated clockwise and anti-clockwise, and further capacitive and inductive sensors are used to segregate glass and plastic waste. All this operations are controlled by Arduino Uno controller [11].

In this paper, the waste is segregated into three main segments, dry, wet, and metal with a flap mechanism like if the waste is pushed into the flap system, the system starts by the detection of IR proximity sensor, and then the object is sensed by metal detection sensor to detect whether the object is metallic; and further if the object is non-metal, a capacitive sensor is used to distinguish between wet and dry wet. The flap then drops the object into the appropriate container after detection [12].

In this paper, the waste is separated by the capacitance value and IR light absorption capability of the material. The system starts by opening the garbage bin using a proximity sensor. The main garbage container has a two clab plate arrangement aligned at 45° to horizontal. So when a wet or dry waste is dropped in the bin, it first calculates the capacitance, since the capacitance values increase considerably due to high permittivity in wet waste. Further if the object is plastic, it is detected by the light absorption capability of the object. The bin inside the garbage can is rotated accordingly to drop the object in the respective bin [13].

This paper proposes a conveyor belt-like mechanism to separate waste, which contains two conveyor belts (main and sub-conveyor). The waste is initially placed on the main conveyor belt which is detected using a ultrasonic sensor, and it is sensed using a metal detector to classify it as metal or non-metal. If it is identified as metal, the conveyor belt moves with help of servo motor and drops the metal object onto the sub-conveyor which finally drops the metal waste into the metallic bin, and if it is non-metal, it is further classified into wet or dry waste by using a capacitive sensor [14]. In this paper, the waste segregation is done by using IR sensor and two ultrasonic sensors to classify wet and dry waste. The whole proposed system is powered by a solar system. A GSM module is used to send alert notifications to the authority. All these operations are carried out by Arduino Nano microcontroller [15].

In this paper, the waste is classified by building an image classifier which identifies the kind of garbage being processed by a convolutional neural network (CNN). Four different CNN models are used here like ResNet50, DenseNet169, VGG16, and AlexNet are trained on ImageNet, which extracts different features from the image samples and feed them to the image classifier to distinguish the type of waste [16]. This paper adds on to prove that SVMs can be modelled in a way to classify images where the basic structure can be featured into dimensional histograms. This mechanism can be used as a viable alternative to RBF kernels. A simple recurrence of the supplied input vector can be observed in the process as well. The superior performance of the SVM can be narrowed down to the high-calibre generalisation in high-dimensional regions. The steps involved are separating the hyperplanes and generalising the learning technique to classify said images. Selection of kernels involves using the Gaussian-plane separation formulae and experiments to sort out the histograms obtained. It is recommended to use nonlinear transformation of the input vectors rather than introducing kernels to maximise efficiency [17].

This paper emphasises the need and effectiveness in using lower-level visual characteristics for the image classifying problem. Binary Bayesian classifiers are used to capture the high-detailed complexities from the image supplied with the specified constraint. The main agenda aims to combine multiple two-class classifiers to form a single hierarchical classifier. The Bayesian Framework formally requires the quantisation of vectors used by the machines. The density estimation is done by selection of size of codebook. High-level semantics can be provided for larger databases as well. The vector machine formed comes under the hierarchical SVMs. The proposed methodology can be used to classify both indoors and outdoors as well [18].

The paper solely aims at sub-categorising the different parts of a 'global' image segment into smaller such segments. The study helps prove that a single model data set can indeed provide an effective competition to the performance of a multi-level model and also provide us with superior classifying capability. The classification methodology begins with drawing size proportions and estimating each image region. The work takes inspiration from a model established in 2005 which uses probabilistic latent semantic analysis (pLSA) to set up the learning algorithms and uses the kernel nearest neighbour classifier for classification of results, but the other models usually tend to adopt Scale-Invariant Feature Transform (SIFT) features. The graphical model can be adopted to learn and understand the salient patterns from an already existing class [19].

The paper aims at targeting relatively smaller workloads that can easily fit into the ram of a typical desktop (about 4G to around 48G). Larger scale classifications are proven to be expensive and publicly unavailable, and hence there is a need for the development of a medium-scale classifier. The main criteria to satisfy would be to balance the efficiency of the output while not compromising with the expenses of said machine. A parallel stochastic gradient (AGSD) algorithm is used to train the vector machine. Local Coordinate Coding and Super-vector Coding are implemented to enhance the performance of the feature extraction mechanism. Although the output is not quantitatively connectable to the extremely high performance of the machines available, the AGSD algorithm still manages to provide a very fast convergence rate [20].

The paper focuses on utilising feature mining methods to provide better results on classification of pedestrians and also aims at lessening the burden on core processing power by evolving and deriving a standard approach for histogram classification using ML. This also concludes an in-depth study on the different types of data sets and offers a theoretical explanation about the vast techniques surrounding mining strategies. Feature mining as a whole is set up to serve as the basis for future classification and design systems. Although proving efficient, the framework does have few demerits, the most notable one being that the number of features do not increase beyond a certain limiting point [21].

## 25.3 CLOUD INTEGRATION AND IoT-BASED SOLUTIONS

The paper puts forth a generic image classifier based on randomised decision trees to form up a ML algorithm. The source image is scanned for subwindows whose sizes are randomly chosen between $1 \times 1$ pixels and the minimum differential length along the image. A selection of a larger number of subwindows proves more efficient and yields better performance. The subwindows are trained and resized to build up a collection of extra-trees and these form the basis for historical recognition. The model can be evaluated at the higher-end services where images exhibit higher variability and cluttered backgrounds with noises too [22].

This paper proposes a model where the waste is placed on an upper tray which is the only visible part to the user. An IR sensor is used to detect the waste, the system is activated, and the waste is dropped onto the main tray where moisture sensor is used to distinguish wet and dry waste. The main tray is rotated using a servo motor by applying accurate angular momentum to dispose the waste into the appropriate sub-bin [8].

Waste management is explained with two-stage operations, where the first stage has two subdivisions. The first subdivision consist of an effective route-obtaining system for the bins to collect the garbage and the second subdivision focus on priority of visiting bins based on their level and other parameters. The second model takes into account trash sorting and transferring them into the recovery value centre in order to optimise recovery value and reduce visual pollution [23].

This paper proposes a smart waste management system, in which all the bins' locations are identified by a unique code, and also each bin is fitted with an ultrasonic distance sensor to get the waste levels in each bin. These bins also contain other sensors to detect the presence of hazardous and stinky gases. The real-time knowledge of the bin levels allows us to selectively empty only the necessary bins, and along with cloud computing, it would be possible to find the shortest paths for the waste collections using the Dijkstra. This paper also demonstrated experimentally that the initial additional cost of setting up this smart waste management system is lower compared to the savings because of the optimised routes of the waste collectors [24].

In this paper a smart cloud-based waste management system is proposed, where each bin is given a unique ID, and the weight and level of waste in the bins are determined using SUN load sensor and ultrasonic distance sensors. All these data are collected by a microcontroller and sent to a server using the GPRS module. Using Ant colony optimisation, the shortest route is found [25].

This paper proposes a cloud-based waste management system. An Arduino microcontroller is connected to ultrasonic distance sensors to find the level of the waste in each bin. It is connected to the cloud with the help of NodeMCU and ThingSpeak platform. The authors have used a Global Positioning System (GPS) module to get the exact latitude and longitude of the bin, which is also sent to the cloud. The level of the bin is visually shown to the waste collectors through a mobile application platform, and Google Maps is used to find the shortest route for the waste collectors to the bin which is filled [26].

This paper proposes a cloud-based smart waste management system for smart cities. This system consists of smart bins which contain sensors to read the volume of waste and the weight of the waste in the bins and are connected to the cloud infrastructure with the help of GSM or GPRS technology. Also each type of waste has a different priority according to which the transportation is called. This enables options to optimise the transportation allocation and selectively emptying of the bins is made possible [27].

This paper proposes a waste management model with rewards systems and a data-centric collection service. Each citizen is involved in this process using a custom application that allows them to check their rewards for recycling wastes, service history, and lists of municipal notifications. Smart bags with QR codes are used for door-to-door garbage collection, and smart bins are scattered all over the city whose waste level is monitored and sent to the cloud. Garbage collection service companies can use this data to allocate optimal transportation for collecting the wastes. These bins allow the citizens to be authentic before disposing of the waste using QR codes and rewards are provided for the citizens [28].

This research paper proposes a waste management model with help of IoT and ML for a goal of creating clean, pollution-free cities. Individual trash cans could be transformed into a network of intelligent, interconnected items using IoT. In the provided system, a dumper truck database has been created in order to collect all the information such as dumper truck ID, meeting date, and meeting time of waste collection. The waste management system and all operations of the truck driver are monitored using ML, which enables timely rubbish collection and automates vehicle tracking through a database using the GPS [29].

The paper proposes a proper effective methodology of collection of wastes irrespective of their types. The sole purpose of said paper would be to map out an efficient way to organise the waste-trucks' routes, while also setting up certain limitations as to the number of times a truck passes through a particular street and this is done using the compatibility of the open-sourced Net2Plan-GIS framework. This paper would certainly aid numerous environmental and economic bases like fuel consumption and vehicular pollution [30].

This research paper proposes a way to classify the waste by photographing it and sending the image to a server, where it would be automatically identified and sorting would be done locally according to the response sent back by the server. This would enhance waste collection planning.

In order to segregate e-waste, a novel Region-Based Convolutional Neural Network (R-CNN)-based algorithm is proposed. A more light-weight and faster algorithm is used to segregate multiple objects simultaneously. This algorithm also enables us to get the size and category of waste, hence aiding the waste collection planning. The selected e-waste categories have recognition and classification accuracy ranging from 90% to 97%. Reference [31] discusses how it is possible to automatically identify and classify the waste's size and type from the submitted photographs.

This paper proposes a smart waste management system using smart dust bins. There are two bins whose waste levels are monitored using ultrasonic distance sensors. IR sensor and servo motor mechanism are used to control the opening and closing of the bin. Both bins in a location are interconnected using Arduino microcontroller; only when the first bin gets filled, the next bin will be allowed to open. And the SMS alert is sent to the waste collectors using the GSM module connected to the Arduino. This interconnected network of bins allows the waste collectors to empty the bin only when all the bins get filled, saving fuel and transportation costs [32]. This paper proposes a cloud computing-based approach with image classification to counteract illegal dumping in smart cities. It uses the data from the existing security camera by connecting them to the cloud. An image classification algorithm is used to classify the different wastes into specific products like washing machines. When a suspect item is identified in at least two consecutive frames of the same scene, an alert is triggered along with time stamp, date and time, geo-location and video are sent to the operators to take action [33–36].

The paper talks about the benefits of segregating recyclable waste. It suggests CNN to automate the process of segregating waste, which makes the process efficient. ResNet50, DenseNet169, VGG16, and AlexNet are the pre-trained CNNs. Pre-trained models are used to overcome the time taken to configure and optimise new networks [37–41].

This paper talks about the use of deep learning neural networks. For $28 \times 28$ images, there are 784 neurons used as inputs. The output of this will be an input image based on the trained network's weights and biases. Caffe, a deep learning framework, is used for implementation.

Deep learning networks have only recently presented good results; hence, a shared framework and models are required for further development in this domain [42–44].

In this paper, CNN is used to perform the segregation task. The goal of this design is to reduce human intervention in the process (Adaptive and Interactive Modelling system). This study applies ML strategies for segregating materials using an induction algorithm. It was shown that recursive splitting techniques show promising results, and these will be compared with results from neural network representations in an attempt to maximise processing efficiency [45–47].

## 25.4   CONCLUSION

Thus, the research papers under smart waste management are studied and reviewed, which comprises two main blocks, effective waste segregation and IoT integration for effective disposal, where the waste is mostly separated into biodegradable, non-biodegradable, wet, dry, metal, and non-metal. The waste is distinguished by using ML technique and sensors like inductive, capacitive, and moisture sensors. The segregation mechanisms used are the conveyor belt and flap mechanism which is used to drop the waste into the appropriate bin. Even though the waste is separated and segregated into separate bins, the IoT part which alerts the nearest bin collector can avoid the bin overflowing situation.

## REFERENCES

1. Nakandhrakumar, R.S., Rameshkumar, P., Parthasarathy, V. and Rao, B.T., 2021 WITHDRAWN: Internet of Things (IoT) based system development for robotic waste segregation management. *Materials Today: Proceedings*. doi: 10.1016/j.matpr.2021.02.473
2. Kumar, B.S., Varalakshmi, N., Lokeshwari, S.S., Rohit, K. and Sahana, D.N., 2017, December. Eco-friendly IOT based waste segregation and management. In *2017 International Conference on Electrical, Electronics, Communication, Computer, and Optimization Techniques (ICEECCOT)* (pp. 297–299). IEEE. doi: 10.1109/ICEECCOT.2017.8284686
3. Zhou, X., Cui, N., Li, Z., Liang, F. and Huang, T.S., 2009, September. Hierarchical Gaussianization for image classification. In *2009 IEEE 12th International Conference on Computer* Vision (pp. 1971–1977). IEEE. doi: 10.1109/ICCV.2009.5459435.
4. Aazam, M., St-Hilaire, M., Lung, C.H. and Lambadaris, I., 2016, October. Cloud-based smart waste management for smart cities. In *2016 IEEE 21st international workshop on computer aided modelling and design of communication links and networks (CAMAD)* (pp. 188–193). IEEE. doi: 10.1109/CAMAD. 2016.7790356.
5. Sheng, T.J., Islam, M.S., Misran, N., Baharuddin, M.H., Arshad, H., Islam, M.R., Chowdhury, M.E., Rmili, H. and Islam, M.T., 2020. An internet of things based smart waste management system using LoRa and tensorflow deep learning model. *IEEE Access*, 8, pp. 148793–148811.
6. Sivakumar, N., Kunwar, A.R., Patel, S.K., Kumar, S. and Mala, S.P., 2016, May. Design and development of an automatic clustered, assorted trash segregation system. In *2016 IEEE International Conference on Recent Trends in Electronics, Information & Communication Technology (RTEICT)* (pp. 409–413). IEEE. doi: 10.1109/RTEICT.2016.7807852.

7. Sarker, S., Rahman, M.S., Islam, M.J., Sikder, D. and Alam, A., 2020, June. Energy saving smart waste segregation and notification system. In *2020 IEEE Region 10 Symposium (TENSYMP)* (pp. 275–278). IEEE. doi: 10.1109/TENSYMP50017.2020.9230949.

8. Jayson, M., Hiremath, S. and Lakshmi, H.R., 2018, February. SmartBin – Automatic waste segregation and collection. In *2018 Second International Conference on Advances in Electronics, Computers and Communications (ICAECC)* (pp. 1–4). IEEE. doi: 10.1109/ICAECC.2018.8479531.

9. Nandhini, S., Mrinal, S.S., Balachandran, N., Suryanarayana, K. and Ram, D.H., 2019, April. Electronically assisted automatic waste segregation. In *2019 3rd International Conference on Trends in Electronics and Informatics (ICOEI)* (pp. 846–850). IEEE. doi: 10.1109/ICOEI.2019.8862666.

10. Kabra, N., Tirthkar, P., Umak, P. and Deokar, P., 2018. Automatic Waste Segregation Using Embedded System. *International Journal of Innovations in Engineering Research and Technology*, pp. 1–3.

11. Rafeeq, M. and Alam, S., 2016, October. Automation of plastic, metal and glass waste materials segregation using Arduino in scrap industry. In *2016 International Conference on Communication and Electronics Systems (ICCES)* (pp. 1–5). IEEE. doi: 10.1109/CESYS.2016.7889840

12. Chandramohan, A., Mendonca, J., Shankar, N.R., Baheti, N.U., Krishnan, N.K. and Suma, M.S., 2014, April. Automated waste segregator. In *2014 Texas Instruments India Educators' Conference (TIIEC)* (pp. 1–6). IEEE. doi: 10.1109/TIIEC.2014.009

13. Pereira, W., Parulekar, S., Phaltankar, S. and Kamble, V., 2019, February. Smart bin (waste segregation and optimisation). In *2019 Amity International Conference on Artificial Intelligence (AICAI)* (pp. 274–279). IEEE. doi: 10.1109/AICAI.2019.8701350

14. Gupta, N.S., Deepthi, V., Kunnath, M., Rejeth, P.S., Badsha, T.S. and Nikhil, B.C., 2018, June. Automatic waste segregation. In *2018 Second International Conference on Intelligent Computing and Control Systems (ICICCS)* (pp. 1688–1692). IEEE. doi: 10.1109/ICCONS.2018.8663148

15. Al Rakib, M.A., Rana, M.S., Rahman, M.M. and Abbas, F.I., 2021. Dry and wet waste segregation and management system. *European Journal of Engineering and Technology Research*, 6(5), pp. 129–133. doi: 10.24018/ejeng.2021.6.5.2531

16. Susanth, G.S., Livingston, L.J. and Livingston, L.A., 2021. Garbage waste segregation using deep learning techniques. In *IOP Conference Series: Materials Science and Engineering* (Vol. 1012, No. 1, p. 012040). IOP Publishing.

17. Chapelle, O., Haffner, P. and Vapnik, V.N., 1999. Support vector machines for histogram-based image classification. *IEEE Transactions on Neural Networks*, 10(5), pp. 1055–1064. doi: 10.1109/72.788646.

18. Vailaya, A., Figueiredo, M.A., Jain, A.K. and Zhang, H.J., 2001. Image classification for content-based indexing. *IEEE Transactions on Image Processing*, 10(1), pp. 117–130. doi: 10.1109/83.892448.

19. Chong, W., Blei, D. and Li, F.F., 2009, June. Simultaneous image classification and annotation. In *2009 IEEE Conference on Computer Vision and Pattern Recognition* (pp. 1903–1910). IEEE. doi: 10.1109/CVPR.2009.5206800.

20. Lin, Y., Lv, F., Zhu, S., Yang, M., Cour, T., Yu, K., Cao, L. and Huang, T., 2011, June. Large-scale image classification: fast feature extraction and SVM training. In *CVPR 2011* (pp. 1689–1696). IEEE. doi: 10.1109/CVPR.2011.5995477.

21. Dollár, P., Tu, Z., Tao, H. and Belongie, S., 2007, June. Feature mining for image classification. In *2007 IEEE Conference on Computer Vision and Pattern Recognition* (pp. 1–8). IEEE. doi: 10.1109/CVPR.2007.383046.

22. Marée, R., Geurts, P., Piater, J. and Wehenkel, L., 2005, June. Random subwindows for robust image classification. In *2005 IEEE Computer Society Conference on Computer Vision and Pattern Recognition (CVPR'05)* (Vol. 1, pp. 34–40). IEEE. doi: 10.1109/CVPR.2005.287.

23. Salehi-Amiri, A., Akbapour, N., Hajiaghaei-Keshteli, M., Gajpal, Y. and Jabbarzadeh, A., 2022. Designing an effective two-stage, sustainable, and IoT based waste management system. *Renewable and Sustainable Energy Reviews*, 157, p. 112031.

24. Misra, D., Das, G., Chakrabortty, T. and Das, D., 2018. An IoT-based waste management system monitored by cloud. *Journal of Material Cycles and Waste Management*, 20(3), pp. 1574–1582.

25. Sharmin, S. and Al-Amin, S.T., 2016, November. A cloud-based dynamic waste management system for smart cities. In *Proceedings of the 7th Annual Symposium on Computing for Development* (pp. 1–4). doi: 10.1145/3001913.3006629"

26. Chaudhari, S.S. and Bhole, V.Y., 2018, January. Solid waste collection as a service using IoT-solution for smart cities. In *2018 International Conference on Smart City and Emerging Technology (ICSCET)* (pp. 1–5). IEEE. doi: 10.1109/ICSCET.2018.8537326.

27. Abdullah, N., Alwesabi, O.A. and Abdullah, R., 2019. IoT-based smart waste management system in a smart city. In Saeed, F., Gazem, N., Mohammed, F., Busalim, A. (eds) *Recent Trends in Data Science*

and Soft Computing. *IRCTC 2018. Advances in Intelligent Systems and Computing* (Vol. 843). Springer, Cham.

28. Pelonero, L., Fornaia, A. and Tramontana, E., 2020, September. From smart city to smart citizen: rewarding waste recycle by designing a data-centric IoT based garbage collection service. In *2020 IEEE International Conference on Smart Computing (SMARTCOMP)* (pp. 380–385). IEEE. doi: 10.1109 /SMARTCOMP50058.2020.00081.

29. Khan, R., Kumar, S., Srivastava, A.K., Dhingra, N., Gupta, M., Bhati, N. and Kumari, P., 2021. Machine learning and IoT-based waste management model. *Computational Intelligence and Neuroscience*, *2021*, p. 5942574.

30. Bueno-Delgado, M.V., Romero-Gázquez, J.L., Jiménez, P. and Pavón-Mariño, P., 2019. Optimal path planning for selective waste collection in smart cities. *Sensors*, *19*(9), p. 1973.

31. Nowakowski, P. and Pamuła, T., 2020. Application of deep learning object classifier to improve e-waste collection planning. *Waste Management*, 109, pp. 1–9.

32. Rohit, G.S., Chandra, M.B., Saha, S. and Das, D., 2018, April. Smart dual dustbin model for waste management in smart cities. In *2018 3rd International Conference for Convergence in Technology (I2CT)* (pp. 1–5). IEEE. doi: 10.1109/I2CT.2018.8529600.

33. Coccoli, M., De Francesco, V., Fusco, A. and Maresca, P., 2022. A cloud-based cognitive computing solution with interoperable applications to counteract illegal dumping in smart cities. *Multimedia Tools and Applications*, *81*, pp. 95–113. doi: 10.1007/s11042-021-11238-8

34. Susanth, G.S., Livingston, L.J. and Livingston, L.A., 2021. Garbage waste segregation using deep learning techniques. In *IOP Conference Series: Materials Science and Engineering* (Vol. 1012, No. 1, p. 012040). IOP Publishing.

35. Sudha, S., Vidhyalakshmi, M., Pavithra, K., Sangeetha, K. and Swaathi, V., 2016, July. An automatic classification method for environment: Friendly waste segregation using deep learning. In *2016 IEEE Technological Innovations in ICT for Agriculture and Rural Development (TIAR)* (pp. 65–70). IEEE.

36. Flores, M.G. and Tan, J., 2019. Literature review of automated waste segregation system using machine learning: A comprehensive analysis. *International Journal of Simulation: Systems, Science and Technology*, pp. 15.1–15.7.

37. Manu, D., Shorabh, S.G., Swathika, O.G., Umashankar, S. and Tejaswi, P., 2022, May. Design and realization of smart energy management system for Standalone PV system. In *IOP Conference Series: Earth and Environmental Science* (Vol. 1026, No. 1, p. 012027). IOP Publishing.

38. Swathika, O.G., Karthikeyan, K., Subramaniam, U., Hemapala, K.U. and Bhaskar, S.M., 2022, May. Energy efficient outdoor lighting system design: Case study of IT campus. In *IOP Conference Series: Earth and Environmental Science* (Vol. 1026, No. 1, p. 012029). IOP Publishing.

39. Sujeeth, S. and Swathika, O.G., 2018, January. IoT based automated protection and control of DC microgrids. In *2018 2nd International Conference on Inventive Systems and Control (ICISC)* (pp. 1422–1426). IEEE.

40. Patel, A., Swathika, O.V., Subramaniam, U., Babu, T.S., Tripathi, A., Nag, S., Karthick, A. and Muhibbullah, M., 2022. A practical approach for predicting power in a small-scale off-grid photovoltaic system using machine learning algorithms. *International Journal of Photoenergy*, *2022*, p. 9194537.

41. Odiyur Vathanam, G.S., Kalyanasundaram, K., Elavarasan, R.M., Hussain Khahro, S., Subramaniam, U., Pugazhendhi, R., Ramesh, M. and Gopalakrishnan, R.M., 2021. A review on effective use of daylight harvesting using intelligent lighting control systems for sustainable office buildings in India. *Sustainability*, *13*(9), p. 4973.

42. Swathika, O.V. and Hemapala, K.T.M.U., 2019. IOT based energy management system for standalone PV systems. *Journal of Electrical Engineering & Technology*, *14*(5), pp. 1811–1821.

43. Swathika, O.V. and Hemapala, K.T.M.U., 2019, January. IoT-based adaptive protection of microgrid. In *International Conference on Artificial Intelligence, Smart Grid and Smart City Applications* (pp. 123–130). Springer, Cham.

44. Kumar, G.N. and Swathika, O.G., 2022. 19 AI Applications to. *Smart Buildings Digitalization: IoT and Energy Efficient Smart Buildings Architecture and Applications*, p. 283.

45. Swathika, O.G., 2022. 5 IoT-Based Smart. *Smart Buildings Digitalization: IoT and Energy Efficient Smart Buildings Architecture and Applications*, p. 57.

46. Lal, P., Ananthakrishnan, V., Swathika, O.G., Gutha, N.K. and Hency, V.B., 2022. 14 IoT-Based Smart Health. *Smart Buildings Digitalization: Case Studies on Data Centers and Automation*, p. 149.

47. Chowdhury, S., Saha, K.D., Sarkar, C.M. and Swathika, O.G., 2022. IoT-based data collection platform for smart buildings. In *Smart Buildings Digitalization* (pp. 71–79). CRC Press.

# 26 Cyber Security in Smart Energy Networks

Sanjeevikumar Padmanaban, Mostafa Azimi Nasab,
Tina Samavat, Mohammad Zand, Morteza Azimi Nasab,
and Erfan Hashemi
CTiF Global Capsule

## CONTENTS

26.1 Introduction ........................................................................................................ 310
26.2 Communication Network Architecture in Smart Grids ..................................... 315
    26.2.1 Producing Section ................................................................................ 315
    26.2.2 Transfer Section ................................................................................... 315
    26.2.3 Distribution Section ............................................................................. 315
    26.2.4 Consumer Section ................................................................................ 316
    26.2.5 Market Section ..................................................................................... 316
    26.2.6 Service Providers Section .................................................................... 316
    26.2.7 Operation Section ................................................................................ 316
26.3 Wireless Sensor Network.................................................................................... 318
    26.3.1 Zigbee .................................................................................................. 319
    26.3.2 Low-Power Wi-Fi ................................................................................ 319
26.4 Standards and Protocols of Communication in Smart Networks....................... 319
    26.4.1 Protocol DNP3...................................................................................... 319
    26.4.2 Protocol IEC61850............................................................................... 320
26.5 Smart Network Security Goals ........................................................................... 320
    26.5.1 Availability ........................................................................................... 320
    26.5.2 Integrity ............................................................................................... 320
    26.5.3 Confidentiality ..................................................................................... 320
26.6 The Most Important Cyber-Attacks in Smart Networks .................................... 320
26.7 Intelligent Measuring Equipment and Home Networks .................................... 321
26.8 Security and Privacy of WSN-Based Consumer Applications........................... 321
26.9 Conclusion .......................................................................................................... 323
References................................................................................................................... 323

### Abstract

The novel electricity industry, known as smart grids, integrates the power generation and information transmission system with telecommunication networks to form a two-way infrastructure of electricity and information, which is a prerequisite for an intelligence network. The smart grid structure will be very efficient and cost-effective in terms of expense and management due to the presence of new telecommunication systems and technologies. However, although having many benefits, the integration of these two structures has disadvantages in terms of system security and protection.

Concerning the information security of telecommunication systems of the smart grid, it must be said that the electrical network, which has a minor connection to the Internet, is exposed to considerable threats. These risks include security attacks by enemy gatherings and hackers to disrupt generating, transmission, and distribution of electricity or to manipulate and corrupt data sent over the smart grid. The various layers of cyber security must be designed to minimize the chance of attacks. So, all connections to an Internet network need to be extremely secure. It is not necessary to detect an attacker's intrusion on the network connection port of the Internet. However, through the network, particularly, in the wireless data transmission environment, this system must be able to detect intrusion. All components of systems and networks in a smart grid must be assessed when designing security. At the same time, in the event of an accident, the system must be able to respond appropriately as soon as a disturbance occurs. So, in such networks, the speed of detection, report, decision-making, and response is a vital characteristic of security.

## 26.1  INTRODUCTION

Smart grids, considered by many to be the greatest technological revolution since the invention of the Internet, play a major role in today's society. Governments worldwide invest huge budgets in research, development, and implementation of smart grids with many diverse goals. For instance, smart grids have the potential to reduce carbon dioxide emissions by employing renewable energy sources, energy storage, and hybrid electric vehicles. They can also increase the reliability of electricity sources by monitoring and controlling the generation, transmission, and distribution of power grids by advancing real-time measurements. In addition, they can make power stations and power transmission infrastructure more efficient and apply dynamic pricing and load response strategies. Apart from these benefits, the products and services that the smart grid can provide are unimaginable.

Today, the power grid is a system that supports the operation of electricity generation, transmission, and distribution. Power flows mainly from central generating stations (high-voltage generators) to medium- and low-voltage consumers. Power generation and consumption need to be balanced to prerequisite the stability of the entire network. For example, the energy of consumers who possess a renewable energy supply may flow to the grid. Due to the unpredictable character of renewable sources, it leads to a complex network structure. This makes their integration a challenge and requires optimizing the old electrical infrastructure.

The global energy infrastructure is expected to undergo a similar change to the telecommunications and media industry in the coming decade. The smart grid combines electronic power infrastructure with modern digitally distributed computing facilities and communications networks. It is a set of dependent and complex systems whose main task is to deliver reliable and efficient power. This is made possible by an extensive awareness, peak reduction in load response schemes, and comprehensive integration of intermittent renewable energy sources through real-time control and power storage.

It can be acknowledged that today, smart energy networks in any country are considered vital highways for sustainable development. Proper and cheap supply and distribution of energy to its consumers, which are the country's major industries, give the ability to increase national production by producing products. Energy distribution has undergone major changes today due to new requirements and new methods of energy production, such as distributed generation. Information network infrastructure is considered a requirement of the energy distribution network, and monitoring of all network components in a centralized manner, the ability to execute commands on all sections, and rapid reporting of potential problems are among the first requirements. The connection of power network infrastructure, its dynamic properties, and optimal management strategies (such as load response and intelligent measuring devices) are properly combined. However, such a large volume of communications and data transmission have created many security vulnerabilities that pose a risk to the power system and jeopardize the system's reliability, which is the main goal of the power network in improving the performance of this indicator. For example, Ref. [1] has shown that intruders' intrusion into the power grid can lead to dangerous consequences, including loss of information on the part of the consumer, successive breakdowns, widespread blackouts, and infrastructure failures.

High reliance on the information network certainly puts the smart grid vulnerable to communications and network systems. In traditional power grids, grid control systems were kept isolated from insecure environments such as the Internet. However, in smart grids, security attacks on the grid infrastructure can easily be carried out from different parts of the infrastructure. For example, the adversary does not need access to enclosed places or systems (such as narrators, substations, or command centers) to damage the power delivery process and can easily attack regardless of the distance between attaches and targets.

In recent years, many cases of cyber-attacks raise the question of whether they can justify the security vulnerabilities and the devastating effects these attacks have on the power grid infrastructure. In this section, real examples of vulnerabilities and cyber-attacks will be presented. These cases double the importance of the issue and emphasize the concept of protecting the infrastructure against cyber-attacks.

- The virus *stuxnet* was discovered which infiltrates the Windows operating system, Siemens industrial equipment and the software tried to destabilize the operation of the power system. This type of cyber-attack, which manifests itself in the form of viruses in the industrial equipment of power plants, poses serious threats to both the cyber system and the physical system of the power grid [2].
- On August 11, 1983, large areas of the Midwest and Northeast of the United States and Ontario in Canada experienced a massive blackout that lasted even a day in some parts. The blackout affected about 93 million people and 63 MW of power in several parts of the United States. Although this historical blackout is not directly related to the malicious operations of cyber terrorists, it was caused by defects and errors in the software programs of the cyber system.
- In September 2003, Italy and some parts of Switzerland experienced their greatest blackout, which affected about 96 million people. After 18 hours in Italy, the blackout ended but caused huge financial and economic losses. This blackout was caused by human error and a lack of effective communication between network operators.
- Another major blackout in southwestern Europe occurred on November 4, 2006, due to human error. Inefficient and inadequate network communication was the main reason for this disaster [3].
- On January 25, 2003, MS SQL Server 2000 worm attacked the Davis–Besse Nuclear Power Plant, which caused data overload in the site network; therefore, the presented computers could not communicate efficiently. The main impact of this action was the unavailability of safety parameters.
- In March 1983, the US Department of Defense launched a test cyber-attack that killed a narrator himself. The mentioned experiment was performed in a laboratory unit that resulted in the hacking of a power plant control system and the vibration of a narrator. This type of attack, which was coordinated on a large scale, could damage electricity infrastructure for months.

Security Consultant *Winkler Winkler* and his group were hired by a company active in power electricity to investigate the vulnerability of the network computer system covered by that company. Using social engineering and destructive browsers, they could access the power plant control network in 1 day, enabling them to monitor power generation and distribution. In addition to accessing the system, the group was able to obtain some records. Figure 26.1 shows the importance of smart grid security from the perspective of companies active in the field of energy services. According to Electricity News [4], smart grids have expanded rapidly in recent years, but their security still stands as a growing concern. *Modular* Research Institute has provided the following statistics and results in its online surveys of some companies active in the field of energy services. How important do you think the issue of security in smart grids is at the moment? Figure 26.2 shows position of smart

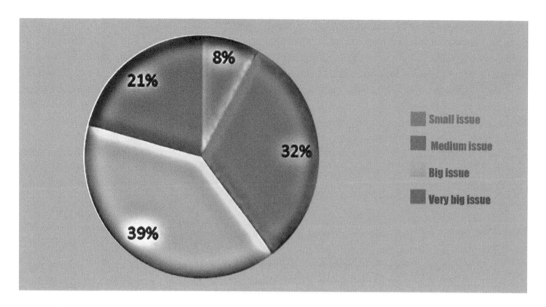

**FIGURE 26.1**  The importance of smart grid security from the perspective of companies active in the field of energy services.

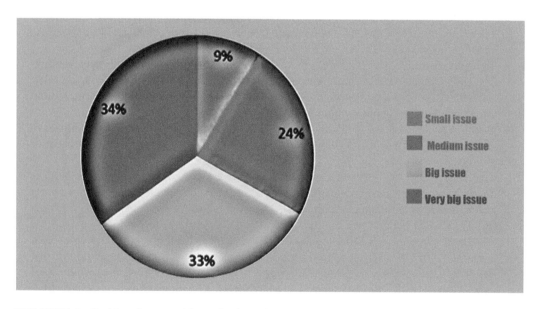

**FIGURE 26.2**  Position of smart grid security from the perspective of companies active in the field of energy services.

grid security from the perspective of companies active in the field of energy services. In your opinion (Figure 26.3), in the next 5 years, what will be the position of the smart network security issue?

- Figure 26.4 shows the best way to reduce security concerns from the perspective of companies active in the field of energy services. Are smart grids vulnerable to security threats?
- Figure 26.5 represents the readiness of companies active in the field of energy services to implement cyber security. How prepared is your organization to invest in cyber security technology over the next 5 years?

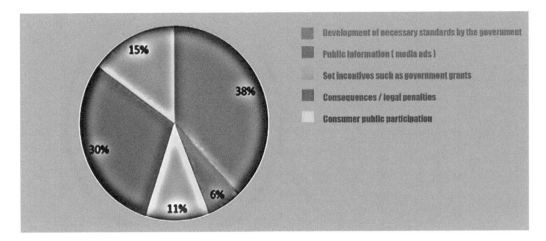

**FIGURE 26.3**   The best way to reduce security concerns from the perspective of companies active in the field of energy services.

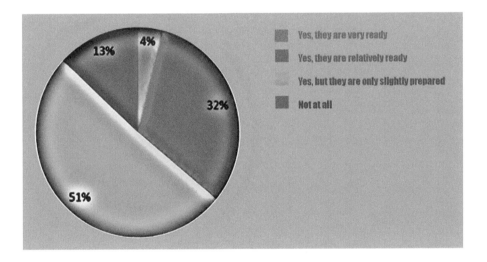

**FIGURE 26.4**   The readiness of companies active in the field of energy services to implement cyber security.

- Figure 26.6 shows the most important areas of smart grid security from the perspective of companies active in the field of energy services. Which area of smart grids will be the most invested next year?

Also, security officials of energy service companies were asked to classify and rate their level of concern about the number of messages caused by security breaches in the above-mentioned areas using a scale of 1–9. Figure 26.7 shows the level of concern in different parts of the smart grid from the perspective of energy companies. Level 1 indicates low-intensity anxiety, and level 9 indicates extremely high-intensity anxiety.

As can be seen from the above description, some of the serious vulnerabilities in smart grid start from the cyber part of the grid. For this reason, the security of smart grid infrastructure must distinguish between the categories of a cyber-attack by terrorists and spy networks, dissatisfied employees, personal errors, equipment breakdowns, and natural disasters and make each one transparent. To protect critical smart grid infrastructure, anomaly detection can play an effective role

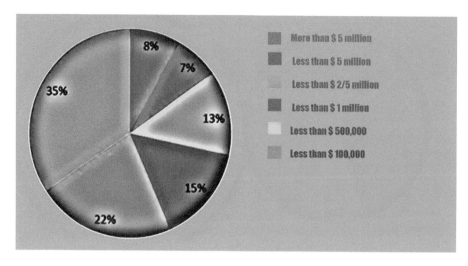

**FIGURE 26.5** How prepared is your organization to invest in cyber security technology.

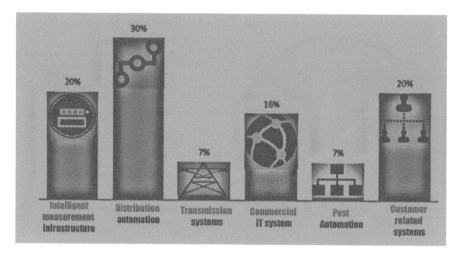

**FIGURE 26.6** The most important areas of smart grid security from the perspective of companies active in the field of energy services.

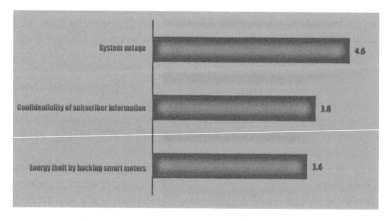

**FIGURE 26.7** Level of concern in different parts of the smart grid from the perspective of energy companies.

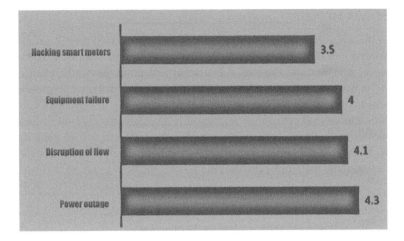

**FIGURE 26.8** Level of severity of the consequences of security breaches from the perspective of companies operating in the field of energy.

in identifying malicious information on the network. The cyber-attack incidents discussed in this section have prompted researchers and electrical industry engineers for future research in this area. Since security challenges arise mainly from malicious attacks on the communication network, it is necessary first to review the basic but essential concepts of computer networks that are the basis of communication and intelligence to deal with the threats that can occur in the smart network. In Figure 26.8, the level of severity of the consequences of security breaches from the perspective of companies operating in the field of energy is shown. Now we can take the discussion to the smart grid and recount the standards and criteria provided.

## 26.2 COMMUNICATION NETWORK ARCHITECTURE IN SMART GRIDS

In this section, after reviewing how cyber-attacks are formed and enumerating the most important ones, the security goals and smart network infrastructure to create communication platforms are discussed. According to the conceptual model of the American National Institute of Standards and Technology, the smart grid consists of seven logical and working spaces:

### 26.2.1 PRODUCING SECTION

Traditional large-scale power generation units, such as nuclear power plants, thermal power plants, wind farms, and solar power plants, are considered components of this section. This production capacity is injected into the downstream network by transmission lines, so the interference of this space with the space of the transmission sector is the most important relationship of this sector.

### 26.2.2 TRANSFER SECTION

Local transmission operators (or system-independent operators) are responsible for the safe operation of the transmission space. Power consumption is at the core of the process, minimizing system losses by reducing the voltage level.

### 26.2.3 DISTRIBUTION SECTION

This workspace is the interface between the transfer and the subscriber. Measuring equipment, loads, distributed generation sources, and microgrids are the main parts of this section.

### 26.2.4  Consumer Section

As the name of this section indicates, the loads fed with the produced energy are classified in this unit, mainly divided into three categories: household, commercial, and industrial. With the emergence of the smart grid concept, the role of consumer and producer can be changed (*prosumer*), and the state of the network determines this role. Issues such as distributed generation, storage, and energy management are driving the change. The key to connecting this group of subscribers to the smart grid lies in a microgrid concept. A microgrid is an interconnected set of loads and distributed generation sources that operate as an integrated network system despite predefined definitions and limitations. The microgrid can be connected to or disconnected from the network and, in both cases, support subscribers with high reliability.

### 26.2.5  Market Section

Market management, retail, collecting companies, implementing the electricity market, and providing ancillary services are some of the important challenges and issues in this sector. Unit participation programs, coordination of distributed generation units, and power distribution between different microgrid units are important issues that should be managed well. Under these circumstances, there will be no more talk of monopoly because the owners of distributed generation resources, as a market player, can enter and compete in this competitive environment.

### 26.2.6  Service Providers Section

Solving issues related to the dynamics of the market system, along with ensuring the safe operation of the basic infrastructure of the network, is one of the commitments of the directors of this section. Customer management, installation of smart equipment, and intelligent management in reaction to the price signals of the load response program are among the conventional programs in this section.

### 26.2.7  Operation Section

This unit is responsible for the safe and secure operation of the power grid. Energy management systems provide effective operation at the transmission level, while distribution management systems provide it at the distribution level. Applications in this area include power grid monitoring, control, protection, and related information analysis. Among the local networks (Figure 26.9), some require two-way communication and power transmission, while others highlight the integration of information and intelligent network management in a centralized manner. In Ref. [4], more detailed information on this classification can be found. Also, to connect all these classes, the smart grid must be hierarchical and widely distributed. Figures 26.9 and 26.10 show the seven sections of the smart grid and its hierarchical structure for connecting these sections, respectively. The communication network consists of nodes that can be gateways in the local network or routers with high bandwidth to send messages in each of the infrastructures and floors of the smart network. Here, conventional wired communication technologies, such as fiber optic technology, can be used to speed up data transmission across different classes. For example, the supervisory control and data acquisition (SCADA) system, which is used to monitor the status of the power grid in the distribution, transmission, and operation sectors, uses this technology [5].

All power quality signals are transmitted from local networks in the distribution and transmission layers to the operating layer through the main network to be centrally processed. A local area network is used for interlayer communication. This network consists of *ad hoc* nodes containing measuring devices, sensors, or intelligent electronic devices. They usually have limited bandwidth and can make limited calculations for protection and monitoring discussions. These nodes are not limited to the use of wired networks. Many of them are expected to use wireless sensor networks.

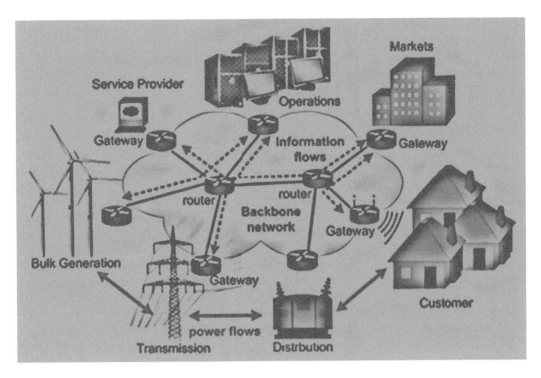

**FIGURE 26.9**  Seven logical networks of smart grids based on NIST.

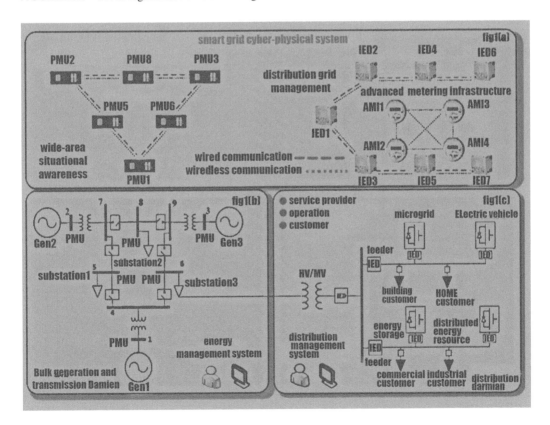

**FIGURE 26.10**  The hierarchical structure of the smart grid to connect its logical spaces.

The advantages of using a wireless network are mentioned in the references. Figure 26.10 shows the hierarchical structure of the smart grid to connect its logical spaces. These advantages include easy access to network information, more dynamism, reasonable price, and less complexity. In the next section, wireless sensor networks (WSNs) will be introduced.

## 26.3 WIRELESS SENSOR NETWORK

WSNs are composed of small and inexpensive mechanical micro-electric systems that, with the help of several sensors, are able to collect the size of the environment and use their limited capacity to process and store these measurements and then, transmit this data through its transceiver devices. Sensor nodes perform this operation with their limited batteries. Although these sensors can also obtain energy from the environment, this energy will be relatively small.

The smart grid integrates Information and Communication Technology into its operations, intending to increase electrical services' reliability, security, and efficiency and reduce greenhouse gas emissions. Figure 26.11 shows the layered structure of the intelligent network. The last layer is the electrical infrastructure, which includes conventional network power receiving equipment. At the top of the electrical infrastructure, there is a communication layer that includes low and high bandwidth standards such as *Zigbee*, *WIMAX*, and *i-Fi*. At the top of the communication, the computational layer is responsible for data collection, storage, processing, and decision-making and is key to smart grid applications. The top layer is smart grid applications, including remote measurement, equipment coordination, smart grid monitoring, plug-in Hybrid Electric Vehicle coordination, and more. The security layer is related to all four of these layers, and security in each layer must be done carefully.

Recently, WSNs have entered home applications, creating more opportunities to integrate homes with smart grids. In this context, the concept of smart homes, focusing on energy efficiency and smart grid operation, has doubled in importance [6]. In Ref. [7], for example, the microprocessor provides a simple interface for the consumer to check his/her monthly bills and the consumption of electrical equipment, thereby improving the home electricity meter. Consumer consumption is

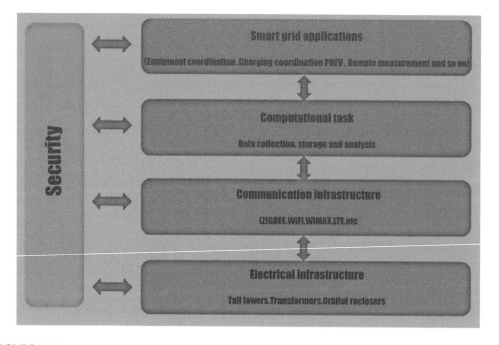

**FIGURE 26.11**   Smart network layer structure.

an important part of the smart grid that benefits from two-way information and energy flow. The flow of major information is controlled by smart meters. While sending consumption information to the electricity company to pay the bill, smart meters also give consumers different hourly price information.

They can use various wireless communication technologies as long as they do not interfere with power efficiency. Short-range communication technologies are preferred due to their dense layout, cost, and energy issues. WSNs may communicate via *Zigbee* or Wi-Fi.

### 26.3.1 ZIGBEE

Recently, several smart devices and home automation devices have been able to communicate via Zigbee.

Smart meter vendors have also developed smart meters that can be used with Zigbee and enable connectivity in smart meters, home appliances, and home automation tools. Communication via Zigbee is to enhance remote metering and Advanced Metering Infrastructure (AMI) and Smart Energy Profile mode and connect industries and home appliances. In short, Zigbee has been widely used in consumer electrical equipment and smart meters. Using Zigbee for connecting the sensor nodes increases the WSN coordination with other appliances in the home. This technology is based on the IEEE 4.15.802 standard.

### 26.3.2 LOW-POWER WI-FI

Conventional Wi-Fi technology has been widely used in home and commercial applications to provide relatively high-speed data communication, and the IEEE 11.802 standard is also intended for Wi-Fi. In smart grids, the goal is for *Wi-Fi* to become part of Home Area Networks (*HAN*s), Neighborhood Area Networks (*NAN*s), and Field Area Networks.

Wi-FI has more range than Zigbee. For example, the approximate range of 500 m outside the home environment has made it an accessible solution for networks beyond distribution networks in the smart grid. Using *Wi-Fi* in *NAN*s, *HAN*s further increases interoperability with other parts, so Wi-Fi is considered as a promising standard for smart grids [8]. Despite the advantages, its high-power consumption is a disadvantage, except in recent Wi-Fi chips with very low power, which *Wi-Fi* can also participate in Wireless Area Networks (WANs). Wi-Fi has been committed to operating like a Zigbee for several years, with data rates of about 1– 2 *Mbps* and a range of 10–70 m outdoor of a home [8]. The applications of Wi-Fi based sensors in smart grids are recently studied and a comparison of low-power Wi-Fi and Zigbee is given in Ref. [9].

## 26.4  STANDARDS AND PROTOCOLS OF COMMUNICATION IN SMART NETWORKS

Power grid protocols have evolved from their previous proprietary to their present standardized forms. The following is a brief discussion of the four major protocols in the smart grid: Protocol DNP3, which is commonly used in North America, Protocol IEC618260, recently standardized by the International Electrotechnical Commission (*IEC*) for advanced substation automation, and Protocols MODBUS and 37IEEE std.c.118, which deal with information sent by phasor measurement units (PMUs).

### 26.4.1  PROTOCOL DNP3

General Electric implemented this protocol in 1550. Originally intended for SCADA systems, it is now used in many electrical, water, and security infrastructures in North and South America, South

Africa, Australia, and Asia. It was originally designed in four layers. These layers were the physical, data link, transfer, and application layers [10]. The physical layer is based on the serial communication protocol referred to in the 422 –RS-232, RS, and 485 –RS standards. Today, this protocol is mapped to the Transmission Control Protocol/Internet Protocol (TCP/IP) model to benefit from the latest technology [11].

### 26.4.2  Protocol IEC61850

This standard has recently been proposed by *IEC* for the automation of posts for network communication [12]. Unlike 3*DNP*, this standard uses different types of protocols such as *UDP/IP, TCP/IP,* and *ADM* for sensitive and real-time communication [13]. In addition, it specifies schedules and time constraints on data transfer and the exchange of information in posts. It is noteworthy that the IEC 61850 standard is intended for replacement with the 3*DNP* protocol in substations and is limited to substation automation only, but has the potential to be used in all power grid communication infrastructures in the near future.

## 26.5  SMART NETWORK SECURITY GOALS

Following the items mentioned in the previous sections, experts in the field of cyber security in *NIST* have recently published a practical guide on smart network security, which is briefly explained below.

### 26.5.1  Availability

Accessing and using the information in the most secure way possible is one of the most important issues in a smart network. Unavailability, disconnection, and non-use of information lead to the depletion of the power system.

### 26.5.2  Integrity

Dealing with changing and destroying information preserves the originality of the data and does not negate it.

### 26.5.3  Confidentiality

Maintaining permissible restrictions on access to and disclosure of the information is an important point in protecting subscribers' privacy.

This feature prevents unauthorized disclosure of information that is not in the public domain. Of the three mentioned cases, the sections with the consumers play an important role and are more effective. However, with respect to transmission and production, the first two cases are more important.

## 26.6  THE MOST IMPORTANT CYBER-ATTACKS IN SMART NETWORKS

Studying the cyber security of the smart grid is considered necessary because of the network's vulnerabilities (areas where security attackers are likely to attack and irreparable economic and social damages occur if they malfunction), and the duties of security managers will become easier. Security issues are mostly studied in the field of control equipment and equipment responsible for collecting and sending data, including remote control terminals, smart meters, PMUs, and programmable logic controllers.

Cyber security issues can be explored from another perspective. When computer infrastructure and equipment software topics are studied, topics such as network routers, firewalls, encryption,

**TABLE 26.1**
**The Most Important Cyber-Attacks in the Smart Grid and Related Research**

| Reference | Type | Reference | Type |
|---|---|---|---|
| [5,6] | Man-in-the-Middle | [7–13] | ARP Spoofing |
| [6] | Precision Insider | [11] | Eavesdropping |
| [6] | Rogue Software | [2,5] | Malformed Packet |
| [11,13] | Denial of Service | [2,5] | Database Attack |

intrusion analysis, and cyber-attack scenarios become the first headlines in this area. One of the most important and dangerous scenarios for a security hacker attack is implementing a middle-man attack. The purpose of the attack, as previously described, was to intercept a data center and local equipment sending data and to manipulate information to achieve its security objectives.

As it turns out, this attack disrupts the integrity and privacy that are among the security goals of the smart grid. Because smart meters deal directly with subscriber information, the importance of these items is doubly important for security purposes. Such an attack can be very dangerous and cause irreparable damage.

A denial-of-service (*DoS*) attack also tries to cause the operator to not perform optimally by overloading the capacity of computer equipment and overflowing the data sent by the network. In this case of accessibility, the first target of smart grid security is attacked.

Table 26.1 lists the various references that have studied the most important smart grid cyber-attacks.

## 26.7   INTELLIGENT MEASURING EQUIPMENT AND HOME NETWORKS

AMI is an important system in smart grids. It provides the infrastructure for communication between the home network and manufacturers and is constantly connected to smart meters to monitor consumption management and issues such as load responsiveness.

The National Institute of Standards and Technology (NIST) report lists nearly ten AMI-related approaches, of which only two are important, related to the goals of smart grid integrity and privacy. Figure 26.12 illustrates these concepts. In this figure, the following approaches can be seen:

1. Exchange of information such as read values and maintenance status between smart meters and manufacturers
2. The connection between the electricity market and the control and analysis centers with the smart meters available at the subscribers' place.

Real-Time Pricing and load response can be mentioned among the connections between smart meters and the electricity market. Unlike the level of distribution and transmission, where the speed of sending and receiving information is of particular importance, in both cases, we are faced with time intervals ranging from a few minutes to several hours [13]. Figure 26.12 shows the two important communication approaches of AMI infrastructure in smart grid. As a result, the security goals of integrity and confidentiality over accessibility have attracted more attention in this area, as there is a great deal of confidential information in this area, including the subscribers' personal use and financial matters.

## 26.8   SECURITY AND PRIVACY OF WSN-BASED CONSUMER APPLICATIONS

The power grid has become smarter with the advent of information and communications technology (ICT), while the development of ICTs may increase the vulnerability of the smart grid to cyber-

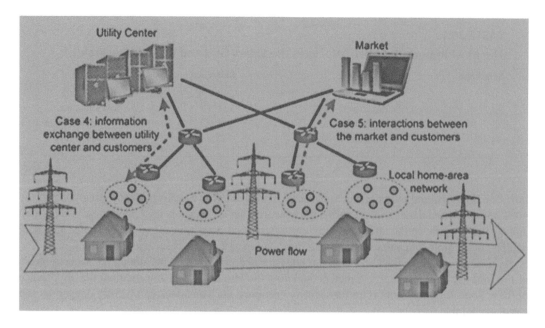

**FIGURE 26.12** Two important communication approaches of AMI infrastructure in smart grid.

attacks. In particular, the development of hundreds of small sensors, which have been expanded ad hoc, has made the protection of smart grid data a major challenge.

The security of a WSN is related to a combination of several criteria such as availability, validation, integrity, and novelty [14]. Availability means that network services are available without interruption under any circumstances, even when an error is detected. Authority controls the access of sensors to which unauthorized sensors cannot send or receive data. Validation ensures the accuracy of the data, which prevents fake messages from being sent by incorrect nodes. Flawless means that a message on its way to the destination has not been modified. Finally, it recently ensures that old messages received from an attacker are not answered. Maintaining these security factors in WSN is challenging for several reasons:

- WSNs use a large number of sensor nodes that are deployed in unprotected environments.
- Sensor nodes have limited processing and storage capabilities.
- Sensor nodes have limited energy.

In WSNs, where wireless communication devices expose sensor nodes to eavesdropping or jamming, advanced signal processing methods cannot be used, because the use of these methods increases costs and power consumption. Sensor nodes may be deployed in an unprotected environment where it is easier to steal and intercept a node or sensor than any other network. In addition, popular encryption methods cannot be used due to the limited processing and storage capabilities of sensor nodes. Public key cryptography requires cost-saving calculations, while symmetric key cryptography requires good key distribution methods, which are challenging in WSN. The limited batteries of the sensor nodes expose them to the dangers of non-service (*DoS*), which can easily drain the battery of the sensor nodes.

In short, the risks associated with WSNs can target different layers of communication, such as the physical, communication, routing, or transmission layers. The defense mechanism against these risks usually increases computational and communication costs. However, these mechanisms are essential for smart grid equipment.

In fact, smart grid security is a broad topic that covers consumer equipment and smart grid accessories [11]. This section will focus on home appliance equipment, which includes smart meters and home power controllers.

Smart meters deliver consumer electricity data to the electricity company. Modifying this data by malicious attackers may result in incorrect billing, inaccurate load figures, incorrect forecasts, and incorrect pricing decisions. For example, smart meters may be a target for the risk of changing Internet loads, which could jeopardize the electricity company's load control signal signals [15]. If these hazards occur through many smart meters, it may jeopardize network stability. In addition, improper configuration of consumer equipment in an interconnected smart grid can provide the basis for the risks of data alteration [16]. This type of risk works in two ways:

1. Modified consumption data can be generated and sent to the electricity company.
2. Modified control signals from the power company may be sent to consumers.

These types of hazards occur more easily in consumer equipment, and the power company has little or no control over this equipment. Attackers may extract data containing key information used to determine the validity of the network from the device memory and embed software that can be extended to other devices in the AMI [17].

Privacy is just as important as security and needs to be handled carefully for WSN-based consumer applications on the smart grid. Privacy refers to a person's information and personal communication or any personal data such as physical, psychological, or economic conditions that a person is reluctant to share with others. Finally, privacy is a person's right to communicate without inspection or monitoring [18]. In Refs. [19–22], the authors have shown that access to detailed information on household activities, such as the presence or absence of a person, sleep cycle, meal times, and baths, will be possible by accessing microdata on electricity consumption. If such data are published, it can pose risks to the consumer and be profitable. For example, this information provides a basis for analyzing the performance of an electric vehicle that can lead to damage to the manufacturer [23–26].

## 26.9   CONCLUSION

In the intelligent network, there is a close connection between all performances. In addition, the information received from each section, even the smallest sections, can significantly impact the performance of the entire system and the decisions of the relevant units. That is why the security of this network is so important. The disadvantage of these networks is the existence of multiple sensors and connections to the Internet to establish a stable connection, which is the gateway for hackers and attackers. This chapter reviews the various components, how the network is vulnerable to profiteers, and how to deal with it. In addition to the extensions that lead to network instability, consumer data may be compromised if high-quality power consumption data is made available to malicious users.

## REFERENCES

[1] Zand, M., Nasab, M. A., Hatami, A., Kargar, M., & Chamorro, H. R. (2020, August). Using adaptive fuzzy logic for intelligent energy management in hybrid vehicles. In *2020 28th Iranian Conference on Electrical Engineering (ICEE)* (pp. 1–7). IEEE.
[2] Asadi, A. H. K., Jahangiri, A., Zand, M., Eskandari, M., Nasab, M. A., & Meyar-Naimi, H. (2022, January). Optimal design of high density HTS-SMES step-shaped cross-sectional solenoid to mechanical stress reduction. In *2022 International Conference on Protection and Automation of Power Systems (IPAPS)* (Vol. 16, pp. 1–6). IEEE.
[3] Ahmadi-Nezamabad, H., Zand, M., Alizadeh, A., Vosoogh, M., & Nojavan, S. (2019). Multi-objective optimization based robust scheduling of electric vehicles aggregator. *Sustainable Cities and Society*, 47, 101494.

[4] Zand, M., Nasab, M. A., Sanjeevikumar, P., Maroti, P. K., & Holm-Nielsen, J. B. (2020). Energy management strategy for solid-state transformer-based solar charging station for electric vehicles in smart grids. *IET Renewable Power Generation*, 14(18), 3843–3852.

[5] Ghasemi, M., Akbari, E., Zand, M., Hadipour, M., Ghavidel, S., & Li, L. (2019). An efficient modified HPSO-TVAC-based dynamic economic dispatch of generating units. *Electric Power Components and Systems*, 47(19–20), 1826–1840.

[6] Nasri, S., Nowdeh, S. A., Davoudkhani, I. F., Moghaddam, M. J. H., Kalam, A., Shahrokhi, S., & Zand, M. (2021). Maximum power point tracking of photovoltaic renewable energy system using a new method based on turbulent flow of water-based optimization (TFWO) under Partial shading conditions. In *Fundamentals and Innovations in Solar Energy* (pp. 285–310). Springer, Singapore.

[7] Dashtaki, M. A., Nafisi, H., Khorsandi, A., Hojabri, M., & Pouresmaeil, E. (2021). Dual two-level voltage source inverter virtual inertia emulation: A comparative study. *Energies*, 14(4), 1160.

[8] Rohani, A., Joorabian, M., Abasi, M., & Zand, M. (2019). Three-phase amplitude adaptive notch filter control design of DSTATCOM under unbalanced/distorted utility voltage conditions. *Journal of Intelligent & Fuzzy Systems*, 37(1), 847–865.

[9] Porcu, D., Castro, S., Otura, B., Encinar, P., Chochliouros, I., Ciornei, I., ... & Bachoumis, A. (2022). "Demonstration of 5G solutions for smart energy grids of the future: a perspective of the Smart5Grid project. *Energies*, 15(3), 839.

[10] Zand, M., Nasab, M. A., Neghabi, O., Khalili, M., & Goli, A. (2019, December). Fault locating transmission lines with thyristor-controlled series capacitors by fuzzy logic method. In 2020 14th International Conference on Protection and Automation of Power Systems (IPAPS) (pp. 62–70). IEEE. doi: 10.1109/IPAPS49326.2019.9069389.

[11] Zand, Z., Hayati, M., & Karimi, G. (2020, August). Short-channel effects improvement of carbon nanotube field effect transistors. In 2020 28th Iranian Conference on Electrical Engineering (ICEE) (pp. 1–6). IEEE, doi: 10.1109/ICEE50131.2020.9260850.

[12] Tightiz, L., Nasab, M. A., Yang, H., & Addeh, A. (2020). An intelligent system based on optimized ANFIS and association rules for power transformer fault diagnosis. *ISA Transactions*, 103, 63–74.

[13] Zand, M., Neghabi, O., Nasab, M. A., Eskandari, M., & Abedini, M. (2020, December). A hybrid scheme for fault locating in transmission lines compensated by the TCSC. In *2020 15th International Conference on Protection and Automation of Power Systems (IPAPS)* (pp. 130–135). IEEE.

[14] Zand, M., Nasab, M. A., Khoobani, M., Jahangiri, A., Hosseinian, S. H., & Kimiai, A. H. (2021, February). Robust speed control for induction motor drives using STSM control. In *2021 12th Power Electronics, Drive Systems, and Technologies Conference (PEDSTC)* (pp. 1–6). IEEE. doi: 10.1109/PEDSTC52094.2021.9405912.

[15] Sanjeevikumar, P., Zand, M., Nasab, M. A., Hanif, M. A., & Bhaskar, M. S. (2021, May). Spider community optimization algorithm to determine UPFC optimal size and location for improve dynamic stability. In *2021 IEEE 12th Energy Conversion Congress & Exposition-Asia (ECCE-Asia)* (pp. 2318–2323). IEEE. doi: 10.1109/ECCE-Asia49820.2021.9479149

[16] Azimi Nasab, M., Zand, M., Eskandari, M., Sanjeevikumar, P., & Siano, P. (2021). Optimal planning of electrical appliance of residential units in a smart home network using cloud services. *Smart Cities*, 4(3), 1173–1195.

[17] Dashtaki, M. A., Nafisi, H., Khorsandi, A., Hojabri, M., & Pouresmaeil, E. (2021). Dual two-level voltage source inverter virtual inertia emulation: A comparative study. *Energies*, 14(4), 1160.

[18] Nasab, M. A., Zand, M., Padmanaban, S., Bhaskar, M. S., & Guerrero, J. M. (2022). An efficient, robust optimization model for the unit commitment considering renewable uncertainty and pumped-storage hydropower. *Computers and Electrical Engineering*, 100, 107846.

[19] Dashtaki, M. A., Nafisi, H., Pouresmaeil, E., & Khorsandi, A. (2020, February). Virtual inertia implementation in dual two-level voltage source inverters. In *2020 11th Power Electronics, Drive Systems, and Technologies Conference (PEDSTC)* (pp. 1–6). IEEE.

[20] Azimi Nasab, M., Zand, M., Padmanaban, S., & Khan, B. (2021). Simultaneous long-term planning of flexible electric vehicle photovoltaic charging stations in terms of load response and technical and economic indicators. *World Electric Vehicle Journal*, 12(4), 190.

[21] Zand, M., Nasab, M. A., Padmanaban, S., & Khoobani, M. (2022). Big data for SMART sensor and intelligent electronic devices–building application. In *Smart Buildings Digitalization* (pp. 11–28). CRC Press.

[22] Laayati, O., Bouzi, M., & Chebak, A. (2022). Smart energy management system: Design of a monitoring and peak load forecasting system for an experimental open-pit mine. *Applied System Innovation* 5(1), 18.

[23] Padmanaban, S., Khalili, M., Nasab, M. A., Zand, M., Shamim, A. G., & Khan, B. (2022). Determination of power transformers health index using parameters affecting the transformer's life. *IETE Journal of Research*, 1–22.

[24] Nasab, M. A., Zand, M., Hatami, A., Nikoukar, F., Padmanaban, S., & Kimiai, A. H. (2022, January). A hybrid scheme for fault locating for transmission lines with TCSC. In *2022 International Conference on Protection and Automation of Power Systems (IPAPS)* (Vol. 16, pp. 1–10). IEEE.

[25] Sanjeevikumar, P., Samavat, T., Nasab, M. A., Zand, M., & Khoobani, M. (2022). Machine learning-based hybrid demand-side controller for renewable energy management. In *Sustainable Developments by Artificial Intelligence and Machine Learning for Renewable Energies* (pp. 291–307). Elsevier.

[26] Khalili, M., Ali Dashtaki, M., Nasab, M. A., Reza Hanif, H., Padmanaban, S., & Khan, B. (2022). Optimal instantaneous prediction of voltage instability due to transient faults in power networks taking into account the dynamic effect of generators. *Cogent Engineering*, 9(1), 2072568.

# Index

adaptive traffic control 181, 186
ambulance management system 111, 114, 115, 116, 118
ambulance tracking system 111, 113, 114
anomaly detection 9, 29, 30, 32, 35, 38, 40, 144, 210, 313
Arduino 11, 43, 44, 45, 47, 48, 49, 63, 67, 68, 111, 114, 115,
    118, 123, 125, 126, 128, 129, 138, 140, 141, 145,
    146, 177, 179, 183, 207, 209, 211, 215, 216, 217,
    218, 221, 223, 223, 226, 227, 228, 229, 231, 232,
    233, 234, 236, 245, 247, 249, 289, 290, 292,
    293, 294, 296, 299, 300, 301, 302, 304, 306
artificial neural network 4, 32, 66, 74, 75, 76, 79, 84, 92,
    94, 95, 144, 148, 183, 184
auto-regressive integrated moving rate 75, 76, 78, 83, 84,
    90, 96

back-propagation neural network (BPNN) 75, 79
bioenergy plants 273, 280
bluetooth 215, 217, 218, 219, 220, 221, 226, 227, 232, 243,
    244, 246, 247, 249, 255
boost converter 257, 258, 260, 261, 265, 270

cloud 9, 10, 11, 16, 58, 65, 66, 67, 69, 70, 74, 91, 93, 113,
    114, 115, 118, 119, 120, 123, 124, 125, 127, 129,
    130, 136, 137, 145, 146, 160, 171, 172, 173, 209,
    217, 219, 229, 230, 244, 245, 246, 247, 250, 251,
    252, 254, 284, 299, 300, 303, 304, 305, 306,
    307, 324
cluster 29, 30, 31, 32, 33, 34, 35, 36, 38, 39, 40, 84, 94, 172,
    187, 305
coconut harvesting 97, 98, 100, 101, 109, 110
connected and autonomous vehicle (CAV) 182
convolutional neural networks 62, 67, 75, 80, 84, 85, 90,
    94, 125, 181, 183, 185, 206, 211, 219, 220, 230,
    302, 304, 305
cooling degree days (CDD) 157, 160, 163, 164, 165, 166,
    167, 169
COVID-19 2, 112, 122, 140, 243, 245, 246, 254, 255, 289
customer lifetime value (CLV) 273, 281
cyber security 309, 310, 312, 313, 314, 320

dashboard 10, 68, 70, 71, 72, 127, 129, 130, 137, 138, 139,
    182, 186, 246, 286
data limitations 189, 198
dataset 66, 75, 76, 79, 80, 81, 82, 83, 85, 87, 88, 89, 90, 91,
    92, 93, 94, 109, 144, 176, 206, 210
DHT11 69, 72, 124, 127, 128, 129, 130, 133, 134, 137, 138,
    145, 220
DNP3 309, 319
donation 17, 18, 19, 21, 27
door detection module 171, 176, 177

earthquake alarm 151, 152, 153, 154, 155
electric power 167, 205, 208, 324
embedded system 146, 183, 228, 299, 300, 306
encoders 29, 32, 34, 35, 38, 39, 40, 147
energy demand 157, 158, 159, 162, 163, 165, 167, 169, 257,
    273, 276, 278, 279, 284

energy supply 157, 166, 167, 232, 278, 310
erasable programmable read-only memory 216
ESP8266 12, 114, 115, 118, 124, 125, 126, 127, 129, 131,
    132, 133, 134, 136, 137, 138, 139, 144, 145, 176,
    211, 220, 229, 231, 233, 234, 238, 241

face detection module (FDM) 171, 173, 174
face recognition module (FRM) 171, 175, 176
firebase 9, 12, 13, 14, 15, 26, 123, 124, 127, 128, 130, 131,
    138, 231, 236, 237, 240, 241

garbage monitoring system 51
GPS 10, 11, 15, 16, 66, 67, 68, 70, 72, 111, 112, 113, 114,
    115, 118, 119, 120, 121, 122, 124, 126, 172,
    245, 304

Hamilton–Jacobi–Bellman (HJB) 182
health monitoring 113, 143, 144, 145, 147, 148, 160, 243,
    254, 255
heating degree days (HDD) 157, 160, 163, 164, 165, 166,
    167, 169
home automation system 215, 216, 217, 218, 219, 220, 228,
    229, 230

IEC61850 309
IIoT 123, 124, 137, 139
intelligent system 16, 58, 62, 63, 73, 95, 97, 99, 107, 108,
    109, 110, 148, 182, 187, 218, 229, 307, 324

Kalman filter 183

leakage detector 41, 48, 49
light fidelity 216, 217, 228
long short-term memory (LSTM) 75, 80, 81, 84, 85, 90,
    92, 95, 206, 216
long-range radio 216, 217
LPG 41, 42, 43, 44, 45, 47, 48, 49

machine learning 35, 53, 62, 66, 67, 68, 73, 75, 76, 84, 91,
    94, 95, 97, 110, 124, 143, 144, 146, 147, 148,
    149, 155, 173, 174, 179, 181, 183, 187, 205, 206,
    211, 212, 218, 219, 245, 273, 274, 277, 287, 288,
    300, 307, 325
message queuing telemetry transport (MQTT) 145, 148,
    216, 217, 218, 228
microcontroller 10, 43, 44, 45, 46, 49, 67, 113, 115, 118,
    123, 124, 125, 127, 132, 133, 134, 136, 138, 139,
    144, 145, 146, 177, 207, 208, 210, 211, 216, 218,
    220, 228, 231, 232, 233, 234, 290, 300, 301,
    302, 303, 304
monitoring system 6, 9, 10, 11, 14, 15, 16, 51, 65, 66, 68,
    71, 73, 113, 122, 143, 144, 146, 147, 148, 149,
    172, 179, 187, 208, 211, 212, 241, 243, 254,
    255, 296
multilayer perceptron 75, 80, 84, 91, 92, 95, 147

neural network 4, 29, 30, 32, 38, 54, 62, 63, 66, 67, 74, 75, 76, 79, 80, 83, 84, 92, 94, 95, 125, 144, 147, 148, 181, 182, 183, 184, 185, 206, 211, 216, 217, 220, 281, 302, 304, 305, 306

NodeMCU 10, 11, 16, 111, 114, 115, 118, 123, 124, 125, 127, 128, 129, 130, 131, 132, 133, 134, 136, 137, 138, 139, 173, 176, 177, 220, 231, 233, 234, 236, 241, 247, 249, 304

normal standard error 75

nuclear power plant 123, 124, 136, 137, 139, 145, 311, 315

optimized scheduling 273, 281

organic light-emitting diode (OLED) 73, 216, 220

piezoelectric sensor 151, 153, 155

power plant 123, 124, 136, 137, 139, 145, 273, 276, 277, 280, 281, 283, 286, 311, 315

power theft 205, 206, 207, 208, 209, 210, 211, 212, 213

predictive analytics 65, 67, 70, 72, 74, 124

programmable logic controller 146, 210, 216, 299, 320

protection 1, 5, 41, 43, 46, 93, 145, 146, 149, 212, 287, 307, 309, 316, 322, 323, 324, 325

pseudo code 97, 107

random forest 66, 75, 76, 79, 147, 206, 211

remote monitoring 53, 143, 145, 233, 255

RFID 111, 112, 113, 114, 115, 118, 119, 122, 125, 172, 300

robot 3, 110, 123, 124, 125, 126, 127, 128, 129, 130, 132, 135, 136, 137, 138, 139, 140, 141, 187, 228, 229, 299, 301, 305

root mean square error 30, 75, 84, 92

serial monitor 68, 127, 231, 236, 238

silhouette score 29, 30, 38

smart energy network 309, 310

smart grid 143, 147, 149, 206, 207, 208, 210, 211, 212, 213, 242, 273, 274, 277, 283, 285, 287, 288, 297, 307, 309, 310, 311, 312, 313, 314, 315, 316, 317, 318, 319, 320, 321, 322, 323, 324

smart healthcare 1, 2, 3, 4, 5

smart waste management system 55, 56, 299, 303, 304, 305, 306

static electro-mechanical 143

sum of squares 29, 30

support vector machines (SVM) 35, 75, 76, 79, 83, 84, 89, 90, 92, 94, 144, 183, 184, 206, 282, 300, 302, 306

support vector regression (SVR) 75, 76, 79, 84, 89, 94

theft alert 205

ThingSpeak 9, 10, 12, 14, 15, 111, 114, 115, 118, 119, 120, 121, 123, 124, 127, 128, 129, 130, 137, 138, 145, 146, 231, 232, 233, 234, 238, 239, 241, 244, 246, 247, 304

threat score 75, 84, 91

TinkerCAD 226, 289, 292, 293

toddler security 187

transformer 143, 144, 145, 146, 147, 148, 149, 209, 210, 215, 221, 296, 324, 325

vehicle monitoring system 9, 11, 14, 15, 16, 113, 122

virtual automation tool 289

waste segregation 51, 53, 59, 60, 62, 299, 300, 301, 302, 305, 306, 307

wildfire detection 65, 73

wind turbine 220, 273, 277, 278, 280, 281, 286

wireless fidelity (Wi-Fi) 2, 4, 12, 58, 70, 114, 115, 118, 125, 126, 131, 140, 144, 145, 173, 176, 179, 183, 208, 210, 216, 217, 218, 219, 220, 229, 231, 232, 233, 234, 238, 246, 309, 319